D1264931

Statistical Downscaling and Bias Correction for Climate Research

Statistical downscaling and bias correction are becoming standard tools in climate impact studies. This book provides a comprehensive reference to widely used approaches, and additionally covers the relevant user context and technical background, as well as a synthesis and guidelines for practitioners. It presents the main approaches including statistical downscaling, bias correction and weather generators, along with their underlying assumptions, skill and limitations. Relevant background information on user needs and observational and climate model uncertainties is complemented by concise introductions to the most important concepts in statistical and dynamical modelling. A substantial part is dedicated to the evaluation of regional climate projections and their value in different user contexts. Detailed guidelines for the application of downscaling and the use of downscaled information in practice complete the volume. Its modular approach makes the book accessible for developers and practitioners, graduate students and experienced researchers, as well as impact modellers and decision makers.

Douglas Maraun is an associate professor and Head of the Regional Climate Modelling Group at the Wegener Center for Climate and Global Change at the University of Graz. His research interests include the processes governing local extreme events, assessing uncertainties of regional climate projections, and statistical post-processing of climate simulations for adaptation planning. He chaired the VALUE network which carried out the most comprehensive inter-comparison and evaluation of different downscaling approaches, and is involved in steering activities of the international downscaling initiative CORDEX.

Martin Widmann is a senior lecturer and climate scientist in the School of Geography, Earth and Environmental Sciences at the University of Birmingham. His current main research area is regional climate change, in particular the development and validation of statistical downscaling methods. He was one of the first to apply bias correction in a climate change context, and recently co-chaired the VALUE network. His other field of research is past climates, in particular the development of data assimilation methods to combine climate simulations with empirical knowledge from proxy data.

Statistical Downscaling and Bias Correction for Climate Research

DOUGLAS MARAUN

Karl-Franzens-Universität Graz Austria

MARTIN WIDMANN

University of Birmingham

CAMBRIDGE
UNIVERSITY PRESS

CAMBRIDGE
UNIVERSITY PRESS

University Printing House, Cambridge CB2 8BS, United Kingdom

One Liberty Plaza, 20th Floor, New York, NY 10006, USA

477 Williamstown Road, Port Melbourne, VIC 3207, Australia

314-321, 3rd Floor, Plot 3, Splendor Forum, Jasola District Centre,
New Delhi - 110025, India

79 Anson Road, #06-04/06, Singapore 079906

Cambridge University Press is part of the University of Cambridge.

It furthers the University's mission by disseminating knowledge in the pursuit of
education, learning and research at the highest international levels of excellence.

www.cambridge.org
Information on this title: www.cambridge.org/9781107066052
DOI: 10.1017/9781107588783

First published 2018

Printed in the United Kingdom by TJ International Ltd. Padstow Cornwall

A catalogue record for this publication is available from the British Library

ISBN 978-1-107-06605-2 Hardback

to our families

Contents

Preface

Statistical downscaling and bias correction are becoming a core element of climate impact studies. They are often intended to and have the potential to inform costly and far-reaching real-world adaptation decisions. The international statistical downscaling community, however, is not organised to meet this challenge, partly because the field is inherently interdisciplinary. Major methodological contributions come from climatologists, impact modellers – in particular hydrologists – and statisticians, all with their different scientific backgrounds, experiences and interests. No consensus exists on the appropriate use and evaluation of different methods; the underlying assumptions are often not explicitly spelled out, rarely are they tested. The downscaling language is, not surprisingly, far from being unified and varies from community to community and region to region.

A number of review articles have been published (e.g. Hewitson and Crane 1996, Zorita and von Storch 1997, Wilby and Wigley 1997, Onof et al. 2000, Fowler et al. 2007, Maraun et al. 2010b, Wilks 2010, Teutschbein and Seibert 2012, Maraun 2016), but some are becoming outdated. They are mostly narrow in scope and in general serve as a literature overview rather than an in-depth introduction to the subject. The IPCC has published a guidelines document on statistical downscaling (Wilby et al. 2004), and one textbook exists on empirical statistical downscaling by Benestad et al. (2008), as well as some book chapters, for example, in Willems et al. (2012). These contributions, however, are limited to some of the approaches in use and provide a mostly technical view of the subject. Other books like the recent contribution by Wilby (2017) cover a broad overview of climate change and society but only briefly lay out the concepts and methods of statistical downscaling. A book that presents the full range of statistical downscaling approaches in some depth and puts these methods into a broader context was missing. Thus, developers and users of downscaling or PhD students starting to work in the field were essentially forced to read review papers, individual papers or book chapters. This book attempts to close this gap.

The first aim of the book is to introduce the main approaches of statistical downscaling – namely perfect prognosis, model output statistics (which is often simply a bias correction), weather generators and some hybrid approaches. We present the most widely used methods that have been developed within these classes and discuss the underlying assumptions and how their structure affects their skill.

The second aim of the book is to provide readers with the necessary background knowledge. We review subjects such as regional climate and climate change itself, the needs of users of climate information, the necessary basics of statistical and dynamical modelling and the uncertainties of climate projections.

The third aim of the book is to present guidance for practical applications both for downscalers and users of downscaled information. We therefore discuss a framework to evaluate downscaling approaches, review the most comprehensive evaluation studies and synthesise the discussions of the book into a list of guidelines. The main focus of the book is on climate change studies, but of course many of the concepts are applicable to seasonal or decadal climate predictions as well.

Given the scope of the book, we have tried to write it such that it is accessible to different audiences: first, to experienced users and developers who need a reference or who are interested in the broader context of downscaling. Second, to researchers starting to work in the field, such as PhD students or advanced MSc students, who may look for an introduction to the different approaches and their performance but also require a concise overview of the relevant background knowledge in, for example, statistics. And third, to users of downscaling information, such as impact modellers or climate service providers, who require an overview of statistical downscaling and who seek guidance on the limitations and applicability of the different approaches in a decision-making context.

The book is divided into three parts: Part I provides the broad context and background, with more generally accessible chapters on, for example, user needs or climate model uncertainties and two more technical chapters on statistical and dynamical modelling. The latter two chapters require some background in undergraduate maths and statistics but are not required to follow the main ideas in the other parts of the book. Part II of the book introduces the different statistical downscaling approaches and their limitations – it is mainly a reference. Part III discusses the performance of statistical downscaling, links to the ongoing debate about the limitations of regional climate modelling and provides practical guidelines. Readers who are mainly interested in practical applications may start reading the introduction and could then jump directly to Chapter 18. They will then be directed to the different chapters for more in-depth discussions.

With the book, we also attempt to unify the statistical downscaling language. Even though the terminology may be scattered and sometimes misleading in climate change research, a more or less well-defined language exists in the numerical weather-prediction community. At first, the terms used in that community might sound unfamiliar for someone from the climate community. But the use of, for example, 'model output statistics' as a broader term for bias correction techniques has been tested for several years in international initiatives and has been proven useful. We therefore use this language, with some climate-specific adjustments, throughout this book. But we refer to widely used terms where they are suitable. The aim is to use familiar language as much as possible whilst being as precise as necessary. For the mathematical parts of the book, we have decided to stay as close as possible to common notation. That is, our notation is local and differs from chapter to chapter. For instance, in the statistical chapters x refers to a predictor, whereas in the dynamical modelling chapter x refers to a space coordinate.

We hope that this book contributes to integrating the community, stimulates discussions within and beyond the community and fosters the improvement and development of statistical downscaling.

Acknowledgements

This book summarises much of what we have learned about downscaling over the last 10 years. We have acquired much of this knowledge through discussions with colleagues. These discussions helped sharpen our arguments, inspired further research or changed our prior beliefs. Without these discussions, this book would not have been possible. The first of these discussions started at the international workshop on statistical downscaling at the University of East Anglia in May 2009. Since then several initiatives have provided platforms for scientific debate: the European VALUE initiative, funded as EU COST Action ES1102, the international CORDEX-ESD activities and related workshops and, unfortunately for a short time only, the US NCPP initiative on developing a downscaling vocabulary. We are greatly thankful for all the discussions at the meetings, the numerous tele-conferences and the joint writing of papers. To name but a few, we would like to thank, in alphabetical order, Joe Barsugli, Rasmus Benestad, Maria Laura Bettoli, Richard Chandler, Jens Christensen, Jonathan Eden, Jesús Fernandez, Andreas Fischer, Tilmann Gneitung, Galia Guentchev, José Gutierrez, Bill Gutowski, Stefan Hagemann, Alex Hall, Elke Hertig, Bruce Hewitson, Heike Hübener, Radan Huth, Chris Jack, Ian Jolliffe, Sven Kotlarski, Linda Mearns, Christel Prudhomme, Ingo Richter, Ole Rössler, Mathias Rotach, Ted Shepherd, Pedro Soares, Thordis Thorarinsdottir, Heimo Truhetz, Claudia Volosciuk, Mathieu Vrac, Daniel Walton, Rob Wilby and Renate Wilcke.

In particular, we would like to thank our colleagues from the University of Cantabria in Santander and their spin-off company Predictia: without the endurance and professionalism of Joaquin Bedia, Daniel San Martín, José Gutierrez and Sixto Herrera, the VALUE portal would not exist, and many of the VALUE results would not be available. Many of the plots in this book are based on portal content.

When writing the book we discussed its content with several colleagues. They gave important input and commented on the manuscript. These are Emanuele Bevacqua, Michela Biasutti, Jan Haerter, Stefan Hagemann, Clara Hohmann, Ed Maurer, Thomas Mendlik, Christian Onof, Christian Pagé, Marie Piazza, Victor Venema, Giuseppe Zappa and Eduardo Zorita. We thank Chris Jack and Bertrand Timbal for providing links to data sets and Ian Phillips for proofreading the document.

We thank Susan Francis, Zoë Pruce and the team from Cambridge University Press for their continual support and advice. Sharelatex provided an excellent online platform to jointly write the book. We acknowledge funding of the Volkswagen Foundation (grants 85423 and 85425).

Finally, D.M. is deeply thankful for the support and patience of Heike Marie and Jesse Alexander. M.W. thanks family and friends for their support.

1 Introduction

Climate is changing and will continue to change. Societies and ecosystems are affected by and often depend on climate and its variability. Already in 1992, the United Nations Framework Convention on Climate Change stated that all parties shall "cooperate in preparing for adaptation to the impacts of climate change" (United Nations 1992). Over the last decades, several countries have developed national adaptation strategies. The EU strategy on adaptation to climate change (European Commission 2013), for instance, acknowledges the need to take adaptation measures at all levels ranging from national to regional and local levels. The Global Framework for Climate Services (GFCS), established in 2009, sets out to develop and communicate climate information to "enable better management of the risks of climate variability and change and adaptation to climate change" (http://www.wmo.int/gfcs/vision). In short, there is an urgent demand for scientifically credible climate change information, in particular at the regional scale (Hewitt et al. 2012). One approach to obtain information about regional climate change is downscaling of global climate projections. In fact, a plethora of different data products have already been made available via internet portals.

Yet the provision of regional climate change information is one of the big challenges in climate science (Schiermeier 2010) and still a subject of essentially basic research (Hewitson et al. 2014). A *Nature* editorial prominently pointed out that "certainty is what current-generation regional studies cannot yet provide" (Nature 2010). Kundzewicz and Stakhiv (2010) argue that climate models have originally been developed to guide mitigation decisions. They could provide a broad picture of global climate change but would not yet be skillful to serve as input for regional adaptation planning. Kerr (2011*b*) brings forward a range of arguments which have been issued against current downscaling practice, and, in a later piece (Kerr 2011*a*), discusses the challenges of providing actionable climate information.

Against this background, the book at hand attempts to provide a reference for a range of approaches and methods often summarised as statistical downscaling. At the same time, the book aims to put the more technical issues of statistical downscaling into the broader context of user needs, regional climate modelling uncertainties and limitations, and good scientific practice. To begin with, we would like to sketch the scientific idea of statistical downscaling and then give some guidance on how to best approach this book.

Figure 1.1 Landmask and elevation model of a typical state-of-the-art GCM. The horizontal resolution is approximately $1.13° \times 1.13°$. Adapted from Figure 1.4 (bottom panel AR4), Solomon et al. (2007).

1.1   Statistical Downscaling and Bias Correction in a Nutshell

On the 18th of July 2009, heavy rains fell in the city of Graz, Austria. The soil was still saturated from a wet spell in late June, such that the city's streams burst their banks, and several districts were flooded. Hydrologists, engineers and city planners might all be interested in the risk of such a flooding to happen again: it depends on the precipitation history over the preceding weeks, on the intensity of the rainfall event and on its spatial-temporal distribution. But all these users of climate information are more and more concerned not only with risk in present climate but also with potential changes of risk in a warmer future climate.

Much of our knowledge about future climate change stems from projections with global general circulation models (GCMs). For instance, the ensemble simulations carried out within the coupled model intercomparison project (CMIP, Meehl et al. 2007a, Taylor et al. 2012) have been the backbone of many prominent messages published in the last Intergovernmental Panel on Climate Change (IPCC) assessment reports (e.g. Meehl et al. 2007b, Collins et al. 2013). But even state-of-the-art GCMs still have a rather coarse resolution (Figure 1.1). As a consequence, regional-scale topography and meteorological processes, in particular those responsible for many types of extreme events, are not represented by these models.

The idea of downscaling is to bridge the gap between the large spatial scales represented by GCMs to the smaller scales required for assessing regional climate change and its impacts. Two major types of downscaling exist: in dynamical downscaling, a high-resolution regional climate model (RCM) is nested into the GCM over the domain of interest (Rummukainen 2010). In statistical downscaling, empirical links between the large-scale and local-scale climate are identified and applied to climate model output.

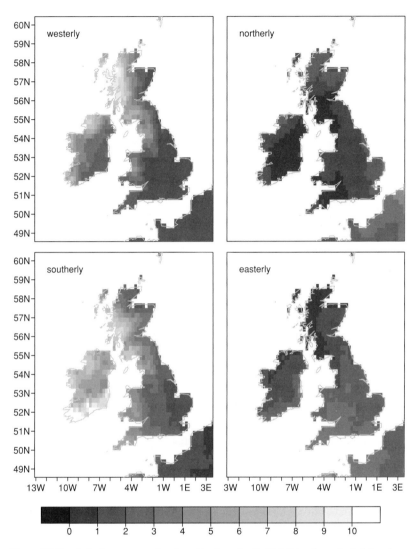

Figure 1.2 Precipitation composites [mm/day] on the British Isles for selected Lamb weather types (Lamb 1972), which describe the main atmospheric circulation patterns over the British Isles. Based on E-OBS daily data (Haylock et al. 2008) and the Lamb weather types from the Climatic Research Unit (Jones et al. 2013) for the period 1950 to 2016.

Figure 1.2 illustrates such an empirical relationship for the British Isles. The panels show the average precipitation which falls under four different situations of the large-scale atmospheric circulation: in case of a westerly flow, the highest precipitation is expected along the west coast of Ireland and Great Britain. Highest intensities occur in particular in the western Scottish Highlands – the South East of England is typically dry under such conditions. Northerly airflow instead brings cooler air, which typically carries less moisture. Precipitation intensities are thus lower – with relatively high values in the exposed regions of Northern Scotland, the North East of Ireland and East

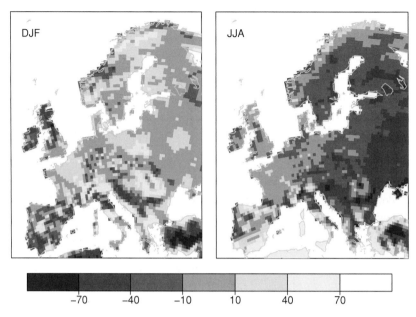

Figure 1.3 Relative precipitation bias [%] of the RCM RACMO2 (van Meijgaard et al. 2008), driven by the GCM EC-EARTH (Hazeleger et al. 2010) compared to E-OBS data (Haylock et al. 2008), for the period 1971–2000. Left: winter; right: summer.

Anglia. If the flow arrives from the south, precipitation is highest in the South of Ireland but also along the west coast of Great Britain. Finally, easterly flow brings higher intensities to the East coast and to the hills of South-West England and Wales, the first orographic barrier in the south of Great Britain. Other such situations – called weather types in meteorology – would also be associated with typical precipitation patterns. Thus, knowing the large-scale flow, one can predict the regional distribution of precipitation, including the effects of regional orography. Applying this empirical relationship to the large-scale circulation simulated by a GCM would thus downscale the GCM and generate regional-scale precipitation fields. In statistical downscaling jargon, this would be a weather-type-based perfect prognosis model. Under the assumption that the empirical link between large-scale circulation and local precipitation remains valid in a future climate, one could apply the model to generate regional precipitation projections.

But even if a GCM would resolve the climate processes relevant for a particular user, the simulated climate would typically still substantially deviate from real-world climate. In fact, even after dynamical downscaling the simulated regional climate is in general biased compared to observations. Figure 1.3 shows the relative error between simulated and observed mean winter (left) and summer (right) precipitation climate. The simulation has been conducted with the RCM RACMO2, driven by the GCM EC-EARTH, two well-performing climate models. In some parts of Europe, the relative error is below 10%, but in many regions it exceeds plus or minus 70%. Impact modellers often cannot use such simulations directly; they demand some form of statistical post-processing to adjust the model output towards observations. Again, one could establish an empirical link: here the ratio between the simulated and the observed mean precipitation. Apply-ing this scaling factor to the simulation, one would "remove" the model bias. This bias correction procedure is a simple form of model output statistics. Under the assumption that the correction function is applicable in a future climate, one could post-process future precipitation projections.

The terms "statistical downscaling" and "bias correction" are used differently in dif-ferent communities and countries. Many US researchers use the terms essentially inter-changeably. In other countries, climatologists often reserve the term "statistical down-scaling" or even "empirical statistical downscaling" for the first approach, which we call perfect prognosis. The term "bias correction" is used by dynamical downscalers and hydrologists exclusively for the second approach, but some users of empirical statistical downscaling would claim that also their approaches are bias correcting. Recently, some authors began to argue that bias correction, as it does only post-processes model output, should better be called bias adjustment. And being slightly meticulous, one could even argue that many statistical methods from either approach do not generate time series rep-resenting local climate – that is, they are not really downscaling. We therefore decided to follow a semi-pragmatic approach. In general, and in particular when we compare the two approaches, we use the terminology originally proposed by Klein and Glahn 1974: we call the first approach perfect prognosis (PP) and the second model output statistics (MOS). The key advantage is that these terms are precisely defined and at the same time get more and more used across disciplines. But since most MOS approaches in this book are mere bias corrections, we often use this simpler term. The term "statistical downscaling" is used rather colloquially to subsume both approaches.

1.2 How to Read This Book

We hope the book will prove useful for different audiences, each with its specific back-grounds and needs. One could approach the book simply by reading it in the given order or use selected chapters as reference. In particular, these would be the technical chapters on statistical methods (Chapter 6) and dynamical modelling (Chapter 8), the different downscaling approaches (Chapters 11–14), and finally the evaluation and performance (Chapters 15 and 16).

Readers who are new to the field or who are mainly interested in using and inter-preting downscaling results may instead start reading the book from Chapter 18. This chapter provides a condensed summary of how to best apply statistical downscaling in practice: what are important issues to be considered? Which methods are useful in which context? How could one deal with uncertainties? In each section, the reader is then directed towards more in-depth discussions in the preceding parts of the book. We will sketch these in the following.

Part I provides both a broader context and the necessary technical background. In Chapter 2 we introduce climate and weather phenomena governing regional climates. After a historical overview (Chapter 3), we discuss the main assumptions, requirements and concepts of downscaling (Chapter 4). User needs are reviewed and discussed in

Chapter 5. Chapters 6 to 9 provide background in statistical modelling, a summary of observational data and dynamical climate modelling and a discussion of their limitations and uncertainties.

Part II is the core of the book and introduces the overall structure of downscaling methods (Chapter 10), as well as the major approaches PP, MOS, weather generators and combinations of these approaches (Chapters 11–14). Each of these chapters provides an overview of widely used methods as well as their structural limitations and the assumptions underlying their use.

Part III discusses the performance of statistical downscaling and regional modelling and its use in practice. A general framework for the evaluation of regional climate projections is presented in Chapter 15 and a synthesis of actual performance in Chapter 16. The ongoing debate of the applicability of downscaling is critically reviewed in Chapter 17. Finally, Chapter 18 provides a synthesis of the book and guidelines for the use of downscaling in practical applications.

Part I

Background and Fundamentals

2 Regional Climate

Before discussing regional climate modelling in more detail, it is sensible to briefly sketch regional climate itself, and the factors controlling regional climate and climate change. We use the term rather loosely, spanning a range of scales below the continental and synoptic scale. Some may prefer to distinguish between regional and local climate – we decided to use only one term – which scales we refer to should become clear from the context. In fact, we will discuss that it is essential to define the relevant scales for any user case individually. In some situations, a region might be a whole country, in other situations a district or even just a specific valley.

Regional climate is determined by a vast number of climatic processes spanning global to local scales. To successfully model regional climate change, it is essential to successfully model the processes relevant for the specific application. We will therefore sketch these processes in the following, considering the Alps as a showcase.

2.1 Large- and Planetary-Scale Processes

The global temperature field is to a first order determined by the influence of latitude and land–sea distribution on the radiative balance, and the different thermal capacity of land and ocean. The Alps are located in the mid-latitudes, in a temperate climate (Figure 2.1, left). The mid-latitudes are a region with a strong meridional temperature gradient and high baroclinic instability, which controls the position of the jet stream and fuels the North Atlantic storm track (Hoskins and Valdes 1990, Lynch and Cassado 2006). Jet streaks, regions of acceleration and divergence in the jet stream, control the genesis of cyclones. The upper-level winds also steer the path of cyclones. Conversely, the passage of cyclones along the storm track drives the jet stream, intricately linking the two phenomena (Woollings 2010). The Alps are located just south of the climatological mean position of the polar front and polar jet in both summer and winter. Cyclones and anticyclones advect different air masses to central Europe. For instance, arctic continental air brings cold and dry air, tropical maritime air is warm and moist, tropical continental air is hot and dry. The jet stream itself follows planetary-scale Rossby waves meandering around the globe. If the meanders have a large amplitude, cold air is transported far south and warm air far north (see Figure 2.2). Such situations are often persistent in time, associated with blocking events: high-pressure systems that block the passage of cyclones and cause heat waves and drought in summer and cold spells in winter.

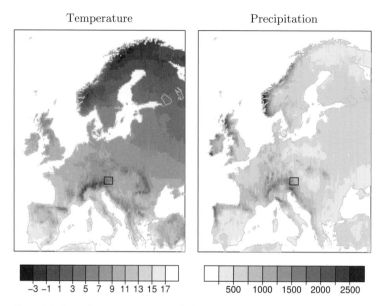

Figure 2.1 Annual mean temperature [°C] and annual mean precipitation [mm] in Europe for the reference period 1971–2000, calculated from the EOBS data set, version 14.0 (Haylock et al. 2008). See Chapter 7 for a discussion of gridded data sets.

If the meanders are weak, the flow is mainly zonal, corresponding to a positive phase of the North Atlantic Oscillation (NAO) – the pressure gradient between the Icelandic low and the Azores high. The mean position of Rossby wave meanders – and hence the position of the stationary mid-latitude high- and low-pressure systems – is controlled by the land–sea contrast and major mountain ranges such as the Rocky Mountains or the Greenland ice shield but also the sharp sea surface temperature gradient in the north Atlantic and the north-eastward orientation of the eastern North American coastline (Hoskins and Karoly 1981, Held et al. 2002, Minobe et al. 2008, Brayshaw et al. 2015).

Mid-latitude winter climate is influenced by processes in the stratosphere (Limpasuvan et al. 2004). During the polar night, a strong circumpolar vortex of westerly winds rotates around the pole. In the Northern Hemisphere, tropospheric Rossby waves penetrating into the stratosphere may slow down these winds and ultimately break down the vortex. As a result, stratospheric temperatures may suddenly rise by several tens of degrees. The breakdown of the polar vortex weakens the polar jet (Baldwin and Dunkerton 2001) and favours cold air outbreaks into the mid-latitudes (Thompson et al. 2002). El Niño/Southern Oscillation influences European winter climate, arguably via a stratospheric teleconnection (Ineson and Scaife 2009).

Severe weather often happens along the fronts between different air masses. Winter storms typically pass north of the Alps, supported by the orientation of the main ridge of the Alps, which spans an arc from south-west to north-east. As a result, precipitation is typically higher along the north-western flank (Figure 2.1, right). The northwest of the Alps is still under maritime influence from the Atlantic, the south from the

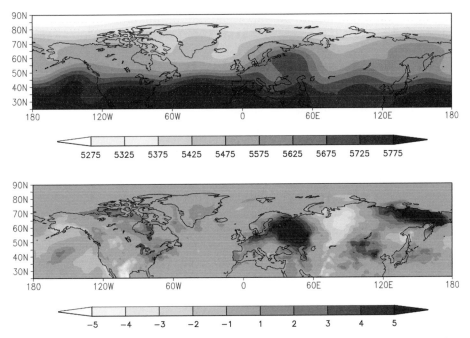

Figure 2.2 Extreme events of July 2010. Monthly mean 500hPa geopotential height [gpm] (top) and maximum temperature anomaly [K] (bottom). The polar jetstream, in the position of the strong geopotential height gradient, meanders strongly as a Rossby wave with a particularly high amplitude over eastern Europe. Hot air is carried northwards in eastern Europe and eastern Siberia, causing, e.g., the severe Russian heat wave. Likewise cold air is transported southwards, causing a cold wave over central Siberia. Based on ERA-Interim (Dee et al. 2011*b*), adapted from a figure created with KNMI Climate Explorer (https://climexp.knmi.nl).

Mediterranean. The Alps themselves contribute to lee cyclogenesis in the gulf of Genoa (Barry 2008). Some of these Genoa lows, called Vb cyclones, travel across Italy and the Adriatic towards eastern Europe. They are responsible for the precipitation maxima in the Julian and Dinaric Alps, north-east of the Adriatic.

In other regions of the world, similar large-scale processes influence the particular climate. Tropical and sub-tropical climates are controlled by the inter-tropical convergence zone (ITCZ), the Hadley and Walker circulation and the Monsoons (Goosse 2015). Along the ITCZ, moist air converges and deep, organised convection occurs. Along the tropopause, the air travels polewards, cools, is deflected by the Coriolis force and sinks over the subtropics, where the Earth's major deserts are located. Temperature contrasts between the oceans and land induce the continental-scale Monsoonal circulation. During summer, the winds blow onshore, carry moisture and cause heavy rains. The orographic forcing of the Himalaya is responsible for the highest rainfall totals worldwide. During winter, the dry winds blow offshore. Specific to the climate within about $5°$ of the equator is the vanishing Coriolis force: here, no tropical cyclones can form. Oceans are a major source of moisture, and their high heat capacity dampens seasonal temperature variations and thus shapes maritime climates.

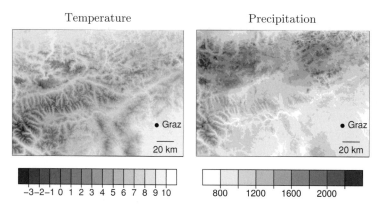

Temperature   Precipitation

• Graz   • Graz

20 km   20 km

–3 –2 –1 0 1 2 3 4 5 6 7 8 9 10   800   1200   1600   2000

Figure 2.3 Annual mean temperature [°C] and annual mean precipitation [mm] in the eastern Alps of Austria (marked by the box in Figure 2.1), for the reference period 1971–2000. Based on the the SPARTACUS (temperature, Hiebl and Frei 2016) and GPARD (precipitation, Hofstätter et al. 2013) data sets. See Chapter 7 for a critical discussion of such high-resolution gridded data sets.

2.2   Regional to Microphysical Processes

As a major mountain range, the Alps modulate climate themselves (Barry 2008). Temperatures typically decrease with altitude. Along the windward side, orography induces and amplifies precipitation, as in the examples discussed earlier. In the wind shadow, precipitation is reduced. Orographic precipitation is directly related to Föhn winds: latent heat is realised by the precipitation, such that the downslope winds in the lee side are warmer than the up-slope winds. The influence of orography becomes evident in particular at smaller scales (Figure 2.3). In particular, inner alpine valleys show up as much warmer and drier than the surrounding peaks, often even drier as the foothills of the Alps.

At even finer scales, local phenomena create distinct microclimates (Barry 2008). Daytime heating and nighttime cooling of mountain slopes cause upslope and upvalley winds during the day and downslope and downvalley winds during nights. Especially during winter, cold air flows down into the valleys, causing temperature inversions. Given that the surrounding mountains effectively shield the valleys from wind and turbulent mixing, these inversions are often persistent. Local feedbacks additionally modulate the climate. Snow cover, for instance, increases the albedo, reducing local temperatures and thus sustaining the snow cover. The duration of snow cover will also depend on the aspect of a given slope: a north face might have a snow pack throughout the year, whereas a south-facing slope at the same altitude might be snow free in summer. Climate, in particular in complex terrain, thus varies at very small scales.

Soil moisture strongly modulates summer temperature and precipitation at continental scales (Seneviratne et al. 2010). For instance, during the 2003 European heatwave, a soil moisture deficit reduced latent cooling by evapotranspiration and thereby strongly amplified the event (Fischer et al. 2007). Soil moisture can influence precipitation directly via moisture recycling, or indirectly, for example, by increasing convective instability (Schär et al. 1999, Seneviratne et al. 2010). In arid climates, soil moisture feedbacks can even influence the large-scale atmospheric circulation and amplify persistent drought conditions (Giannini et al. 2003).

In particular, precipitation emerges from the interplay of local and even microphysical processes (Rogers and Yau 1996, Pruppacher et al. 1998). Once a layer of moist air is lifted, it cools adiabatically until it is super-saturated and condensation sets in. Cloud droplets form and grow by diffusion and collisions. The growth is critically accelerated by mixing with dry air, which is entrained into the clouds. Latent heat release provides further energy to fuel the updrafts, which in turn hold the droplets in the cloud and sustain their growth. At some point, the droplets have grown into heavy raindrops, and fall out of cloud. If they do not fully evaporate, they reach the ground as rain. Snow develops in a similar way. The lifting occurs either by orographic forcing, by large-scale instabilities, for example, along weather fronts, by convergence of air or by local heating. Key to local convective precipitation is convective instability: often, the atmosphere is stratified such that it is stable as long as it is dry. At some level after condensation has set in, the latent heat release destabilises the air and enables free convection. The energy required to lift the moist air to this level of free convection is called convective inhibition. The energy which is available to drive convection is called convective available potential energy and is determined by the moisture and temperature stratification.

In other regions, other regional-scale processes may be important. In particular, coasts – either along the oceans and seas but also along major lakes – have distinct regional climates. Here, the high heat capacity of water dampens diurnal temperature variations, and the land–sea temperature contrast drives land breezes during the day and sea breezes during night. These wind systems often reach some tens of kilometres inland. Onshore winds are slowed down by increased surface friction, causing convergence, cloud formation and precipitation a couple of kilometres inland.

2.3 Regional Climate Variability and Change

Many of the factors determining regional climate may respond to global climate change. The Hadley cell, for example, is projected to weaken in the Northern Hemisphere. The relative importance of different drivers is still a matter of debate (Held and Soden 2006, Vecchi et al. 2006, Seo et al. 2014, He and Soden 2015, Merlis 2015). As a result of increased subtropical static stability, however, the Hadley cell is expected to also expand polewards (Lu et al. 2007). Polar amplification, that is, the increased warming in polar regions due to the ice-albedo feedback, will reduce the surface temperature gradient between poles and equator. Moreover, the projected slow-down of the meridional overturning circulation in the Atlantic ocean will increase the meridional sea surface temperature gradient in the Northern Hemisphere (Woollings et al. 2012). In contrast, the enhanced warming of the upper tropical troposphere, called tropical amplification, will increase the upper-level meridional temperature gradient. These competing changes will in turn change the position and intensity of the storm track as well as the character of planetary waves (Butler et al. 2010, Harvey et al. 2015).

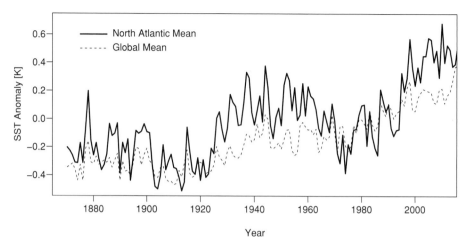

Figure 2.4 Area average annual mean sea surface temperature anomalies (reference 1971–2000) for the North Atlantic (0°N–60°N, 75°W–7.5°W) and global oceans. In addition to an overall warming, the Atlantic exhibits strong multidecadal variability with a timescale of some 60 years, called the Atlantic multidecadal oscillation (HadISST data accessed via https://climexp.knmi.nl).

At local scales, feedbacks may modulate local warming. For instance, increasing temperatures may dry out the soil and in turn amplify regional temperatures. Similarly, warming at high elevations will decrease snow cover, reduce the albedo and thus amplify the warming (Hall et al. 2008). Changes in regional climate may be distinguished into dynamic and thermodynamic changes (Shepherd 2014). The former refers to changes in the atmospheric circulation, the latter to non-circulation changes such as an increased water vapour content in a warmer atmosphere. Which of these factors dominates depends on the region, scale and variable considered (Kendon et al. 2010, Woollings 2010). To realistically simulate regional climate and to credibly project changes in regional climate, all climatic phenomena relevant in a specific context have to be adequately represented by the chosen climate models. We will therefore discuss the performance of both global and regional climate models in Section 8.5.

Changes in regional climate are, however, not only controlled by external forcings. In fact, internal models of variability, caused by feedbacks between different components of the climate system, cause substantial fluctuations of climate on interannual to multidecadal, perhaps centennial timescales (Latif and Park 2012). Prominent examples are El Niño Southern Oscillation (ENSO; Philander 1990), the Indian Ocean dipole (IOD; Saji et al. 1999), Pacific Decadal Oscillation (PDO; Mantua et al. 1997) or the Atlantic multidecadal oscillation (AMO; Schlesinger and Ramankutty 1994), see Figure 2.4. For instance, the AMO explains a major part of the long-term droughts in the Sahel (Giannini et al. 2003). Central and western European annual mean temperatures have been simulated to change by more than half a degree Celsius from a negative to a positive AMO phase (Knight et al. 2006). A similar result has been found in observed spring and summer temperatures for western and eastern Europe, respectively (Sutton and Dong 2012).

2.4 Further Reading

An introduction to climate and climate dynamics is given in (Goosse 2015). Lynch and Cassado (2006) provide a concise and easily accessible introduction to atmospheric dynamics. The books by Whiteman (2000) and Barry (2008) focus on mountain meteorology and climate. Barry and Blanken (2016) present local and microclimate and their relation to large-scale climate. An excellent review of the dynamical influences on European climate is given by Woollings (2010), on the role of internal climate variability by Latif and Park (2012).

3   History of Downscaling

3.1   Downscaling in Weather Forecasting

The first downscaling methods had been invented already in the late 1940s (Klein 1948) and became operational in the early days of numerical weather prediction at the end of the 1950s. Back then, operational numerical weather prediction models were by far too coarse to predict local weather, and furthermore they did not forecast all variables of interest but only a few such as pressure and temperature. At that time a considerable network of observed weather time series was available already. Klein et al. (1959) employed this data to infer statistical relationships between the observed large-scale circulation – for those variables that were simulated by the models – and the observed local-scale weather variables of interest. The statistical model was then applied to downscale the actual numerical forecast of the large-scale circulation to a forecast of the local weather. The key assumption of this approach is that the large-scale predictor has been perfectly forecasted by the numerical model, hence the approach itself has been coined perfect prognosis (PP). After some years, a considerable database of past forecasts had been archived. Analyses of this data revealed that numerical forecasts even of the large-scale weather were of course not perfect but showed systematic deviations compared to observations. Yet this database also became key to mitigate this problem: Glahn and Lowry (1972) developed a new approach that – during calibration – did not take the predictors from observations but from the archived numerical forecasts. For a new weather prediction, the inferred statistical link is then applied to the new numerical forecast. As this approach is basically a post-processing of numerical model data, it has been coined model output statistics (MOS). The key advantage of MOS is that it contains by construction a bias correction of the numerical model. Current weather prediction systems employ complex MOS approaches with several predictors that are continually recalibrated to provide the highest predictive skill.

In parallel, numerical approaches were developed to improve the resolution and accuracy of forecasts over a target region. The first limited-area model was developed at the US National Meteorological Center (Howcroft 1966, Gerrity and McPherson 1969) and became operational in 1971. This model covered the US, Canada and the Arctic Ocean at a horizontal resolution of 190.5km at 60°N and was driven at the lateral boundaries with input from a Northern Hemisphere numerical weather prediction model. Current weather forecast models are run at much higher resolutions and often provide multiple nesting and adaptive grid resolutions. For instance, the Unified Model of the UK Met

Office (in the year 2017) consists of a global model at a horizontal resolution of 25km. A nested model of 12km resolution covers the North Atlantic and Europe. Finally a second nested model is run over the UK with a resolution of 1.5km over the UK itself and 4km in surrounding areas (Clark 2009).

3.2   Concerns about Climate Change

In 1960, Keeling published his initial multiyear atmospheric carbon dioxide measurements. At the South Pole, he found the first evidence for a systematic increase in carbon dioxide concentrations roughly consistent with estimates of anthropogenic emissions. In the following decade, precipitation over the Sahel gradually decreased, causing severe droughts and famine. These events raised concerns that human activities may change – and may have changed already – climate at a global scale. Scientists started their quest to constrain Arhennius' 1896 first estimate of equilibirum climate sensitivity: how warm would the atmosphere be if the carbon dioxide concentration would be doubled? Manabe and Wetherald published the first climate simulation with a global general circulation model in 1967 and, a couple of years later, the first climate change simulation (Manabe and Wetherald 1975). Their results suggested climate sensitivity to be between approximately 2K and 3K. The First World Climate Conference was held on 12–23 February 1979 in Geneva and led to the establishment of the World Climate Programme and the World Climate Research Programme (WCRP) in 1980. In 1988, the Intergovernmental Panel on Climate Change (IPCC) was finally founded by the World Meteorological Organisation and the United Nations Environment Programme. At its first meeting in Geneva, the panel agreed on three main tasks that led to the formation of three working groups (IPCC 1988): (i) assessment of available scientific information on climate change; (ii) assessment of environmental and socio-economic impacts of climate change; (iii) formulation of response strategies.

3.3   Early Downscaling in Climate Research

In line with the awareness of global climate change, concerns rose also about potential impacts. Decision makers were interested in various issues such as national food security or the assessment of risks associated with nuclear waste repositories. This regional aspect of climate change created a demand for regional climate change studies: how would regional climates respond to global climate change? What would be the impacts of these changes?

A popular approach at those times was to study past warm climates, both in the instrumental record and in paleo data, as analogues for a warmer future climate (e.g. Wigley et al. 1986). Yet Crowley (1990) pointed out that the forcings that caused high global temperatures in the past are in general different from the radiative forcing of anthropogenic greenhouse gases. Thus, the response of the atmospheric circulation, in particular at regional scales, in a future climate is likely to be different compared to past climates. In parallel to the purely empirical analogue approaches, researchers therefore started to pursue model-based approaches.

One of the first studies to quantify regional impacts of climate change has been carried out by Schwarz (1966). In his study, Schwarz used three approaches to assess possible impacts of changes in climate on the water supply in the north-eastern US: a case study approach, a qualitative assessment of the system's sensitivity and a quantitative sensitivity study based on a simulation approach. For the latter analysis, he employed a stochastic stream flow generator developed by the US Army (US Army Corps of Engineers 1971). Schwarz generated stream flow time series for different possible future climates varying in mean, standard deviation, skewness and autocorrelation. In the 1970s, the global response of the climate system to increasing greenhouse gases was barely known, let alone possible regional changes in the water cycle. Consequently, Schwarz considered both positive and negative changes in all parameters. A similar study, but now explicitly anticipating a global warming, has been carried out by Mearns et al. (1984) for potential changes in extreme temperatures at Des Moines in the US corn belt. Such so-called "change factor" approaches are still widely used, for example, in the UK Climate Projections project (Murphy et al. 2009).

In the 1980s, more and more climate change simulations from global climate models became available, and researchers started to develop tools to make direct use of these data sets for regional impact studies. The simplest approaches were interpolations of GCM surface variables to local scales (e.g. Cohen and Allsopp 1988). Similarly, Wigley et al. (1990) derived empirical relationships between observed grid-scale surface variables and local surface variables and transferred these relationships to climate model simulations. Grotch and MacCracken (1991), however, demonstrated that GCMs do not accurately simulate climatic fields below a minimum skillful scale which is considerably larger than the horizontal model resolution. Also Giorgi et al. (1991) showed that the climate change signal simulated with a high-resolution regional climate model differs substantially from GCM simulations interpreted at the grid box scales. Thus, Giorgi et al. (1991) and von Storch et al. (1993) argued that grid box data cannot be used to directly downscale to finer scales.

Several statistical and dynamical approaches for regional climate simulations have since then been developed that are still in use. Two strands in statistical downscaling have initially been followed. In the first type, a large-scale GCM simulated time series is downscaled to a local time series, that is, the local weather variability is synchronised with the large-scale weather. The other type of simulating regional climate change does not employ the full climate model time series but only the simulated long-term climate change signal.

The first statistical downscaling study was probably the analysis by Kim et al. (1984). The authors inferred a statistical relationship between the variability of a climate variable (temperature and precipitation) averaged over a large area and its local variability. Subsequently, this relationship was transferred to climate change simulations. The authors coined this approach the "climate inversion", as its aim was to invert the averaging from local to large scales. Gates (1985) published a first conceptual discussion and arguably was the first to introduce the term "downscale". Karl et al. (1990) recognised that the problem of downscaling climate simulations is conceptually similar to the PP and MOS approaches in numerical weather prediction. Although they did not apply these concepts according to their strict meaning, Karl et al. (1990) formulated two key points in downscaling: first they suggested to use free-atmospheric variables as predictors, as these are not dominated by local-scale surface boundary conditions. Second they emphasise the difficulties of extrapolating the predictor/predictand relationships into unobserved climatic regimes. Von Storch et al. (1993) further refined the downscaling concept; they recognised that the simulated predictors should be defined on a scale larger than the minimum skillful scale. This is essentially the PP assumption: the predictors should be accurately simulated by the climate model. They also argued that predictors should explain a large part of the predictand variability.

The second type of statistical downscaling, as stated earlier, employs only the simulated long-term climate change signal. In the simplest implementation, the change signal was simply added to an observational present-day climate record (e.g. Rosenzweig 1985, Santer 1985, Gleick 1986). This approach has become popular as the "delta change method" and is still in use. Wilks (1988) proposed a more sophisticated variant that employed a weather generator. Weather generators had already been in use for several years. The first wet-day generator was developed by Gabriel and Neumann (1962); Katz (1977) and Buishand (1977) constructed precipitation generators, and Richardson (1981) finally published the first full weather generator that simulated random sequences of precipitation, temperature and solar radiation. Wilks (1988) changed the parameters of a weather generator by a GCM-simulated climate change signal to assess possible impacts of climate change on US agriculture. Wilks also realised that local weather is not deterministically predictable from large-scale information, and thus a stochastic approach is required to simulate time series of, for example, local daily precipitation. All delta change or change factor approaches employ climate model data at the grid scale or averages across a relatively small area. These approaches thus implicitly assume that at least the climate change signal is accurately simulated at the scale considered.

In parallel to these developments, mainly hydrologists started to develop approaches that combine the advantages of both statistical downscaling types. Hay et al. (1991), Bárdossy and Plate (1991) and Bárdossy and Plate (1992) conditioned the parameters of a weather generator on a day-by-day basis on weather types describing the large-scale atmospheric circulation, that is, they combined downscaling with simulation approaches. In their studies, the authors also suggest using the statistical models, calibrated to real-world data, to downscale GCM simulations of future climate. The first actual applications of this method to GCM simulations were carried out by Matyasovszky et al. (1993) and Bartholy et al. (1995). Conceptually, these models were the first PP downscaling weather generators.

As an alternative to statistical downscaling, limited-area models have been applied to dynamically downscale GCM simulations. Dickinson et al. (1989) set up the first regional climate modelling system and initially downscaled five GCM simulated winter storms over the western US. Giorgi and Bates (1989) showed, that the limited area model driven with observed boundary conditions simulated a realistic climate. In a follow-up study, Giorgi (1990) downscaled GCM simulations of six Januaries. These studies demonstrated that RCMs produced a stable climate and realistically reproduced the mesoscale atmospheric circulation and orographic effects in the western US. Giorgi et al. (1991) finally conducted the first climate change simulation with a limited-area model.

As a result of this early burst in downscaling research, many of the fundamental concepts in downscaling climate simulations that are still in use today had been laid out already by the early 1990s. The research into the construction of high-resolution climate change scenarios was further stipulated by the IPCC. In its first assessment report from 1990, it was stated that "Finer resolution than used at present is also required for the atmospheric component if regional variations of climate are to be predicted. Present-day climate models do not have sufficient resolution to represent in a meaningful way the climate of specific regions as small as, for example, the majority of individual nations" (McBean et al. 1990, p. 325).

3.4   The 1990s and 2000s

From the mid-1990s onwards, both dynamical and statistical downscaling approaches have been further improved and applied. For reviews on statistical downscaling methods, refer to Hewitson and Crane (1996), Zorita and von Storch (1997), Wilby and Wigley (1997), Xu (1999), Yarnal et al. (2001) Hanssen-Bauer et al. (2005) and Maraun et al. (2010b). For reviews and overviews of dynamical downscaling refer to, for example, Giorgi and Mearns (1999), Laprise (2008) or Rummukainen (2010). Comprehensive overviews of applications are given by the IPCC assessment reports (Giorgi et al. 2001, Christensen et al. 2007) and the reviews by Prudhomme et al. (2002) and Fowler et al. (2007).

The growing awareness that simulations by individual models are affected by model errors (e.g. Giorgi and Bates 1989) resulted in two new research avenues: first, model intercomparison projects were initiated, and second, ways to correct model biases were explored. The first intercomparison study by Christensen et al. (1997) was soon followed by the Project to Intercompare Regional Climate Simulations (PIRCS; Takle et al. 1999) under the auspices of the WCRP. In Europe, the PRUDENCE project (Christensen and Christensen 2007) was launched to compare RCMs. The European STARDEX project pursued a similar aim for statistical downscaling methods, with a focus on extreme events (Haylock et al. 2006, Goodess et al. 2010). These projects were followed by the European ENSEMBLES project (Hewitt 2005, van der Linden and Mitchell 2009). The most visible and directly useful outcome of the ENSEMBLES project is the huge database of RCM simulations over Europe and north-west Africa. The North American equivalent was the North American Regional Climate Change Assessment Program (NARCCAP; Mearns et al. 2009). Apart from providing standardised experiments for model intercomparison, the PRUDENCE, ENSEMBLES and NARCCAP projects were also designed to create multimodel ensembles for sampling model uncertainties. For a list of model intercomparison projects, refer to the WCRP (http://www.wcrp-climate.org/index.php/modelling-wgcm-mip-catalogue). Notably, the United Kingdom undertook a huge national effort to create its own ensemble of regional climate change simulations (Murphy et al. 2009). The RCM simulations further provided change factors for single-station weather generators (Kilsby et al. 2007).

Closely related to model intercomparisons and uncertainty assessments is the evaluation of downscaling skill and added value. Already in early downscaling studies, the simulated present-day climate has been compared with observations (e.g. Dickinson et al. 1989, Giorgi and Bates 1989). Charles et al. (1999) were the first to study downscaling performance in a simulated future climate based on a perfect model or pseudo-reality approach. They calibrated a statistical downscaling model to GCM-simulated present climate and assessed whether the downscaling model was able to correctly represent future climate, simulated by the same GCM, with predictors taken from the same GCM. Denis et al. (2002) developed the "big brother" experiment to isolate RCM downscaling skill: they simulated a reference climate over a large domain with a high-resolution RCM. This reference climate was then degraded to a typical GCM resolution and in turn used to drive the same RCM over a smaller domain. The final RCM downscaling can then be compared with the reference climate (created by the same RCM). Added value was mainly assessed for RCMs (Giorgi and Bates 1989, Castro et al. 2005, Frei et al. 2006).

Wilby et al. (2000) showed that climate model biases caused considerable errors in subsequent simulations of river runoff; but a simple removal of mean biases strongly improved the runoff simulation. Therefore, researchers explored possibilities to bias correct global and regional climate model output. Wood et al. (2002) bias corrected a long-term ensemble forecast with the NCEP global spectral model (GSM), and Hay et al. (2002) bias corrected dynamically downscaled reanalysis data. As transfer function, Wood et al. (2002) used a quantile mapping (Panofsky and Brier 1968) that maps simulated quantiles onto observed quantiles. Hay et al. (2002) post-processed precipitation with a similar approach and bias corrected temperature by simply removing the difference in long-term means.

Around the same time, Widmann et al. (2003) found that numerically simulated precipitation is a skillful predictor of observed precipitation. Specifically, they bias corrected NCEP reanalysis precipitation against observed precipitation. Because in this example simulated precipitation was by construction in synchrony with observed precipitation, the authors were able to use classical forecast verification measures to assess the skill of the bias correction. Salathé (2005) applied bias correction to transient climate change simulations with GCMs. With the increasing availability of global (Meehl et al. 2007a, Taylor et al. 2012) and regional (Hewitt 2005, Christensen and Christensen 2007, Mearns et al. 2009, van der Linden and Mitchell 2009) ensemble simulations, bias correction has become more and more popular (e.g. Maurer 2007, Li et al. 2010, Hagemann et al. 2011, Dosio et al. 2012).

3.5   Recent Developments

A major step in the operationalisation of downscaling was the establishment of the Coordinated Downscaling Experiment (CORDEX: Giorgi et al. 2009) of the WCRP.

The Global Framework for Climate Services (GFCS; Hewitt et al. 2012) was founded in 2012 and further pushed for operational regional climate change products. Additional pressure came from development banks, international aid organisations and national to local governments or funding agencies (Hewitson et al. 2014). As a result, a vast number of bias-corrected national and global climate change projections have been conducted (Maurer 2007, Li et al. 2010, Hagemann et al. 2011, Dosio et al. 2012, Stoner et al. 2013, Girvetz et al. 2013, Hempel et al. 2013, Maurer et al. 2014) and have in turn served as input for impact studies (Gangopadhyay et al. 2011, Girvetz et al. 2013, Hagemann et al. 2013, Warszawski et al. 2014). These results have been the basis for assessment reports (Cayan et al. 2013, World Bank 2013, Georgakakos et al. 2014) and have been made available through online data portals (Worldbank n.d., prepdata n.d.).

In response to the ongoing operationalisation, also critical discussions arose. Several studies highlighted that the signal-to-noise ratio between the projected climate change signal and internal variability could be very low at the regional scale, in particular for precipitation (Hawkins and Sutton 2009, Deser et al. 2012, Maraun 2013b).

Another debate was concerned with the ability of climate models to simulate regional trends. For instance, in western Europe observed temperature trends were not consistent with historical simulations from climate model ensembles (van Oldenborgh et al. 2009, Bhend and Whetton 2013; in fact, simulated trends were weaker than observed trends). In this context, also the issue of added value came up again: does downscaling improve historical GCM trends? Racherla et al. (2012) downscaled a historical GCM simulation with an RCM and compared grid-box trends in both simulations. They concluded that no added value was evident, but Laprise (2014) argued that the model setup, in which internal variability contributed substantially to long-term trends, was ill designed. Pielke and Wilby (2012) argued that downscaling for climate projections was pointless because of limited skill. For a discussion of these issues, refer to Chapter 17.

Around the same time, several problems with the use of bias correction in climate change modelling had been discussed. It had been shown that biases are not time invariant (which is an important assumption in bias correction), but may depend on the state of the climate system (Christensen et al. 2008, Buser et al. 2009, Vannitsem 2011, Boberg and Christensen 2012, Maraun 2012). Quantile mapping was found to modify simulated trends (Hagemann et al. 2011). Some authors argued that these modifications may actually account for state-dependent biases (Boberg and Christensen 2012, Gobiet et al. 2015), whereas others modified quantile mapping to conserve the raw climate model trends (Li et al. 2010, Haerter et al. 2011, Hempel et al. 2013, Pierce et al. 2015). Another issue was related to large-scale circulation errors. Eden et al. (2012) argued that bias correction could only sensibly post-process local errors resulting from parameterisations and the representation of orography but not large-scale errors in the atmospheric circulation. Bias correction problems that may occur in the presence of circulation errors have been reported by Maraun and Widmann (2015), Addor et al. (2016) and Maraun et al. (2017a). Furthermore, Maraun (2013a) demonstrated that bias correction was not able to create sub-grid variability and could therefore not in general be used for downscaling.

Along with the push to use climate model simulations to inform decision making, the interface between climate modellers and users caught more and more attention

(Henderson-Sellers 1996). Researchers addressed the requirements for useful data (Cash et al. 2002), in particular in the presence of deep uncertainties (Lempert et al. 2004). Several researchers concluded that climate models were not fit to provide regional climate information (Kundzewicz and Stakhiv 2010, Wilby 2010, Pielke and Wilby 2012). In response, strategies for impact modelling and robust adaptation have been pursued that do attempt to avoid downscaling (Dessai 2009, Prudhomme et al. 2010, Wilby and Dessai 2010, Brown et al. 2012).

Being aware of both the limitations of downscaling and user needs, Hewitson et al. (2014) highlighted the ethical dimension of providing regional climate information. In an even broader context, Adams et al. (2015) propose an ethical framework for climate services. To ensure that downscaled climate information is credible, Barsugli et al. (2013) and Hewitson et al. (2014) call for a systematic evaluation of statistical downscaling approaches. Such an evaluation framework has recently been developed by the European network VALUE (Maraun et al. 2015). This framework was the basis for a comprehensive evaluation of downscaling performance (Gutiérrez et al. 2017, Hertig et al. 2017, Maraun et al. 2017a, Soares et al. 2017, Widmann et al. 2017). Recently, two issues have been highlighted: downscaling is just one source of regional climate change information, and different sources often contradict each other. As a consequence, there is an urgent need to distil credible and salient climate information from all available sources of information (e.g. WCRP WGRC 2014, Hewitson 2016). In 2013, CORDEX-ESD has been launched to develop a framework for a global statistical downscaling intercomparison and to further coordinate the development of statistical downscaling methods.

3.6 Further Reading

A general historical overview of climate modelling has been published by Edwards (2011). Recently, Rockel (2015) reviewed the history of dynamical downscaling, and Giorgi and Gutowski (2016) presented a review of coordinated regional climate modelling experiments.

4   Rationale of Downscaling

In this chapter, we discuss the basic ideas, assumptions and concepts underlying downscaling. The concept itself is considered in Section 4.1. In different user contexts, different aspects of the climate system – expressed in statistical terms – will be relevant. We introduce these aspects in Section 4.2. Each downscaling model is based on a set of assumptions; these are presented in Section 4.3. But also the downscaled model itself has to fulfill specific requirements, as will be discussed in Section 4.4. In Section 4.5 we will discuss remaining issues such as added value.

4.1   What Is Downscaling?

As already introduced in Chapter 1, the main rationale and purpose of downscaling is to bridge the gap from the large spatial scales represented by GCMs to the smaller scales required for assessing regional climate change and its impacts. Dynamical downscaling employs regional climate models (RCMs) to simulate the atmosphere and its coupling with the land-surface at a higher resolution, but over a limited domain (Rummukainen 2010). Boundary conditions are taken from the driving GCM. Statistical downscaling derives empirical links between large and local scales and applies these to climate model output. The two main variants of statistical downscaling have already been sketched in Chapter 1; they will be introduced in more detail in Part II. For now only the basic difference is important: so-called perfect prognosis statistical models – essentially all regression and weather type methods – are calibrated against observed large-scale predictors and local-scale predictands. Under climate change, the statistical model is applied to predictors from a GCM. So-called model output statistics methods – essentially all bias correction methods – calibrate a transfer function between climate model simulations and observations in present climate, and apply this transfer function to future climate model simulations. Given that bias correction is often applied to RCMs rather than directly to GCMs, we will in the following sections discuss not only statistical but briefly also dynamical downscaling.

Generally speaking, downscaling uses additional information from regional or local scales that is not present in GCMs to derive information about regional climate and climate change, conditional on the driving GCM. RCMs resolve regional-scale processes and use regional information on the orography; statistical downscaling uses information on observed climate at selected locations.[1]

---

[1] This is why statistical downscaling is often called empirical statistical downscaling (Benestad et al. 2008).

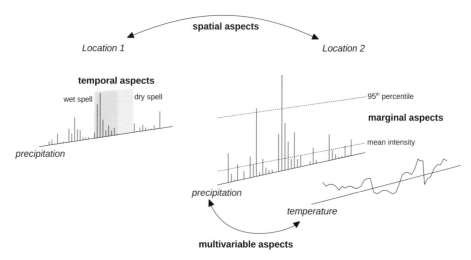

Figure 4.1 Aspects of the multivariate climate distribution.

4.2 Statistical Aspects of Regional Climate

Downscaled climate change projections are not future observations. They are based on dynamical and statistical models, that is, they are inherently approximate and imperfect representations of reality. In particular, every model is designed to represent some aspects of reality only, that is, a model is designed for a specific purpose. This holds especially for statistical downscaling models, which represent only a very limited section of the climate system. For a given user problem, a specific downscaling model might thus be perfectly sensible or completely inappropriate. To clarify how a downscaling model may meet the needs of a specific user and to develop a sensible evaluation framework, it is therefore useful to consider which aspects of the climate system the model represents and how well it represents them.

Climate in a wider sense is defined as the state of the climate system. This state is often described statistically (Planton 2013). A full statistical description not only comprises long-term climatic means but – in line with the argument made above – a range of other statistical aspects that characterise the full multivariate climate distribution: what is the typical variability around the mean? What are typical extreme events? What are characteristic time and space scales? In other words: the climate of a region of interest may be decomposed into its marginal, temporal, spatial and multivariable aspects (see Figure 4.1). We will refer to these aspects throughout the book as statistical climate aspects.

4.2.1 Marginal Aspects

Marginal aspects comprise all unconditional aspects such as the mean and the variance but also wet-day probabilities or a high percentile (Figure 4.2; see also the definition of

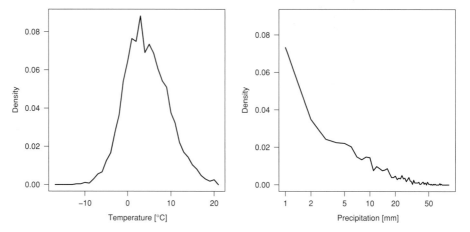

Figure 4.2 Marginal aspects. Histograms of observed winter temperature (left) and summer precipitation (right) at Graz, Austria (15°27'00"E, 45°49'52"N). Temperature is well approximated by a Gaussian distribution but exhibits heavy tails. Precipitation is highly skewed. Based on ECA-D data (Klein Tank et al. 2002), for the period 1979–2008.

marginal distributions in Section 6.1.5). At a daily resolution, the marginal distribution is spanned by the sequence of weather events such as the passage of cyclones and air masses, potentially modified by local feedbacks and forcings.

Temperature roughly follows a Gaussian distribution (Section 6.1.6). In summer, heat waves typically result in positive skewness, that is, stronger extreme events occur as would be expected from a Gaussian distribution. Conversely, cold air outbreaks in winter typically cause a negative skewness towards low temperature extremes. The marginal distribution of precipitation is more complex. First of all, precipitation is an intermittent process with dry and wet days. On a rain day, the precipitation distribution is heavily skewed, with many moderate events and few heavy events.

4.2.2   Temporal Aspects

In general meteorological processes exhibit some temporal dependence (Figure 4.3). Partly this dependence stems from real memory in the considered variables; for instance, soil moisture feedbacks may link temperatures on subsequent days, and similarly precipitation recycling may lead to a series of rainfall events. Often, however, the dependence is imprinted by slowly varying large-scale conditions, such as the passage of cyclones or the persistence of a blocking high. To distinguish such aspects of persistence from systematic variations, such as the seasonal cycle, we refer to them as residual temporal dependence. They comprise short-term memory such as the lag-one autocorrelation or wet–dry transition probabilities, but also longer-term memory such as spell length distributions and interannual variability. Residual temporal dependence is usually stronger at larger scales, in particular for processes such as precipitation that exhibit strong variability at small spatial and short timescales.

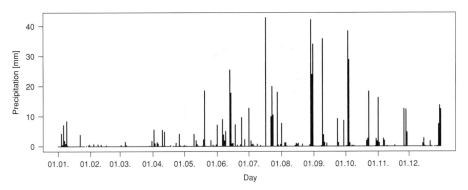

Figure 4.3 Temporal aspects. Observed precipitation during the year 2003 in Graz (15°27'00"E, 45°49'52"N). Precipitation is an intermittent processes, occurring in wet spells, followed by dry spells. Based on ECA-D data (Klein Tank et al. 2002), for the period 1979–2008.

4.2.3   Spatial Aspects

Climate tends to vary smoothly in space; more abrupt changes are mainly imposed by topography: coastlines modify the temperature and precipitation distribution, and mountain ranges create elevation-dependent temperatures, rainshadows, or even large-scale climate divides. We call such spatial variations in climate systematic spatial variations.

Meteorological processes generally extend across space – thunderstorms, cyclones and air masses all have a characteristic spatial scale (Figure 4.4). Thus, the weather at different locations exhibits residual spatial dependence on a wide range of timescales in addition to the systematic spatial variations. Consider, for example, two nearby weather stations along the wind-facing slope of a mountain chain. If a cyclone passes by, it will

Figure 4.4 Spatial aspects. Observed precipitation on 28 Oct. 2000, when the storm Nicole hit western Europe and caused widespread flooding in the UK. Precipitation tends to cluster in space, but with considerable variability. Based on ECA-D data (Klein Tank et al. 2002), for the period 1979–2008.

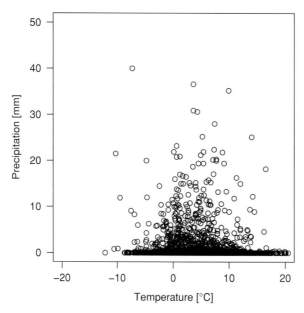

Figure 4.5 Multivariable aspects. Observed scatterplot of winter temperature and precipitation in Graz (15°27'00"E, 45°49'52"N). Based on ECA-D data (Klein Tank et al. 2002), for the period 1979–2008.

likely rain at both locations. Thus, if it rains at one of the stations, the probability of rain is high at the second station as well; sunshine at one and rain at the other is rather unlikely. The relationship is stronger when the stations are closer.

4.2.4   Multivariable Aspects

Multivariable aspects characterise the dependence between different meteorological variables such as temperature and precipitation (Figure 4.5). For instance, sunshine duration and precipitation are in general highly anti-correlated. Multivariable aspects are crucial for many climate impacts. For instance, spring runoff is highest when snow melt – caused by non-freezing temperatures – coincides with heavy precipitation. Similarly, a meteorological drought may be even more severe if it is accompanied by a heat wave. Such multivariate extreme events are called compound events (Leonard et al. 2014).

4.2.5   Systematic Seasonal and Spatial Variations

As discussed earlier, climate varies systematically throughout the year. Such systematic variations are often considered for mean climate only. But of course, any aspect of the climate system may have a seasonal cycle. Extreme precipitation events tend to occur in particular seasons (Maraun et al. 2009), and the spatial-temporal scales of precipitation are typically smaller and shorter in summer than in winter. Similarly, every aspect of the climate system varies systematically in space: temperature decreases with height, and its variability is dampened towards the coast. Precipitation is high on exposed mountain slopes and lower in shielded valleys.

4.3 Requirements for a Downscaling Model

A downscaling model should accurately represent the climatic aspects of interest on the required timescales at the spatial-temporal target resolution. For dynamical models this involves a realistic representation of the relevant processes, either explicitly or by parameterisations (Chapter 8). A statistical model needs to include the relevant predictors representing the regional variability. Moreover, the influence of the predictors on the predictands needs to be realistically represented. These requirements will be discussed in more detail in Part II.

Downscaling in the context of climate change is intended to simulate credible[2] regional responses to large-scale climate change. As such, it is distinct from downscaling in weather forecasting. A sensibly downscaled weather forecast requires predictors that represent the day-to-day variations of weather; predictors representing climate change may be completely irrelevant (gradually changing climate can be accounted for by recalibrating the downscaling system). In contrast, it is essential for downscaling in a climate change context to represent the influence of large-scale climatic changes on the regional variables of interest. This requirement goes along with key assumptions:[3]

- In dynamical downscaling, the response of the relevant regional-scale processes to large-scale changes needs to be credibly represented, either explicitly or in sub-grid parameterisations.
- In statistical downscaling, predictors representing the influence of a changing climate on the regional variable of interest need to be included. But moreover, the downscaling model has to be designed in such a way that the influence of the predictors on the variable of interest on long timescales is represented and may be extrapolated to the projected different climate.

To construct models that fulfil these assumptions and to evaluate whether these assumptions hold is far from simple, especially when characteristics different from mean temperature are sought. We will discuss these assumptions in more detail in Part II.

4.4 Requirements for the Downscaled Model

As stated earlier, downscaling output is always conditional on the driving model. Then obviously also the quality of a downscaled climate projection depends on the driving model: garbage in, garbage out (Hall 2014).

---

[2] The term 'credible' will be defined in Section 5.2.
[3] These assumptions are often loosely referred to as stationarity assumptions.

Dynamical downscaling takes its input from the lateral boundary conditions. Consequently it is required that the relevant input fields at the boundaries are well represented, including their response to global climate change. What is relevant depends on the domain size and the spatial-temporal variability of the processes within the domain. A brief review of climate model performance is given in Section 8.5.

Perfect prognosis (PP) downscaling methods are calibrated in the real world and then applied to climate model output. The predictors are therefore required to be perfectly simulated – this is the so-called perfect prognosis assumption. Perfect means that the predictors are realistically and bias free simulated in present climate and that their response to large-scale climate change is credibly simulated. Therefore, the processes controlling the variability of the predictors have to be realistically represented in the model, either explicitly resolved or parameterised. The PP approach does in general not require the driving model to correctly simulate regional processes – only to the extent that the regional processes feed back into the relevant large-scale processes. In fact, perfect prognosis is designed to represent mesoscale processes by statistical relationships. We will discuss the PP assumption in more detail in Part II.

For bias correction the situation is different: in climate change modelling, these methods do not intend to bypass a cascade of processes from synoptic to local scales but rather to post-process regional climate model simulations. The post-processing intends to correct (rather than to bypass) mis-represented processes. Thus, the assumption underlying bias correction is that the driving model – GCM and RCM – realistically represents the full process chain down to the target variable, apart from a potentially time-varying bias, including the response of the involved processes to climate change. In weather forecasting, MOS systems similar to PP are in operation that represent missing (or bypass not adequately simulated) processes.

4.5   Further Issues

4.5.1   Added Value and Representativeness

A key question is whether downscaling or bias correction adds value, that is, whether it improves simulations with a coarser dynamical model. Added value can be separated into two contributions (Luca et al. 2012, Di Luca et al. 2015).

First, value may be added by reducing model bias caused by, for example, a coarse model topography or a poor representation of resolved or parameterised processes. For instance, a mountain chain is typically rather flat in a GCM and thus does not influence the large-scale flow as strongly as in reality. Simulated precipitation fields will then be systematically wrong, even when aggregated to the scale of the driving model. Downscaling can in principle reduce such model errors even at the large scale.

But second, a GCM may perfectly well describe climate beyond its native resolution. It would, however, in general not correctly represent local climate. For instance, the GCM area average daily precipitation will have characteristics different from point-scale precipitation: intensities will be lower, temporal persistence will be higher and even the climate change signal might be different. These differences are not a model error or a lack of skill but a matter of representativeness (Klein Tank et al. 2009, Maraun and Widmann 2015): processes have different characteristics on different scales. Another aspect of added value is therefore bridging this scale gap, that is, generating simulations which are representative of the local target climate. Some authors refer to this aspect as added detail (Rummukainen 2010).

In a climate change context, adding value additionally means that downscaling is intended to produce a more credible response to climate change than the coarser resolution GCM – again, either by improving the GCM signal at the resolved scale or by introducing sub-grid variations of the climate change signal. A typical example is the snow-albedo feedback: warming in high altitudes is amplified, because it reduces snow cover, which in turn decreases the surface albedo and thereby increases the warming (Mountain Research Initiative EDW Working Group 2015). A GCM, which does not resolve regional orography, will simulate a warming consistent with its elevation. An RCM better resolves the regional orography and will show a systematically varying warming signal (Salathé et al. 2008, Walton et al. 2015). The higher the resolution, the better the elevation-dependent warming will be represented. Thus, a key element of added value is to modify the climate change signal where necessary in a physically credible way.

4.5.2   Consistency with the Driving Model

Downscaling may intrinsically break consistency with the driving model. The closer the output is intended to match observed regional climate, the stronger the inconsistency will be.[4]

Consider a typical dynamical downscaling case. The GCM simulates a specific large-scale atmospheric flow, conditional on the coarse model topography. In a region of complex terrain, which is not resolved by the GCM, the simulated surface climate will poorly match observations. A free-running RCM would adjust the large-scale flow inside its domain to be consistent with the better-resolved orography – the RCM flow would then be inconsistent with the flow of the driving model but be closer to observations. PP downscaling is calibrated to observed data, which are trivially consistent. Applying the calibrated model to GCM simulated predictors will thus produce a solution which is consistent with the driving GCM; but as the GCM is likely biased, the downscaling will not fully match observations.

By contrast, bias correction is designed to "correct" biases: the statistical post-processing transforms some statistical aspects – for example, the long-term mean – of the simulated time series to closely match the corresponding observed aspects, and thereby immediately introduces inconsistencies with the driving GCM. For instance, if precipitation is too low, because too little moisture is advected, the moisture budget of the GCM will be inconsistent with the bias corrected precipitation fields.

---

[4] Already the discussion illustrates the point: in general, the RCM climate change signal, averaged to the GCM resolution, will be different from that of the GCM.

Here, a careful balance is required that ultimately calls for the selection of suitable GCMs: consider again the example of a GCM that does not resolve El Niño/Southern Oscillation. A PP downscaling consistent with the driving GCM will not be sensible; it would not represent observed regional-scale climate. But also a bias correction would not be sensible: one may transform the GCM output such that some aspects perfectly match observations. Still, one would not trust a future simulation, because the relevant processes are not represented. In any case, the stronger the inconsistency between the driving GCM and the downscaling, the lower the confidence in future projections.

4.5.3   Downscaling and Interpolation

Downscaling is not an interpolation of the driving model. Consider precipitation in a 200km GCM grid box over Graz. A simple (e.g. bilinear) interpolation would produce a time-varying but spatially smooth field, with magnitudes in between those of the neighbouring GCM grid boxes. Systematic local effects – high values in the windward side of a mountain, low values in the wind shadow – would be missed. Observational data sets are often interpolated with more sophisticated algorithms that include covariates – for example, elevation, distance from the sea, slope – which systematically modulate the observed climatological fields. But GCMs resolve neither these sub-grid covariates nor their influence on sub-grid climate. Therefore, downscaling is also not "zooming" into a GCM.

4.6   Further Reading

For reviews of value added by downscaling, refer to Deser et al. (2011) and Di Luca et al. (2015). The papers by Salathé et al. (2008), Walton et al. (2015) and Meredith et al. (2015a) highlight the added value by representing relevant processes to credibly capture the climate change signal.

5   User Needs

Regional climate modelling might be purely discovery driven. Often, however, it is funded and designed to supply users with information. In this chapter, we will lay out the needs of different users in the broader context of climate change adaptation.

5.1   Context

The VALUE network (Maraun et al. 2015) has carried out a survey on requirements by users of regional climate change data (Roessler et al. 2017). Information on temperature, precipitation, wind speed, relative humidity and global radiation are demanded by more than 50% of the surveyed users (which were dominated by impact modellers from the water sector). Around 40% demand fields of data at a spatial resolution of 1km or finer, about one third at an hourly temporal resolution, and another 50% at daily resolution. Often users essentially demand "future observations". As will become clear from the discussions throughout this book, robust information at such a high resolution in both space and time can in general not be provided in the form of time series.

In recent years, however, users are in practice often additionally confronted with a very different problem, which Barsugli et al. (2013) coined the practitioner's dilemma: web-based portals now provide access to a proliferation of high-resolution climate projections. These products are based on different methods and assumptions and often provide contradictory results (Hewitson et al. 2014), but users do not have guidance to select appropriate data sets and use them wisely. According to Barsugli et al. (2013, p. 424), "products are sometimes selected on the basis of availability, convenience of format, and familiarity with the provider." Obviously, there is a huge discrepancy between the expressed user needs and the actual choice of data.

At the interface between providers and users of regional climate information, also a provider's dilemma exists: demands by national and local governments, climate services providers, development banks or international aid organisations pressure the scientific community to operationalise the provision of regional climate projections – but, as will be discussed throughout this book, the science required to provide such information is still foundational research (Hewitson et al. 2014). In this context, scientists face questions about their responsibilities when delivering uncertain information which may be used for real-world decisions. Thus, the provision of climate information has

Figure 5.1 Müritz-Elde canal. Photo: Katharina Brückmann, BUND Mecklenburg Vorpommern. Used with permission.

an inherently ethical dimension (Adams et al. 2015). Climate science has in fact been referred to as post-normal: "where the stakes are high, uncertainties large and decisions urgent, and where values are embedded in the way science is done and spoken" (Hulme 2007). Recently, a debate has arisen whether climate models are ready to provide input for impact models and to inform adaptation planning (Kundzewicz and Stakhiv 2010, Wilby 2010, Pielke and Wilby 2012).

Often it might be possible to implement low-regret measures to adapt to climate change – such measures have only small adverse effects and thus do not need to be backed up by accurate climate information. But other cases are not that clear cut. Consider a model ensemble projecting increased flood risk along, say, the Müritz-Elde canal in Northern Germany (see Figure 5.1). A possible adaptation measure might be to reinforce the dams along the canal. The implementation, however, would require cutting down several hundred oak trees. A delicate ecosystem and a century-old heritage would thus be destroyed. Tourists canoeing on the canal may keep off the region. Knowing what is at stake, responsible planners need high-quality climate information to justify possible decisions.

Against this background – high user demands but little guidance to choose appropriate products; pressure to rapidly deliver information, where in fact more research is needed; naive expectations and strong scepticism – it is important for providers and users to acquire a clear understanding of the applicability and limitations of different regional climate change products but also of the real user requirements beyond stereotypical lists of variables and resolutions. In fact, the demands expressed by users are often a wish list rather than absolutely necessary information (Pingel 2012), and the concept of uncertainties is often just a phrase (Groot et al. 2004).

5.2   General Requirements

To begin with, it might be helpful to highlight that users of weather forecasts and climate projections – as a consequence of the different types of information that can be provided – have broadly different needs. Users of weather forecasts are interested in a specific weather state: what are the chances to enjoy a barbecue for their birthday party five days ahead? Users of climate projections, however, require information about possible future climates, that is, the statistics of future weather. Note also that a statement such as "precipitation might increase or decrease by 10%" might be non-trivial and thus be helpful information for a user of future climate information (see Chapter 9), whereas it would be a mere joke for the host of the party.

   In the adaptation example given in the previous section, we stated rather vaguely that decision makers may require high-quality information – it is useful to define a set of criteria to characterise the quality of information more precisely. Hewitson et al. (2014) argue that any decision-relevant regional climate information must meet the criteria of being plausible, defensible and actionable. Similarly, Cash et al. (2002) discuss the requirements for effectively linking knowledge and action and conclude that information needs to be credible, salient and legitimate to be considered by decision makers. The criterion of legitimacy is qualitatively different from the others, as it refers to the information-generating process. According to Cash et al. (2002), information is legitimate if it is perceived as unbiased and meets standards of political and procedural fairness. The other criteria characterise the quality of information. In the following, we define those terms we use in the book.

Credible Information
Model-based information is credible if the model that generates the information fulfils the following two criteria. First, the model has to produce plausible output, that is, output that is not obviously inconsistent with our process understanding. Second, the model has to realistically include all relevant processes. Thus, credibility in a modelling context involves a comprehensive process-oriented evaluation.

   Credibility is a necessary condition for any information to be useful and will play a key role throughout the discussions of this book. All regional climate projections assume that they are based upon credible input from global climate models (see Chapter 8). Without future observations and a complete understanding of climatic processes, it is of course impossible to positively conclude that a model projection is correct. Still, credibility may be established by assessing whether – to the best of our knowledge – all relevant processes and feedbacks dominating climate change are realistically included (see Chapter 15).

Defensible Information
Credible information becomes defensible if the associated uncertainties are comprehensively characterised (Maraun et al. 2015). The uncertainty assessment may allow a user to discriminate the resulting outcome from other potential outcomes. In Chapter 9, we

will discuss that ensembles of climate models are required to assess the uncertainties of climate projections. A climate projection based on a single climate model cannot be defensible.

Salient Information

Information is salient if it is relevant in a specific context (Cash et al. 2002). In a climate modelling context, information is salient if, for example, the required variables are provided at the required resolution, the relevant time horizon is analysed and the output represents the required statistical climate aspects.

Salience is closely linked to the design of a downscaling method – a method that only provides mean values may be appropriate in some applications but might fail to provide salient information when extremes are relevant. In the methodological chapters in Part III, we will therefore discuss how the structure of different types of downscaling methods limits their skill. But the question of salience is also related to the evaluation of climate model data (Chapter 15): does the evaluation address the relevant issues in a given context?

Actionable Information

Defensible and salient information is actionable if it provides evidence strong enough to guide real-world decisions in the context of an accepted level of risk (Hewitson et al. 2014). Both the perception and the tolerance of risk depend on the specific situation; they are subjective and differ from culture to culture and individual to individual. Thus, actionable information is necessarily user specific and context dependent. In short, information is actionable if one would spend their own money on it (Hewitson 2011).

5.3   User Groups

Users of climate model data may, on a very basic level, be categorised into scientific and non-scientific users. The former group comprises climatologists analysing climate projections and various types of impact modellers such as hydrologists, agricultural scientists, public health scientists and scientific engineers. The latter group comprises mainly members of the public administration, local to national to international governance, as well as non-governmental organisations and the private sector. Having different interests and backgrounds, these user groups also have distinct needs.

Boundary organisations are intermediaries to foster the dialog between the scientific community and, in particular, non-scientific stakeholders. Such organisations may be long-established bodies such as the IPCC or advisory boards of national governments and parliaments but also institutions more recently installed within the global framework for climate services. Boundary organisations are themselves not users of climate information but have to ensure that credible, legitimate and salient information is communicated. Their role will be discussed in Chapter 18.

Scientific users

Scientific users of regional climate information have a scientific interest: advancing our understanding of the climate system, its potential impacts or how society might adapt to these impacts. Even if scientists intend to inform decision makers, they – at least ideally – aspire to "speak truth to power" (Hoppe et al. 2013, p. 283).

To optimally accomplish their scientific aims, scientific users need to understand the specific structural skills and limitations of a method as well as the actual performance in a given context.

Often scientific users may want to characterise the uncertainties of potential future impacts and may thus require to sample uncertainties in future climate as well. To avoid driving their impact models with a big ensemble of climate model simulations, they would need a sensible selection of simulations, either spanning the full range of simulated uncertainties or representing some worst-case scenario.

Non-scientific users

Non-scientific users ask for scientific knowledge to guide decisions and to frame policy development. They want "politics on top and science on tap" (Hoppe et al. 2013, p. 283), which is of course fully legitimate in a democratic society which is not to be ruled by technocrats.

In this context, however, the incentive for non-scientific users to understand the information and their scientific basis is very limited, as long as it is credible, salient and legitimate.[1] Thus, non-scientific users typically do not demand detail on the used methods. To use scientific information in a decision-making context, they are more concerned about the robustness of the information: where are scientific messages certain, where are they uncertain? In any case, we will discuss in Chapter 18 that all users of regional climate information should develop a clear understanding of the overall uncertainties.

5.4 User-Specific Needs

The specific user needs depend very much on the type of user and the specific application (Hewitson et al. 2014, Maraun et al. 2015). In some situations, qualitative information about long-term changes might be sufficient, and in some situations ensembles of credible time series with a high spatial-temporal resolution may be required. Depending on the application, some or all of the statistical climate aspects defined in Section 4.2 might be relevant (Maraun et al. 2010*b*). For instance, a water resources manager might be interested in individual peaks of precipitation but also in the temporal variability ranging from short events to long wet or dry spells. The resulting runoff will also depend on the spatial fields of precipitation across the considered catchment, and finally the dependence between precipitation and temperature is important to assess whether precipitation falls as snow or rain.

---

[1] In some situations, of course, policy makers request scientific information not to guide the selection of policy options but rather to support an already-chosen option. In this context, the user is of course more interested in a "useful" scientific message than in a credible and legitimate one.

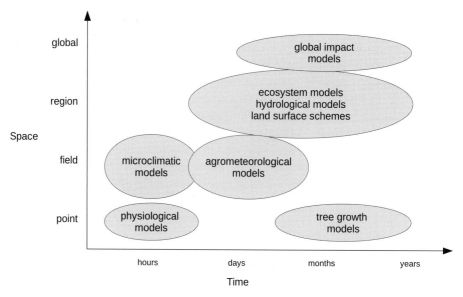

Figure 5.2 Spatial and temporal scales, on which selected impact models are operated. Adapted from Eitzinger and Thaler (2016).

Scientific impact modellers often demand climate model data that has the same characteristics as the observations which have been used to calibrate the impact model (Roessler et al. 2017). Often this data is available at the point scale (from meteorological stations) at a daily resolution. But not all impact models require data of such a high spatial-temporal resolution (see Figure 5.2). For instance, tree growth models may require a high spatial but a much lower temporal resolution.

In general the resolution required for climate downscaling is not necessarily the input resolution of the impact model. Depending on the context, it may be lower or even higher: the required climate model resolution is determined by the relevant impact processes and the climate model resolution that is required to simulate the meteorological and climatological drivers of these processes. A list of selected hydrological processes, which might be relevant for adaptation planning, is shown in Figure 5.3. For instance, to assess the impact of climate change on flash flooding in a small Alpine river catchment, simulations with a high spatial and temporal resolution might be required to capture the influence of local orography on precipitation and to credibly simulate the response of extreme convective precipitation to climate change. But if only the mean discharge of the same river would be of interest, a much lower spatial-temporal resolution might suffice. Here, a 12km daily resolution RCM, interpolated in space and time to the input resolution of the hydrological model, might suffice.[2]

In principle, user needs should be expressed in terms of meteorological phenomena – understanding these phenomena then assists the identification of required variables,

---

[2] Of course, the impact model may also be explicitly run at a lower resolution. But in particular process-based models are often tuned to a specific resolution.

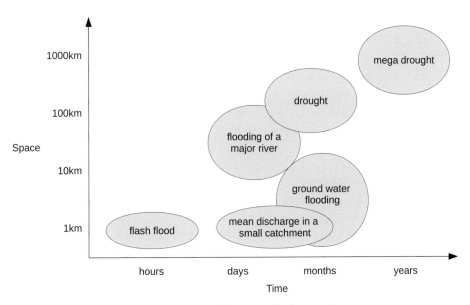

Figure 5.3 Spatial and temporal scales, on which selected hydrological processes occur.

spatial-temporal resolution and the relevant statistical climate aspects (see Section 4.2). For instance, for a particular user, meteorological drought and heat waves might be relevant. Then the important variables are temperature and precipitation, the important aspects are either long spells of daily joint threshold exceedances (low precipitation/high temperature) or joint threshold exceedances of weekly to seasonal averages over a certain region. This conceptualisation is not only important for the selection of suitable downscaling methods (see the discussions on structural skill in the methodological chapters of Part III) but also for a user-focused evaluation (see Chapter 15).

Last but not least, users and providers should be explicit about the time horizon of interest: are long-term projections really of interest, or are seasonal or decadal predictions the relevant products? For instance, a city planner might be interested in the potential flood risk for a century ahead. Whereas a farmer planning for the next season might require a seasonal prediction, the insurance company insuring the farmer might additionally benefit from information about the coming decades to inform their long-term strategy.

Thus, the guiding questions to understanding user needs and to provide credible and salient information in a given context are:

- What is the relevant time horizon?
- What are the relevant impact processes?
- What are the meteorological and climatological processes determining these impact processes?
- What is the required spatial-temporal resolution to simulate these meteorological and climatological processes for present and future climate?

- Which variables are required?
- What is the desired file format?

In some situations, available regional climate projections may not meet the needs of specific users in a specific context: for instance, because the large-scale climate is not credibly simulated or because the required variables are not, or not credibly, simulated. In such situations, close cooperation between users and different scientific disciplines is required; see the discussion in Section 18.5.2.

5.5   Further Reading

Key papers discussing the user context of downscaling are Barsugli et al. (2013) and Hewitson et al. (2014). A key paper discussing the link between research and decision making is Cash et al. (2002). A comprehensive review can be found in Roessler et al. (2017). A recent book providing a user perspective on regional climate change is Wilby (2017).

6 Mathematical and Statistical Methods

Statistical downscaling relies on a wide range of statistical concepts. This chapter presents a concise overview of these concepts. We attempt to explain the different issues as simply as possible but with the necessary detail and background to successfully implement a broad range of statistical downscaling methods. Section 6.1 gives an overview of random variables and probability, including the most relevant probability distributions, and some very basic introduction to the modelling of extreme values. Parameter estimation, that is, basis for calibrating statistical models, is introduced in Section 6.2. Here we focus on the widely used concept of maximum likelihood but give a brief overview of Bayesian estimators as well. Regression models, the backbone of many statistical downscaling models, are presented in Section 6.3, including a discussion of statistical model selection and evaluation. Weather generators are specific stochastic processes – these are laid out briefly in Section 6.4. Finally, predictors are often post-processed based on principal component analysis, and some statistical downscaling methods employ canonical correlation analysis. These pattern methods are introduced in Section 6.5. In the following, we will write scalars in normal font, while vectors and matrices will be written in bold font. We assume that readers have basic knowledge in significance testing – it will not be introduced in this chapter but only occurs in the context of model selection. In fact, significance testing is very useful in many contexts, but it is not essential to implement a statistical downscaling method. Also, even though relevant for some predictor transformations, we do not discuss clustering algorithms.

6.1 Random Variables and Probability Distributions

6.1.1 Events and Random Variables

In synoptic meteorology, the atmospheric circulation is often characterised by discrete weather types. At each day, the actual circulation might randomly assume one of these types. In statistics, the space of all possible weather types is called sample space or event space S. For the considered weather types, the sample space could be, for example, $S = (W, NW, N, NE, E, SE, S, SW)$. Subsets of the event space, for example, SE, or (NW, N, NE) (i.e. all Northerly weather types) are called events. The complement $\neg A$ of event A in S is defined such that $S = A \cup \neg A$ and $A \cap \neg A = \emptyset$. For instance, the complement of the three westerly weather types would be all other weather types from northerly to southerly. Mutually exclusive events are defined such that $A \cap B = \emptyset$, for instance northerly and southerly weather types.

The random occurrences of weather types on a given day can be formally described by random variables. Other random occurrences of events could be, for example, the occurrence of precipitation or the occurrence of a specific intensity of precipitation. A random variable itself is denoted with a capital letter, such as, X. The actual measured outcome – also called realisation or random number – is denoted with lowercase letters, such as, x. The occurrence of a weather type might be described by the random variable X. The actual outcome that occurs is x, for instance, $x = $ SE. Random variables can be categorical (e.g. precipitation occurrence), discrete (e.g. the number of wet days) or continuous (e.g. the intensity of precipitation). The key difference between categorical and discrete random variables is that the latter are ordered and allow for mathematical operations on the sample space (e.g. it is not possible to calculate the mean of wet and dry).

6.1.2   Probability

Attached to a random variable X are the probabilities of possible events A in S to occur. For instance, in the example of weather type occurrence, one could denote the corresponding probabilities as $P_W, \ldots, P_{SW}$. In a classical frequentist perspective, the probability of an event A to occur is defined as the limit of the frequency n_A of measured occurrences relative to the total number of events n that have been measured, for $n \to \infty$:

$$P(A) = \lim_{n \to \infty} \frac{n_A}{n}. \qquad (6.1)$$

This definition is in line with common sense: if no knowledge about the current atmospheric conditions was at hand, the best guess for the occurrence probability of a particular type at a given day would be the relative climatological occurrence frequency of that weather type.

Probabilities satisfy the following axioms. Given events A in an event space S, then

1. $P(A) \geq 0$ (probabilities are positive);
2. $P(S) = 1$ (the probability that any possible event occurs is one);
3. given countably many sets A_i, $i = 1 \ldots$ with $A_i \cap A_j = \emptyset$ for $i \neq j$, then $P(\cup A_i) = \sum P(A_i)$ (when combining disjunct events, their probabilities add up).

From these axioms, the following relations can directly be derived:

- $P(\neg A) = 1 - P(A)$ (the probability that an event occurs is one minus the probability that its complement occurs);
- For arbitrary events A and B, $P(A \cup B) = P(A) + P(B) - P(A \cap B)$ (the probability of one or another event occurring is equal to the sum of the probabilities of either event occurring, minus the probability that their intersection occurs);
- $P(A) \leq 1$ (probabilities are smaller or equal to one).

Conditional Probability

Consider the occurrence of precipitation, say, at the eastern slope of a mountain chain in the mid-latitudes. Without any knowledge about the current atmospheric conditions, one would assume that a wet day occurs with the climatological wet-day probability, given as the relative frequency of wet days over a long period of time. Now assume that the current weather type is known, for instance a westerly or an easterly flow. In the former case, the probability of precipitation would be decreased (as the location of interest would be in the rain shadow), in the latter case increased. The probability that an event A – a wet day – occurs, given that an event B (with $P(B) > 0$) – an easterly weather type – occurs is called conditional probability and given as

$$P(A|B) = P(A \cap B)/P(B). \tag{6.2}$$

For instance, assume that the probability of an easterly flow to occur (no matter whether it rains or not) is $P(B) = 20\%$ and that the probability of precipitation and an easterly flow to occur simultaneously is $P(A \cap B) = 10\%$. Then, in case of an easterly flow the probability of precipitation occurrence would be $P(A|B) = 10\%/20\% = 50\%$. Obviously, the concept of conditional probabilities will be central in statistical downscaling.

Bayes Theorem

According to Equation 6.2, the probability that A and B occur simultaneously, $P(A \cap B) \equiv P(B \cap A)$ can be written either as $P(A|B)P(B)$ or $P(B|A)P(A)$. Thus it directly follows that

$$P(A|B) = \frac{P(B|A)P(A)}{P(B)}. \tag{6.3}$$

This seemingly trivial relationship is known as the Bayes theorem and plays a central role in many statistical problems. It also provides the basis for Bayesian statistics (see Section 6.2.3).

6.1.3 Probability Distributions

The distribution of probabilities over the event space $\mathcal{S}$ is described by probability distributions. Assume a categorical or discrete random variable, for example, the number of wet days within a month. Then one can define a probability mass function

$$f_X(x) = P(X = x) \tag{6.4}$$

that gives the probability of occurrence for all possible (discrete) outcomes x. For continuous random variables, however, no probability mass function can be defined. For instance, the probability that precipitation intensity is exactly 5mm is zero. As an equivalent, one therefore defines the probability density function (PDF) $f_X(x)$ such that

$$P(a \leq X \leq b) = \int_a^b f_X(\xi)d\xi \tag{6.5}$$

A high PDF indicates a region in the event space where events are more probable to occur. For discrete and continuous random variables, one can define the (cumulative) probability distribution (CDF) as

$$F_X(x) = P(X \leq x) = \int_{-\infty}^{x} f_X(\xi)d\xi. \tag{6.6}$$

In case of a discrete random variable, the CDF $F_X(x)$ would be calculated as the sum over the probabilities $P(X = x)$ of all events x_i with $x_i \leq x$. The CDF measures the probability that an event x or a smaller event occurs. If the PDF $f_X(x)$ is continuous at x, it can be derived from the CDF as

$$f_X(x) = \frac{dF_X(x)}{dx}. \tag{6.7}$$

For examples of distributions, see Section 6.1.6.

Moments
Characteristic properties of a probability distribution can be quantified by so-called moments. The nth moment μ_n of a probability density $f_X(x)$ is defined as

$$\mu_n = E[X^n] = \int x^n \cdot f_X(x)dx, \tag{6.8}$$

and the nth central moment μ'_n as

$$\mu'_n = E[(X - \mu)^n] = \int (x - \mu)^n \cdot f_X(x)dx. \tag{6.9}$$

The operator $E(.)$ reads "expected value of". The first moment is the expected value or simply the mean of a distribution,

$$\mu = E[X] = \int x \cdot f(x) \, dx. \tag{6.10}$$

The variance is defined as the second central moment

$$\sigma^2 = \text{Var}(x) = E[(X - \mu)^2] = E[X^2] - \mu^2 \tag{6.11}$$

From the third and fourth central moments, one can derive further properties such as the skewness and kurtosis, a measure of peakness.

Quantiles
The α-quantile x_a of a probability distribution F_X is defined as the value x_α such that

$$F_X(x = x_\alpha) = P(x < x_\alpha) = \alpha. \tag{6.12}$$

For instance, the median of a distribution is the 50% quantile; the probability of not exceeding the median is 50%. Instead of referring to the α-quantile, one often also refers to the 100α percentile. High quantiles in the tail of a distribution are often referred to as return values, the associated probabilities are expressed as return periods (Figure 6.1, see also Section 6.1.6).

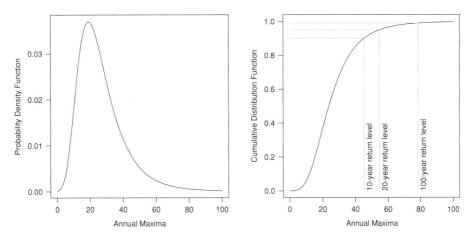

Figure 6.1 Return levels. Probability density function (left) and cumulative distribution function (right) of a GEV distribution, representing annual maxima of an arbitrary variable. The probabilities associated with specific quantiles can be expressed in terms of return periods.

6.1.4 Computations with Random Variables

In many situations, one is interested in functions of random variables, for example, in the calculation of confidence or prediction intervals (see Section 6.2). Here we list some basic rules.

Assume two independent (not necessarily Gaussian, see Section 6.1.6) random variables X and Y with expected values μ_X and μ_Y and variances σ_X^2 and σ_X^2. Then

- $Z = \alpha + X$ has expected value $\mu_Z = \alpha + \mu_X$ and variance $\sigma_Z^2 = \sigma_X^2$;
- $Z = \alpha X$ has expected value $\mu_Z = \alpha \mu_X$ and variance $\sigma_Z^2 = \alpha^2 \sigma_X^2$;
- $Z = X + Y$ has expected value $\mu_Z = \mu_X + \mu_Y$ and variance $\sigma_z^2 = \sigma_X^2 + \sigma_Y^2$.

6.1.5 Multivariate Random Variables and Probability Distributions

Consider a weather station at which both precipitation and temperature are measured. The occurrence of a certain precipitation intensity and temperature can then be described by a two-dimensional random variable. In general, a vector of scalar random variables $\mathbf{X} = (X_1, \dots, X_n)^T$ is called a multivariate random variable or a random vector. Other examples of multivariate random variables are, for example, sequences of random variables such as precipitation at subsequent days or spatial random fields such as precipitation at different sites.

The joint occurrence of two continuous events x and y in a two-dimensional subset $\mathcal{D}$ of the event space is characterised by the joint probability density function f_{XY} such that

$$P(XY \in \mathcal{D}) = \int_{\mathcal{D}} f_{XY}(x, y) dx\, dy. \tag{6.13}$$

Table 6.1 Joint occurrence of a weather type X (describing the main flow direction) and the occurrence of precipitation Y. Table (a) shows the joint probabilities and associated marginals, table (b) the conditional probabilities for precipitation occurrence Y, given that a particular weather type X has occurred.

(a)		X = west	X = north	X = east	X = south	$f_Y(Y)$
$f_{XY}(X, Y)$	Y = dry	0.25	0.15	0	0.2	0.6
	Y = wet	0.15	0.05	0.1	0.1	0.4
$f_X(X)$		0.4	0.2	0.1	0.3	1
(b)						
$f_{Y|X}(X|Y)$	Y = dry	0.625	0.75	0	0.67	
	Y = wet	0.375	0.25	1	0.33	

Correspondingly, the joint probability distribution

$$F_{XY}(x, y) = P(X \le x, Y \le y) \tag{6.14}$$

describes the probability that simultaneously $X \le x$ and $Y \le y$. Joint CDF and PDF are linked by

$$f_{XY}(x, y) = \frac{\partial^2 F_{XY}}{\partial x \partial y}. \tag{6.15}$$

When integrating the joint density function over all other variables, one obtains the marginal density functions associated with a specific variable

$$f_X(x) = \int f_{XY}(x, y)dy \tag{6.16}$$

$$f_Y(y) = \int f_{XY}(x, y)dx \tag{6.17}$$

These describe the occurrence of a variable regardless of which values the other variables assume. Equivalent to conditional probabilities, the conditional density functions are defined as

$$f_{Y|X}(y|x) = f_{XY}(x, y)/f_X(x) \tag{6.18}$$

$$f_{X|Y}(x|y) = f_{XY}(x, y)/f_Y(y) \tag{6.19}$$

and describe the occurrence of y given x and vice versa. Figure 6.2 illustrates joint and marginal densities.

As an example, consider the joint probability mass function that describes the joint occurrence of a certain weather type and the occurrence of precipitation (see Table 6.1). Two random variables X and Y are called independent if

$$f_{XY}(x, y) = f_X(x)f_Y(y), \tag{6.20}$$

that is, if the joint probability density function is simply the product of the marginal density functions. From this definition and Equations 6.18 and 6.19 it follows that in case of independence

$$f_{Y|X}(y|x) = f_Y(y), \quad f_{X|Y}(x|y) = f_X(x), \tag{6.21}$$

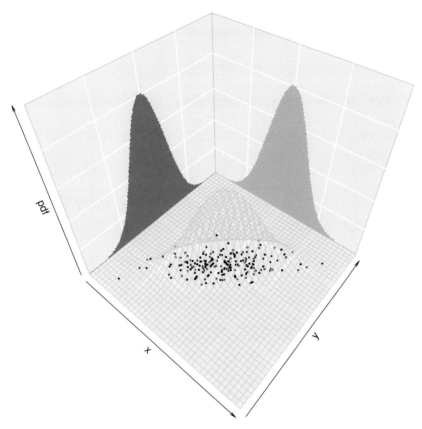

Figure 6.2 Multivariate normal distribution. Grey mesh: joint PDF; (dark) grey areas: marginal distributions; black dots: sample. Adapted from Maraun (2016).

that is, knowledge of X does not help to predict Y and vice versa. A sequence of random variables $X_1, \ldots, X_N$ is called independent and identically distributed (i.i.d.), if the X_i are mutually independent and follow the same probability distribution.

Moments of Multivariate Distributions
The m-dimensional expected value of a random vector $\mathbf{X}$ is given as

$$\mu = \mathrm{E}[\mathbf{X}] = \int_{\mathrm{R}^m} \mathbf{x} \cdot p(\mathbf{x}) \, d\mathbf{x}, \tag{6.22}$$

where the individual elements μ_i of μ are the expected values of the marginal distributions of $\mathbf{X}$. For multivariate distributions, mixed higher-order moments can be defined that include products of different variables. The most relevant is the covariance

$$\mathrm{Cov}(X, Y) = \mathrm{E}[(X - \mu_X)(Y - \mu_Y)] \tag{6.23}$$

that quantifies joint variations of two random variables. The (symmetric) covariance matrix Σ of a multidimensional random variable is constructed from the individual variances $\sigma_{i,i}^2$ along the main diagonal and the individual covariances $\sigma_{i,j}^2 = \sigma_{j,i}^2$ off the diagonals.

As the magnitude of a covariance is difficult to interpret, one defines the (Pearson) correlation by normalising the covariance by the individual variances

$$\text{Corr}(X, Y) = \frac{\text{Cov}(X, Y)}{\sqrt{\text{Var}(X)\text{Var}(Y)}}. \tag{6.24}$$

The correlation measures the strength of the linear relationship between X and Y and assumes values between -1 (perfect anti-correlation) and 1 (perfect correlation). Note that two independent variables have zero correlation, but a zero correlation does not necessarily imply independence.

Correlations are not a sensible measure to quantify the dependence of extreme events. For instance, the extremes of correlated variables do not necessarily exhibit dependence. Vice versa, overall only weakly correlated variables may have a strong tendency for their extremes to occur jointly. Therefore, specific measures have been designed to quantify the dependence of extremes. One such measure is the tail dependence. Given two random variables X and Y following marginal distributions with quantile functions (i.e. inverse cumulative distributions) $q_X(p)$ and $q_Y(p)$, upper-tail dependence is defined as

$$\lambda_u = \lim_{p \to 1} P(Y > q_Y(p) | X > q_x(p)). \tag{6.25}$$

Similarly, a measure for lower-tail dependence can be constructed as

$$\lambda_l = \lim_{p \to 0} P(Y \leq q_Y(p) | X \leq q_x(p)). \tag{6.26}$$

The upper-tail dependence, for example, measures the probability that one variable exceeds a high threshold (associated with the probability p), given that the other variable also exceeds a high threshold (associated with the same probability p), for $p \to 1$. In practical applications, one will of course estimate the tail dependence based on values of p smaller than 1. This estimate will always, also for independence, assume finite values of $(1 - p)^2$. For example, for $p = 0.95$, one will trivially find an estimate of tail dependence of at least 0.25%, which is simply the probability that two independent variables at the same time exceed the 95th percentile.

6.1.6   Examples of Distributions

In the following, a selection of probability distributions will be given that will be referred to in the later sections.

Bernoulli Distribution
The Bernoulli distribution describes the so-called Bernoulli trial, a random experiment with only two possible outcomes:

$$P(X = 0) = p; \qquad P(X = 1) = 1 - p. \tag{6.27}$$

The mean of the Bernoulli distribution is given by $\mu = p$, the variance by $\sigma^2 = p(1 - p)$. A typical example of the Bernoulli distribution is the occurrence of precipitation.

Binomial Distribution

Assume you repeat a Bernoulli trial n times. The Binomial distribution $B(k|p, n)$ gives the probability of having k times 1 (called successes) for n trials. It is defined as

$$B(X = k|p, n) = \begin{cases} \binom{n}{k} p^k (1 - p)^{n-k} & \text{for } 0 \leq k \leq n \\ 0 & \text{otherwise.} \end{cases} \quad (6.28)$$

The mean of the Binomial distribution is given by $\mu = np$, the variance by $\sigma^2 = np(1 - p)$. The length of spells, if one can assume independence between individual events, can be described by the binomial distribution. For high n, where the event space can be assumed to be quasi-continuous, the binomial distribution converges to a normal distribution.

Poisson Distribution

For a low probability of success and $n \to \infty$, the binomial distribution converges to a Poisson distribution. This approximation holds already reasonably well for $p \leq 0.05$ and $n \geq 50$. The Poisson distribution is defined as

$$P_\lambda(X = k) = \frac{\lambda^k}{k!} \exp^{-\lambda} \quad (6.29)$$

with $\lambda = np$. Both mean and variance of the Poisson distribution are given by $\mu = \sigma^2 = \lambda$. A typical application of the Poisson distribution is to model the number of exceedances over a high threshold. For large λ ($\lambda \geq 30$), the Poisson distribution can be approximated by a normal distribution with $\mu = \lambda$ and $\sigma^2 = \lambda$.

Uniform Distribution

The uniform distribution $U(a, b)$ has equal density between two values a and b and zero density outside this interval. Its density function is defined as

$$f_X(X) = \begin{cases} \frac{1}{b-a} & \text{for } x \in [a, b] \\ 0 & \text{otherwise.} \end{cases} \quad (6.30)$$

The uniform distribution plays an important role in modelling occurrence processes but also in the use of copulas to model multivariate distributions (Section 6.1.7).

Normal or Gaussian Distribution

The normal or Gaussian distribution plays a central role in statistical modelling. As a result of the central limit theorem, many processes in nature approximately follow a normal distribution. The PDF of the normal distribution with mean μ and variance σ^2 is defined as

$$f_X(x; \mu, \sigma) = \frac{1}{\sqrt{2\pi}\sigma} \exp\left(-\frac{(x - \mu)^2}{2\sigma^2}\right) = \mathcal{N}(\mu, \sigma^2). \quad (6.31)$$

The normal distribution is fully specified by the first and second moment, and all other moments are zero. Sixty-eight percent of the probability mass lies between $\mu \pm \sigma$, 95% between $\mu \pm 1.96 \cdot \sigma$.

The joint PDF of the d-dimensional multivariate normal distribution $\mathbf{MVN}_d(\boldsymbol{\mu}, \boldsymbol{\Sigma})$ with the d-dimensional vector of expected values $\boldsymbol{\mu}$ and $d \times d$ covariance matrix $\boldsymbol{\Sigma}$ is defined as

$$f_{\mathbf{X}}(\mathbf{X} = \mathbf{x}) = \frac{1}{(2\pi)^{k/2}\det(\boldsymbol{\Sigma})^{1/2}}\exp\left(-\frac{1}{2}(\mathbf{x} - \boldsymbol{\mu})^T \boldsymbol{\Sigma}^{-1}(\mathbf{x} - \boldsymbol{\mu})\right). \tag{6.32}$$

Exponential Distribution
The exponential distribution is a skewed distribution that describes the waiting time between events in a Poisson process (Section 6.4.2). The PDF is defined as

$$f_\lambda(r) = \begin{cases} \lambda e^{-\lambda r} & r \leq 0, \\ 0 & r < 0, \end{cases} \tag{6.33}$$

with rate parameter λ. The mean of the exponential distribution is given by $\mu = 1/\lambda$, the variance by $1/\lambda^2$. See Figure 6.3 for example densities and Figure 6.4 for example realisations.

Gamma Distribution
The gamma distribution is a skewed distribution that is often used to model daily precipitation intensities. It can be parameterised in different ways. One possibility is to define the PDF as

$$f_{\lambda,\gamma}(r) = \frac{\lambda^\gamma}{\Gamma(\gamma)}r^{\gamma-1}e^{-\lambda r}, \qquad \lambda, \gamma > 0, \tag{6.34}$$

with shape parameter γ and rate parameter λ. The mean of the gamma distribution is given by $\mu = \gamma\lambda$, the variance by $\sigma^2 = \gamma\lambda^2$. See Figure 6.3 for example densities.

Extreme Value Distributions
Statistical modelling of extreme events is based on extreme value theory. The key aim is to derive a parametric distribution from high observational values that allows one to specify high return periods as accurately as possible and to extrapolate beyond the range of observed values. Basically two approaches exist that differ in their choice of high values; both approaches, however, can be transferred into each other. For a detailed introduction to extreme value theory refer, for example, to Coles (2001).

The first approach is the so-called block maxima approach that models the magnitude of the highest value in a long sequence. Assume a sequence (commonly referred to as block) of n i.i.d. random variables, $X_1, \ldots, X_n$ with unknown distribution. The maximum of this sequence is given as

$$M_n = \max\{X_1, \ldots, X_n\}. \tag{6.35}$$

A key result from extreme value theory is the so-called extremal types or Fisher-Tippet theorem. It states that, no matter what the distribution of the X_t, for long blocks, the block maxima follow a specific family of distributions. More specifically, if the probability distribution of the (properly rescaled maximum) converges for increasing block

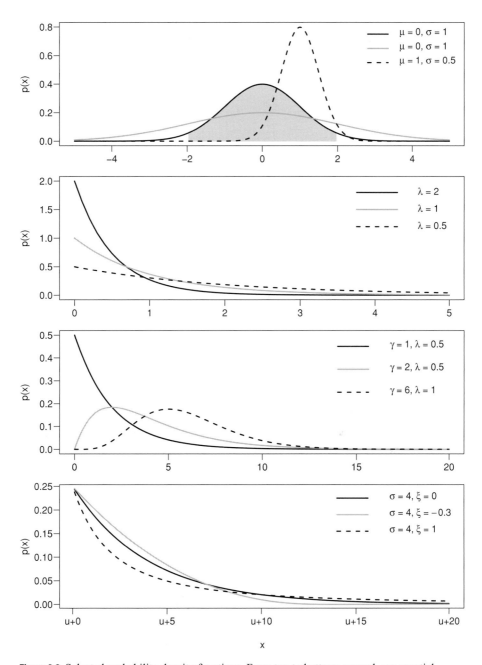

Figure 6.3 Selected probability density functions. From top to bottom: normal, exponential, gamma, generalised Pareto. The grey shading in the top panel marks 95% of the probability mass around the mean. The generalised Pareto distribution with $\gamma = 0$ is an exponential distribution.

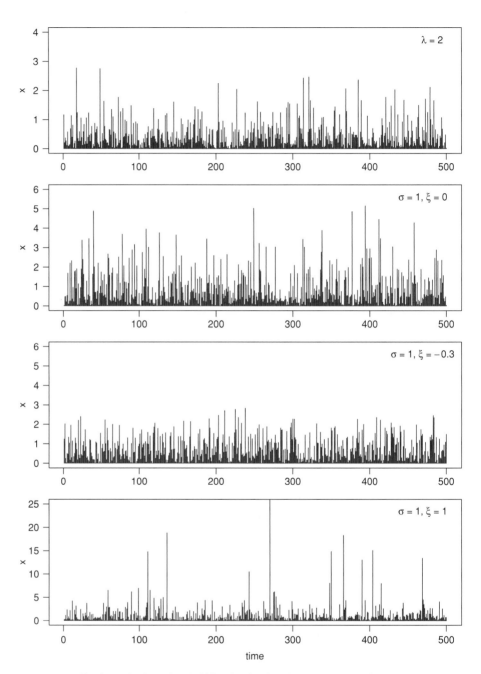

Figure 6.4 Realisations of selected probability density functions. From top to bottom: exponential, generalised Pareto with three different parameter sets. The generalised Pareto distribution with $\gamma = 0$ is an exponential distribution. Note the different y-axis range in the bottom panel.

length ($n \to \infty$) to a limiting distribution $G(z)$, then $G(z)$ belongs to the family of generalised extreme value (GEV) distributions. The cumulative distribution function is given by

$$G(z; \mu, \sigma, \xi) = \exp\left\{-\left[1 + \xi\left(\frac{z - \mu}{\sigma}\right)\right]^{-1/\xi}\right\} \qquad (6.36)$$

defined on $\{z : 1 + \xi(z - \mu)/\sigma > 0\}$. The location parameter μ determines the position, the scale parameter σ (with $\sigma > 0$) the width and the shape parameter ξ the decay of the distribution for large values of z. Three cases are possible, leading to three different distributions: for $\xi < 0$, the tail has a finite upper value (Weibull distribution); for $\xi > 0$, the tail is heavy with a power law decay (Fréchet distribution). In the limit $\xi \to 0$, one obtains a Gumbel distribution with an exponential decay, that is, a light tail (Embrechts et al. 1997, Coles 2001).

The mean of the GEV distribution is given by $\mu + \sigma(\Gamma(1 - \xi) - 1)/\xi$ for $\xi \neq 0$ and $\xi < 1$, the variance by $\sigma^2(\Gamma(1 - 2\xi) - \Gamma(1 - \xi)^2)/\xi^2$ for $\xi \neq 0$ and $\xi < 1/2$. In the Gumbel case ($\xi = 0$), the mean is $\mu + \sigma\gamma$ (with Euler's constant $\gamma \approx 0.577$), the variance $\sigma^2 \frac{\pi^2}{6}$.

The i.i.d. condition can be relaxed such that the Fisher-Tippet theorem holds also for a wide class of stationary but not necessarily independent stochastic processes (Leadbetter et al. 1983). The practical consequence of the Fisher-Tippet theorem is that for long enough blocks, one does not need to care about choosing an appropriate distribution – the GEV is the only possible. To apply the block maxima approach, one has to carefully choose a blocklength that is sufficiently long to justify the choice of the GEV but at the same time as short as possible to use as many data points as possible. A natural choice is annual blocks, but other lengths might be possible or even necessary.

The second approach is the peaks-over-threshold approach, which considers threshold exceedances. The counterpart of the Fisher-Tippet theorem for threshold exceedances is the Pickands-Balkema-de Haan theorem: consider again a sequence of i.i.d. random numbers $X_1, \ldots, X_n$. If the Fisher-Tippet theorem holds for this sequence, then the excess $X - u$ over a large threshold $u \to \infty$ is given by the generalised Pareto (GP) distribution distribution with cumulative distribution function

$$P(X - u < y | X > u) \approx GP(y) = 1 - \left[1 + \frac{\xi y}{\tilde{\sigma}}\right]^{-1/\xi}. \qquad (6.37)$$

The scale parameter $\tilde{\sigma}$ of the peaks-over-threshold approach is related to the scale parameter of the block maxima approach by $\tilde{\sigma} = \sigma + \xi(u - \mu)$. The mean of the GP distribution is given by $\mu + \sigma/(1 - \xi)$ for $\xi < 1$, the variance by $\sigma^2 = \sigma^2/[(1 - \xi)^2(1 - 2\xi)]$ for $\xi < 1/2$. See Figure 6.3 for example densities and Figure 6.4 for example realisations.

It is often argued that the peaks-over-threshold approach makes better use of the data, as the block maxima approach disregards all values within a block but the maximum. A disadvantage, however, is the potential clustering of data that has to be taken into account either by de-clustering (Coles 2001) or by appropriately adjusting the uncertainty estimates of the GP parameters (Smith 1990, Fawcett and Walshaw 2007).

Equivalently to choosing a suitable block length, the peaks-over-threshold approach requires the selection of an appropriate threshold that is high enough to apply the GP distribution but low enough to use as many data points as possible.

From the GEV distribution, one can nicely derive return levels. Consider an annual block length, that is, the GEV describes the magnitude of the highest value within a year. Then the probability p that a maximum larger than a certain value z_P occurs within a year is given by $P(z > z_p) = GEV(z_p) = 1 - p$. The value z_P is called return value, the associated return period is $T_p = 1/p$ years. Thus, within a year, the return value z_p is exceeded with a probability of p; on average, z_p is exceeded once every T_p years. Equivalent calculations are, of course, also possible for the peaks-over-threshold approach.

6.1.7   Copulas

Often it is desired to statistically model the dependence between different variables or variables at different sites. Parametric multivariate distributions are essentially confined to the multivariate normal distribution Equation 6.32. Many meteorological variables such as precipitation, wind or humidity, however, are non-normal. In some situations, transformations of the multivariate normal distribution might provide a solution. For instance, precipitation at multiple sites has successfully been modelled as the third root of a multivariate normal (Yang et al. 2005). Such transformations, however, are not straightforward if the dependence of different variables is sought after, for example, of temperature and precipitation.

Copulas have been proposed as a powerful tool to parametrically model the dependence between different variables with potentially different, non-normal marginal distributions (Figure 6.5). For brief introductions, see Schölzel and Friederichs (2008) and Genest and Favre (2007). The underlying idea is to separately model margins and dependence. Assume we have d variables $X_1, \ldots, X_d$ with different marginal cumulative distribution functions $F_i(x_i)$. According to Sklar's theorem (Sklar 1959), any d-variate cumulative distribution function $F_{1\ldots d}(x_1, \ldots, x_d)$ can be expressed as

$$F_{1\ldots d}(x_1, \ldots, x_d) = C_{1\ldots d}(F_1(x_1), \ldots, F_d(x_d)). \tag{6.38}$$

Note that the cumulative distribution functions $F_i(x_i)$ assume values on $[0, 1]$, that is, they transform the marginal distributions of the variables X_i to variables with univariate margins $U_i = F_i(X_i)$ (see Equation 6.30). The function $C_{1\ldots d}$ is thus simply the multivariate cumulative distribution function of marginally uniform variables $U_i(0, 1)$. It is called copula. The corresponding density $f_{1\ldots d}(x_1, \ldots, x_d)$ is given as

$$f_{1\ldots d}(x_1, \ldots, x_d) = f_1(x_1) \cdots f_d(x_d) c_{1\ldots d}(u_1, \ldots, u_d) \tag{6.39}$$

where $f_i(x_i)$ are the marginal densities of the individual variables and $c_{1\ldots d}$ is the copula density corresponding to $C_{1\ldots d}$. Equation 6.39 illustrates how any multivariate density can be decomposed into its marginals and a copula. A wide range of parametric copulas has been defined. They differ in particular in their behaviour of tail dependence in the lower and upper tails. Many copulas are still limited to three dimensions. The few models available for higher dimensions often still lack the flexibility to individually model

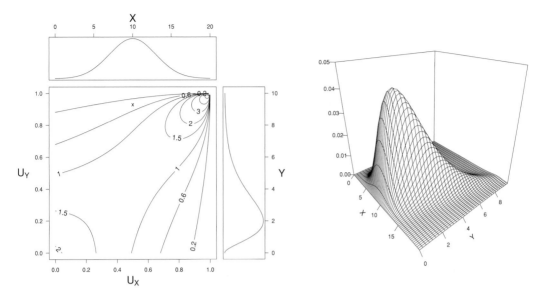

Figure 6.5 Bivariate probability density function modelled with a copula. Left: density with Gaussian and gamma margins, linked via a Survival Clayton copula. Right: resulting bivariate density (courtesy Emanuele Bevacqua).

the dependence between different components. As a way forward, pair-copula constructions have been developed, which decompose a d-dimensional copula into bivariate copulas, some of which are conditional (Joe 1996). For a recent application in regional climate modelling, see Hobaek Haff et al. (2015).

6.2   Parameter Estimation

Statistical models are used to draw conclusions about a a system that is subject to random variations from data generated by the system. For instance, a regression model may be used to understand the link between the large-scale circulation and local precipitation. One would collect a sample of data, adjust the model to the data, and study the behaviour of the model or use it for predictions. The adjustment procedure comprises two issues: first, selecting a suitable model, that is, selecting a suitable regression model and useful predictors. And second, parameter estimation, that is, determining the model parameters for example, (e.g. the intercept and the regression slope) that optimally describe the given data. Along with the parameter estimation, one would aim to quantify the uncertainties of these estimates due to finite sample sizes, such as by confidence intervals. Model selection will be discussed in conjunction with regression models in Section 6.3.

Estimators are often based on the maximum likelihood approach, but some state-of-the-art downscaling methods use Bayesian estimators. We will therefore focus on the maximum likelihood approach in general but provide a brief sketch of Bayesian inference as well. Prior, however, we will give a brief introduction and present some standard estimators. In the statistical downscaling literature, parameter estimation is often called calibration. We will use this term in Part II of the book.

Two well-known estimators are the sample mean and variance. Given a data set x_i, $i = 1 \ldots N$, the sample mean is defined as

$$\hat{\mu} = \bar{x} = \frac{1}{N} \sum_{i=1}^{N} x_i. \tag{6.40}$$

Similarly, the sample variance, the squared standard deviation, is given as

$$\hat{\sigma}^2 = s^2 = \frac{1}{N-1} \sum_{i=1}^{N} (x_i - \bar{x})^2. \tag{6.41}$$

The $\hat{\ }$ indicates that the calculated quantities are estimators, namely of the first and second moment of an assumed distribution. Similarly, estimators of higher moments such as skewness and kurtosis can be defined. For two samples x_i and y_i, an estimator of the (Pearson) correlation is given as

$$r = \frac{\sum_{i=1}^{N} (x_i - \bar{x})(y_i - \bar{y})}{\sqrt{\sum_{i=1}^{N} (x_i - \bar{x})^2} \sqrt{\sum_{i=1}^{N} (y_i - \bar{y})^2}}. \tag{6.42}$$

The Pearson correlation is often not a useful measure of dependence for highly non-normal distributions such as precipitation. In such cases it is better to use the rank-based Spearman correlation coefficient, which is calculated by replacing the actual values x_i and y_i by their ranks.

In the context of parameter estimation, one assumes that the sample x_i is drawn from a random variable X. The estimator μ is a function of X and thus itself a random variable. In fact, any estimator $\hat{\theta}$ for a parameter θ has a distribution with an expected value $E[\hat{\theta}]$ and a variance $VAR[\hat{\theta}]$. The difference between the expected value of the estimator and the true parameter is called bias,

$$B(\hat{\theta}) = E[\hat{\theta}] - \theta. \tag{6.43}$$

Thus, a bias measures a systematic difference between the estimation and the truth.

The assumption that the x_i are sampled from a random variable is crucial to quantify the sampling uncertainty of, say, the estimation of $\hat{\mu}$. Assume, for instance, that the x_i are realisations of a normally distributed random variable X with mean μ and variance σ. The estimator of the mean $\hat{\mu}$ is a linear function of X, and is thus itself a normally distributed random variable. According to the rules given in Section 6.1.4, it has a mean μ and variance $VAR[\hat{\mu}] = 1/N^2 \cdot N\sigma^2 = 1/N\sigma^2$. This distribution describes how the sample mean is expected to vary for new samples of data. The larger the sample size N, the smaller the variance of $\hat{\mu}$, that is, the smaller the sampling uncertainty. Sampling uncertainties are typically expressed by confidence intervals. A $1 - \alpha$ confidence interval spans a range around the estimated parameter such that – if multiple samples were available – for on average $100 \times (1 - \alpha)\%$ of these samples, the true value would lie within the confidence intervals (which are different for each sample). For instance, a 95% confidence interval would cover the true value in 95% of the cases. Confidence intervals for maximum likelihood estimates will be discussed in detail in the next section.

Uncertainty estimates rely on the assumed probability model. Often the tacit assumption for the probability model is that the data is sampled from a normal distribution. If the data is not well described by the chosen model, the uncertainty estimates will in general be unreliable and sometimes too narrow.

6.2.1 Maximum Likelihood Estimation

A very powerful concept in frequentist inference is the maximum likelihood approach (for details see, e.g. Cox and Hinkley 1994, Wilks 2006). Assume that a series of temperature measurements x_i, $i = 1, \ldots, N$ is sampled from a random variable X. Associated with X is a probability density function $f_X(x; \boldsymbol{\theta})$ with unknown parameters $\boldsymbol{\theta} = (\theta_1, \ldots, \theta_K)$. For instance, one could assume that the distribution is normal with parameters $\boldsymbol{\theta} = (\mu, \sigma)$.

One way to estimate the parameters of the distribution would be to simply calculate sample mean and sample variance according to Equations 6.40 and 6.41. Such moment-based estimators can also be defined for non-normal distributions. They link parameters of the distribution to moment estimators. Moment-based estimators, however, cannot easily be used in regression models. A popular estimator in linear regression models is the well-known least-squares estimator. It minimises the squared residuals between observations and statistical model predictions.

A widely used approach with a well-established theory is the maximum likelihood approach. In fact, we will show that the least-squares estimator is a maximum likelihood estimator if the data is sampled from a normal distribution. In the maximum likelihood approach the "best estimate" is defined as those parameters maximising the probability that the observed data would occur. Formally, the likelihood of the parameter vector $\boldsymbol{\theta}$, given a set of observations x_i, is defined as

$$L(\boldsymbol{\theta}|x_1, \ldots, x_N) = \prod_{i=1}^{N} f_X(x_i; \boldsymbol{\theta}). \tag{6.44}$$

Often it is more convenient to consider the log-likelihood (because exponentials, e.g. in the normal distribution, vanish, and products are transformed into sums)

$$\ell(\boldsymbol{\theta}|x_1, \ldots x_N) = \sum_{i=1}^{N} \log f_X(x_i; \boldsymbol{\theta}). \tag{6.45}$$

As the logarithm is a monotonous function, maxima of the log-likelihood coincide with maxima of the likelihood. The maximum likelihood estimator (MLE) is defined as the parameter vector that maximises the (log-)likelihood,

$$\hat{\boldsymbol{\theta}}_{MLE} = \arg(\max_{\boldsymbol{\theta}} \ell(\boldsymbol{\theta}|x_1, \ldots, x_N)). \tag{6.46}$$

It is calculated by finding the roots of the partial derivatives of L or ℓ with respect to the parameters θ_k,

$$\frac{\partial \ell(\boldsymbol{\theta}|x_1,\ldots,x_N)}{\partial \theta_k} \overset{!}{=} 0. \tag{6.47}$$

To illustrate the concept, let us consider a sample x_i of size N, sampled from a normal distribution with unknown parameters μ and σ. The likelihood of some parameters μ and σ, given the sample, is then defined as

$$L(\mu,\sigma|x_1,\ldots x_N) = \prod_{i=1}^{N} \frac{1}{\sqrt{2\pi}\sigma} e^{-\frac{(x_i-\mu)^2}{2\sigma^2}}, \tag{6.48}$$

the log-likelihood as

$$\ell(\mu,\sigma|x_1,\ldots x_N) = -\frac{N}{2}\log(2\pi) - N\log\sigma - \frac{1}{2\sigma^2}\sum_{i=1}^{N}(x_i-\mu)^2. \tag{6.49}$$

The maximum likelihood estimate of μ is then given as

$$\frac{\partial \ell(\mu,\sigma|x_1,\ldots x_n)}{\partial \mu} = -\frac{1}{\sigma^2}\sum_{i=1}^{N}(x_i-\mu) \overset{!}{=} 0. \tag{6.50}$$

Apart from a factor (which can be eliminated by division), this is the least-squares estimator. Rewriting gives $\sum_{i=1}^{n} x_i - N\mu = 0$, or the well-known sample mean $\hat{\mu}_{MLE} = 1/N \sum_{i=1}^{n} x_i$. Similarly, the maximum likelihood for σ is given as

$$\frac{\partial \ell(\mu,\sigma|x_1,\ldots x_N)}{\partial \sigma} = -\frac{N}{\sigma} - \frac{1}{\sigma^3}\sum_{i=1}^{N}(x_i-\mu)^2 \overset{!}{=} 0. \tag{6.51}$$

Solving for σ and inserting $\hat{\mu}_{MLE}$ for μ gives the maximum likelihood estimator

$$\hat{\sigma}^2_{MLE} = \frac{1}{N}\sum_{i=1}^{n}(x_i-\hat{\mu}_{MLE})^2. \tag{6.52}$$

Note that this estimator is different from the sample variance Equation 6.41. In fact, it is a biased estimator for small N.

Some very convenient properties often make the MLE a good choice:

- The MLE can be applied for any distribution (given one has a mathematical model of the density), not only to normal distributions.
- The MLE is asymptotically consistent, that is, for $n \to \infty$ it converges in probability to the true value.
- The MLE is asymptotically efficient, that is, for $n \to \infty$, no unbiased estimator can have a smaller variance.
- The distribution of the MLE is asymptotically normal. This property can be used to derive approximate confidence intervals.
- MLE allows for an easy inclusion of predictors.

In case of a normal distribution, the solution could be derived analytically. In many cases, however, numerical optimisation algorithms are required.

Confidence Intervals

The curvature of the (log-)likelihood function around the MLE provides a measure of the uncertainty of the estimate: if the curvature is weak, the likelihood for different parameter values changes only slowly. These parameter values are thus almost equally likely, resulting in a considerable uncertainty about the parameter values. Yet if the curvature is strong, the likelihood will drop fast for slight changes in parameters. Other parameter values are thus very unlikely, resulting in a very low uncertainty. The $K \times K$ matrix $\mathbf{I}(\theta)$ of curvatures of the log-likelihood with entries $e_{k,l}(\theta) = E\{-\frac{\partial^2}{\partial\theta_k \partial\theta_l}\ell(\theta)\}$,

$$\mathbf{I}(\theta) = \begin{pmatrix} e_{1,1}(\theta) & \cdots & & e_{1,K}(\theta) \\ \vdots & \ddots & e_{k,l}(\theta) & \vdots \\ & e_{l,k}(\theta) & \ddots & \\ e_{K,1}(\theta) & \cdots & & e_{K,K}(\theta) \end{pmatrix},$$

is called the information matrix. For $n \to \infty$, that is, a sufficiently large sample size, the distribution of the K dimensional MLE $\hat{\theta}_{MLE}$ converges to a multivariate normal distribution

$$\hat{\theta}_{MLE} \overset{\cdot}{\sim} \text{MVN}_K(\theta_{MLE}, \mathbf{I}(\theta_{MLE})^{-1}) \tag{6.53}$$

with mean θ_{MLE} and a covariance matrix $\mathbf{I}(\theta)^{-1}$ given by the inverse of the information matrix. The estimate of $\mathbf{I}(\theta)$ is usually approximated during the (often numerical) optimisation procedure.

We denote the entries of the inverted estimated information matrix as $\hat{w}_{k,l}$. The elements in the main diagonal are the squared standard errors SE_k of the estimated parameters, the off-diagonal elements are the corresponding covariances. From these, an approximate $(1 - \alpha)$ confidence interval for $\hat{\theta}_k$ can be calculated

$$\hat{\theta}_k \pm z_{\alpha/2}\sqrt{\hat{w}_{k,k}} = \hat{\theta}_k \pm z_{\alpha/2} \cdot SE_k \tag{6.54}$$

where $z_{\alpha/2}$ is the $(1 - \alpha/2)$ quantile of the standard normal distribution. For instance, with $\alpha = 0.05$ ($z_{\alpha/2} = 1.96$), one could calculate a 95% confidence interval as $\hat{\theta}_k \pm 1.96 \cdot SE_k$. Additionally, the off diagonals of the inverse information matrix provide useful information about correlations between the parameter estimates. The normality assumption holds for many practical applications. But in case of inference for extremal properties, it is preferable to use more precise calculations based on the profile likelihood (e.g. Coles 2001).

6.2.2   Distribution Diagnostics

Once a distribution model has been calibrated to a data set, it should be investigated whether the model adequately describes the data. Depending on which aspects of the data are important, different diagnostics might be useful.

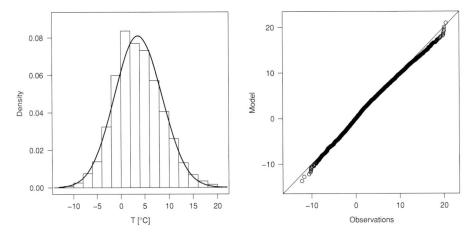

Figure 6.6 Diagnostic plots. Normal distribution used to model daily winter temperature in Graz, Austria, 1979–2008. Left: histogram and density plot. Right: quantile-quantile (QQ) plot. The model does not fully capture the amplitudes of extreme temperatures. Data from ECA-D (Klein Tank et al. 2002).

A first overall comparison between model and observation may be given by comparing the histogram of observations with the modelled density (Figure 6.6, left). The disadvantage of such plots is that they aggregate the observations into distinct bins, and information within a bin is removed. An alternative would be comparing the empirical cumulative distribution function with the modelled cumulative distribution function. A related but even better visualisation is quantile-quantile (QQ) plots (Figure 6.6, right). These plots display modelled quantiles against the corresponding empirical quantiles – the sorted values. Other diagnostics are, for example, probability plots and, in particular for extremes, return-value plots.

6.2.3   Bayesian Estimators

Bayesian statistics relies on some assumptions which are fundamentally different from frequentist statistics. Probability is not defined as the limit of relative frequency for infinite sample sizes but as a measure of plausability – not necessarily based on empirical evidence. Yet maybe even more importantly, Bayesian statistics does not assume a true (and thus fixed) underlying parameter. Instead, a parameter is assumed to be a random variable associated with a distribution. The key benefit of Bayesian inference is to incorporate prior knowledge into the estimation procedure. Whereas frequentist statistics in general relies only on empirical evidence, Bayesian estimators combine knowledge from data with prior information such as expert knowledge.[1]

Mathematically, Bayesian inference starts from the Bayes theorem (Eq. 6.3). Given a sample of data $x_1, \ldots, x_N$ and a statistical model with parameters θ to describe the

---

[1] In fact, one can constrain frequentist estimators, which is similar to including prior information.

data. Then a Bayes estimator for the model parameters given a priori knowledge and the observations is given by the parameters' posterior distribution

$$f(\boldsymbol{\theta}|x_1, \ldots, x_N, \alpha) = \frac{f(x_1, \ldots, x_N|\boldsymbol{\theta})f(\boldsymbol{\theta}|\alpha)}{f(x_1, \ldots, x_N|\alpha)} \sim f(x_1, \ldots, x_N|\boldsymbol{\theta})f(\boldsymbol{\theta}|\alpha). \qquad (6.55)$$

The prior distribution $f(\boldsymbol{\theta}|\alpha)$ is key to Bayesian inference; it enables one to integrate additional knowledge into the estimation. This knowledge could, for example, stem from prior experiments, theoretical considerations or other expert knowledge. The prior distribution is parameterised with so-called hyper-parameters α. The likelihood $f(x_1, \ldots, x_N|\boldsymbol{\theta})$ describes the probability that the observed data would occur, given a vector of parameters $\boldsymbol{\theta}$. Thus, it is essentially the likelihood Equation 6.44. The marginal density that the data occur, regardless of the chosen model, is given as $f(x_1, \ldots, x_N|\alpha) = \int_{\boldsymbol{\theta}} f(x_1, \ldots, x_N|\boldsymbol{\theta})p(\boldsymbol{\theta}|\alpha)d\boldsymbol{\theta}$. It is basically a normalisation constant. For a uniform prior with infinite support, the posterior distribution becomes proportional to the likelihood only. Thus, the maximum likelihood estimator is identical to the most probable Bayesian estimator if no prior information is assumed.

For a long time, a major practical limitation to Bayesian inference was the fact that, in many cases, no analytical solution for the posterior probability density function could be derived. With modern computers, the probability density function is approximated numerically by the so-called Monte-Carlo Markov chain (MCMC) approaches (Gilks et al. 1995).

The Bayesian equivalent to confidence intervals is credible intervals. They can directly be calculated from the posterior distribution by choosing a certain interval. Note the different assumptions on the parameter $\boldsymbol{\theta}$: from a frequentist perspective, a fixed true value exists, and a confidence interval will encompass it in a defined number of times. From a Bayesian perspective, $\boldsymbol{\theta}$ is a random variable and will be inside the credible interval with a defined probability.

6.3 Regression Models

Many statistical downscaling methods are build upon some sort of regression model. Such models relate an input or predictor variable x via some function $f(x, \beta)$ to an output or predictand variable y,

$$y \approx f(x, \beta). \qquad (6.56)$$

Here β is a parameter that is estimated based on pairs of data for x and y. In the following, we introduce the regression models most commonly used in statistical downscaling along with the necessary concepts to develop and test regression models. For an excellent reference, see Davison (2003).

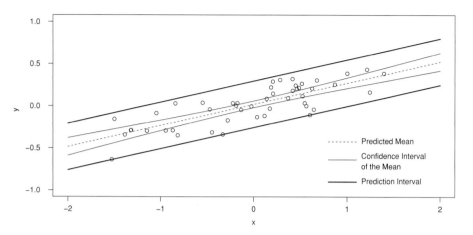

Figure 6.7 Linear regression. Shown are the original data (dots), the predicted mean (dashed line), a 95% confidence interval of the predicted mean (thin solid line) and a 95% prediction interval for possible new data (bold solid line).

6.3.1   Linear Models

The most basic regression models are linear (regression) models (see Figure 6.7). Given N pairs of data (x_i, y_i), the simplest linear model that uses the x_i to predict the y_i is defined as

$$y_i = \beta_0 + \beta_1 x_i + \epsilon_i. \tag{6.57}$$

The relationship between predictors and predictands is thus modelled as a straight line defined by the intercept β_0 and the slope β_1. The residuals between the model prediction and the actual observed y_i are assumed to follow normal distributed noise $\epsilon_i \sim \mathcal{N}(0, \sigma^2)$.

The linear model (Eq. 6.57) can easily be formulated for a vector of M predictors $\mathbf{x}_i^T = (x_{i1}, \ldots, x_{iM})$ and an M-dimensional parameter vector $\beta = (\beta_1, \ldots, \beta_M)$:

$$y_i = \beta_1 x_{i1} + \ldots + \beta_M x_{iM} + \epsilon_i = \mathbf{x}_i^T \beta + \epsilon_i. \tag{6.58}$$

This is the general form of a linear model and is sometimes referred to as multiple linear regression. To keep the notation as concise as possible, the offset in this general formulation is – different to the model Equation 6.57 – captured by β_1; the time series x_{i1} thus consists only of ones. For a concise notation of confidence and prediction intervals, the entries of the predictor vector over all time steps can be written in matrix notation, $\mathbf{X} = (\mathbf{x}_1, \ldots, \mathbf{x}_T)^T$.

Note that Equation 6.57 is not called linear because it links predictors and predictands via a straight line but rather because it is linear in the parameters β_0 and β_1. It is a simple exercise to construct models that are still linear but do not link predictors and predictands via a straight line. During the estimation procedure, the x_i are mere numbers that can be transformed arbitrarily. For instance, one can link x_i^2 and y_i and thus easily construct a squared relationship. To emphasise this point, some authors refer to the model Equation 6.57 as straight-line regression (e.g. Davison 2003).

The parameters of the model can be estimated by the well-known least-squares estimator, that is, by minimising the sum of squared residuals. Since the noise ϵ_i is assumed to be Gaussian, this is a maximum likelihood estimator (see Section 6.2.1) with log likelihood

$$\ell(\beta, \sigma^2) = -\frac{1}{2}\left[n \log \sigma^2 + \frac{1}{\sigma^2}\sum_{i=1}^{T}(y_i - \mathbf{x}_i^T\beta)^2\right] \tag{6.59}$$

The minimum can be found by setting the partial derivatives of $\ell(\beta, \sigma^2)$ with respect to all β_m and – if desired – σ to zero,

$$\frac{\partial \ell(\beta, \sigma^2)}{\partial \beta_m} = 2\sum_{i=1}^{N} x_{im}(y_i - \mathbf{x}_i^T\beta) = 0, \quad m = 1, \ldots, M. \tag{6.60}$$

The resulting maximum likelihood estimate for the parameter vector β reads

$$\hat{\beta} = \frac{\sum_i \mathbf{x}_i^T y_i}{\sum_i \mathbf{x}_i \mathbf{x}_i^T} \tag{6.61}$$

It can further be shown (e.g. Davison 2003) that an unbiased estimator for the variance of the residuals σ^2 can be calculated from the squared residuals between predicted mean and observations,

$$S^2 = \frac{1}{N - M}\sum_{i=1}^{N}(y_i - \mathbf{x}_i^T\hat{\beta})^2. \tag{6.62}$$

Confidence Intervals

Confidence intervals for β can be derived under the assumption of asymptotic normality of the estimator given by Equation 6.61. The estimator is unbiased, so the mean of the distribution of $\hat{\beta}$ is β. The variance of $\hat{\beta}$ can also be calculated from Equation 6.61; with the matrix notation of the predictor vector in time, $\mathbf{X} = (\mathbf{x}_1, \ldots, \mathbf{x}_T)^T$, the variance of $\hat{\beta}$ results as $\sigma^2(\mathbf{X}^T\mathbf{X})^{-1}$. Thus, the mth component of β is distributed according to

$$\hat{\beta}_m \sim \mathcal{N}(\beta_m, \sigma^2 v_{mm}). \tag{6.63}$$

where v_{mm} is the mth diagonal element of $(\mathbf{X}^T\mathbf{X})^{-1}$. The true residual variance σ^2 is of course not known but only estimated by S^2. The estimator, standardised with the estimated mean and variance

$$T = \frac{\hat{\beta}_m - \beta_m}{\sqrt{S^2 v_{mm}}} \sim t_{N-M} \tag{6.64}$$

follows a t distribution with $N - M$ degrees of freedom. Thus, a $(1 - \alpha)$ confidence interval for β_m is given as

$$\hat{\beta}_m \pm S\sqrt{v_{mm}} t_{N-M;\alpha}. \tag{6.65}$$

For sufficiently large samples, the t distribution can be replaced by a standard normal distribution. Typical regression software returns the vector of standard errors $SE_m = S\sqrt{v_{mm}}$. For instance, a 95% confidence interval for β_j is then approximated by $\hat{\beta}_m \pm 1.96 S\sqrt{v_{mm}}$. By error propagation, approximate confidence intervals can also be given for other properties such as the predicted mean $\mathbf{x}_i^T \beta$ at time step i (see Figure 6.7).

Prediction and Prediction Intervals

Given new predictor values $\mathbf{x}_{new}$, the regression function $\mathbf{x}_i^T \beta$ can be used for prediction. As the predictors in general do not explain the full variance of the predictands – unexplained variability is given by ϵ_i – the prediction based on $\mathbf{x}_{new}^T \beta$ does not represent an individual value but rather the mean of a predicted distribution of values that might occur, given x_{new}.

The range of possible values can be quantified by so-called prediction intervals (see Figure 6.7). A $1 - \alpha$ prediction interval is defined such that the actual new value y_{new} will occur within the interval with a probability $1 - \alpha$. To construct a prediction interval, consider first the expected value and the variance of a new predictand value $y_{new} = \mathbf{x}_{new}^T \beta + \epsilon$. They are given by $E(\mathbf{x}_{new}^T \hat{\beta} + \epsilon_{new}) = \mathbf{x}_{new}^T \beta$ and $var(\mathbf{x}_{new}^T \hat{\beta} + \epsilon_{new}) = \sigma^2 \left[\mathbf{x}_{new}^T (\mathbf{X}^T \mathbf{X})^{-1} \mathbf{x}_{new} + 1\right]$, respectively. A $1 - \alpha$ prediction interval is therefore given as

$$\mathbf{x}_{new}^T \hat{\beta}_m \pm S \left[1 + \mathbf{x}_{new}^T (\mathbf{X}^T \mathbf{X})^{-1} \mathbf{x}_{new}\right]^{1/2} t_{N-M,\alpha/2}. \tag{6.66}$$

The fundamental difference between confidence and prediction intervals is that the former provide an uncertainty estimate of a model parameter, that is, the predicted mean, whereas the latter provide a range in which new observations are likely to occur.

6.3.2   Generalised Linear Models

In real-world situations, one often faces at least one of the following problems: often the distribution of regression residuals is not Gaussian; moreover, the predictands cannot always be assumed to depend linearly on the predictors. To address these issues, generalised linear models (McCullagh and Nelder 1983, Dobson 2001) have been developed.

Given data y_i, sampled from a predictand variable Y, and a set of M predictors $x_{i1}, \ldots, x_{iM}$, a generalised linear model is defined as

$$E(Y) = \mu = g^{-1}(\beta_1 x_{i1} + \cdots + \beta_M x_{iM}) = g^{-1}(\mathbf{x}_i^T \beta). \tag{6.67}$$

Here, μ is the expected value of the dependent variable conditional on the predictor vector $\mathbf{x}_i$, and β is a vector of M parameters. The link function $g(.)$ transfers the expected value in such a form that the influence of the predictors can be assumed linear. The distribution of Y, conditional on $\mathbf{x}_i$, is assumed to be from the exponential family. The variance of the predictand is then a function of the predicted mean, $var(Y) = V(\mu)$. Parameter estimation is straightforward via maximum likelihood.

Logistic Regression

A prominent example of a generalised linear model is the logistic regression, which uses a continuous predictor to predict a binary predictand. A typical application is modelling precipitation occurrence. The predictand values are either "dry" or "wet", and the predictors typically describe the large-scale circulation and moisture transport. In the logistic model, "dry" and "wet" are coded as 0 and 1, and the predicted mean is simply the wet-day probability p_i,

$$p_i = \mu = g^{-1}(\mathbf{x}_i^T \beta) = \frac{\exp(\mathbf{x}_i^T \beta)}{1 + \exp(\mathbf{x}_i^T \beta)}. \tag{6.68}$$

The S-shaped function $g(p_i) = \log(\frac{p_i}{1-p_i})$ is called "logit" function and ensures that p_i is restricted to values in $[0, 1]$.

Gamma Model

Another example for a generalised linear model would be a model for daily precipitation intensities. These can often be modelled by a gamma distribution (Katz 1977). Chandler and Wheater (2002) formulated a generalised linear model that predicts the mean μ of the gamma distribution as

$$\mu = g^{-1}(\mathbf{x}_i^T \beta) = \exp(\mathbf{x}_i^T \beta). \tag{6.69}$$

The exponential link function ensures the mean to be positive and transforms the predictand such that the influence of the predictors can be assumed linear. The shape parameter γ is modelled independently of the predictors. According to Section 6.1.6, the variance follows as $\sigma^2 = \mu^2/\gamma$ (in other words: the coefficient of variation σ/μ is modelled as a constant).

6.3.3 Vector-Generalised Linear Models

A further generalisation of generalised linear models is vector generalised linear models (Yee and Wild 1996). These formulate individual linear models for a vector of Q distribution parameters $\boldsymbol{\theta} = (\theta_1, \ldots, \theta_Q)^T$,

$$g_q(\theta_q) = \mathbf{x}_i^T \beta_q. \tag{6.70}$$

For instance, $\boldsymbol{\theta}$ could include the mean $\theta_1 = \mu$ and standard deviation $\theta_2 = \sigma$ of a distribution. These could then depend in different ways on predictors.

6.3.4 Model Selection

In principle, one could improve the fit of any statistical model by simply increasing the model complexity (see Figure 6.8). If the number of model parameters reaches the number of data points, the residuals between model and observation would completely vanish. This perfect model fit, however, is a trivial consequence of overfitting.

Overfitting results from too high a number of parameters, either because too many predictors have been included or because the model structure is too flexible.

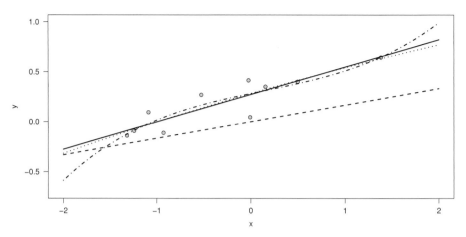

Figure 6.8 Overfitting. Data (circles) simulated from a straight-line regression model with $\beta_0 = 0.2$, $\beta_1 = 0.25$ and residual noise with $\sigma = 0.125$. Shown are models with one predictor x and zero offset (dashed), one predictor x (solid), two predictors x and x^2 (dotted) and three predictors x, x^2 and x^3 (dashed-dotted). The model with zero offset is too simple to describe the data. But in particular the third-order polynomial, even if it well follows the data, will provide poor extrapolations.

Wilks (2006) points out that overfitting is not only a result of including predictors of low predictive power but also happens for informative predictors if the number of data points is too low. Even though many atmospheric processes might influence, say, local precipitation, the number of data points might simply be too low to disentangle these effects. Similarly, the relationship between predictor and predictand might be nonlinear, but the available data might not be sufficient to constrain the shape of the nonlinearity.

Overfitting usually reduces predictive power. Both the predictors and the model structure therefore have to be parsimonious, that is, selected to balance model complexity and model fit. Different model selection criteria and automated procedures have been suggested in the statistical literature (e.g. Cox and Hinkley 1994, Davison 2003) as well as in the atmospheric sciences (e.g. von Storch and Zwiers 1999, Wilks 2006). Here we present some likelihood-based approaches. If no likelihood is specified (e.g. for neural networks), model selection can be carried out, for example, by cross-validation (Section 6.3.6).

Likelihood Ratio Tests

Suppose you have two different statistical models, a complex one and a simplification. For instance, consider a statistical model including a specific predictor and a simplified version that omits this predictor. The simple model $\mathcal{M}_0$ is parameterised by a parameter vector $\Theta_0 = (\theta_1, \ldots, \theta_{Q_0})$ with Q_0 parameters. The complex alternative model $\mathcal{M}_1$ includes additional parameters, resulting in the vector $\Theta_1 = (\theta_1, \ldots, \theta_{Q_0}, \ldots, \theta_{Q_1})$ with $Q_1 > Q_0$ parameters. The simple model is nested into the alternative one, that is, it can be derived from the latter by fixing the parameters $(\theta_{Q_0+1}, \ldots, \theta_{Q_1})$. Given predictor and predictand data, the parameters of both models have been estimated via MLE, resulting in the two maximised likelihoods $L_0(\Theta_0)$ and $L_1(\Theta_1)$ or the corresponding log-likelihoods $\ell_0(\Theta_0)$ and $\ell_1(\Theta_1)$.

The null hypothesis of the likelihood ratio test is that the simplified model $\mathcal{M}_0$ is a valid simplification of $\mathcal{M}_1$. The test statistic is the ratio of likelihoods or, equivalently the difference of the log-likelihoods, also called deviance statistic

$$D = 2\ell_1(\Theta_1) - 2\ell_0(\Theta_0)). \tag{6.71}$$

Under the null hypothesis D is $\chi^2_{Q_1-Q_0}$ distributed with $Q_1 - Q_0$ degrees of freedom. For a chosen significance level α, the critical value of the test statistic is thus given as the $1 - \alpha$ quantile of the $\chi^2_{Q_1-Q_0}$ distribution. The likelihood ratio test requires nested models, so it cannot be used to decide whether a predictor x_1 is better than a predictor x_2.

Information Criteria

An alternative approach is based on information criteria. These are measures that penalise the negative maximised log-likelihood of a model by the number of parameters. The optimal model is then the model that minimises the information criterion. Several criteria have been developed for different situations. Here we present the two most widely used, the Akaike information criterion (AIC; Akaike 1973) and the Bayes information criterion (BIC; Schwarz 1978).

Information criteria can be characterised by their consistency and efficiency. If the true model had a finite dimension and was part of the set of candidate models, an asymptotically consistent criterion would choose the true model for an infinite amount of data. If the true model had an infinite dimension and lay outside the set of candidate models, an asymptotically efficient criterion would choose the candidate model that minimises the mean squared error of prediction.

Given a model with Q parameters Θ; the model parameters have been estimated via MLE from T data points, resulting in the log-likelihood $\ell(\Theta)$. Then the AIC of this model is defined as

$$AIC = -2\ell + 2Q. \tag{6.72}$$

The AIC is asymptotically efficient (Stone 1977) but not consistent. For a finite number of data points, it tends to select too-complex models. The BIC is defined as

$$BIC = -2\ell + Q \log T. \tag{6.73}$$

The penalising term depends on the number of data points and is – for data sets of a reasonable minimum size – greater than in the AIC. The BIC is consistent but not efficient. For finite sample sizes, it tends to select too-parsimonious models.

Forward Selection and Backward Elimination

In many cases, the number of different candidate models is rather high, and an efficient model selection procedure is desirable. The simplest option would be to calibrate all

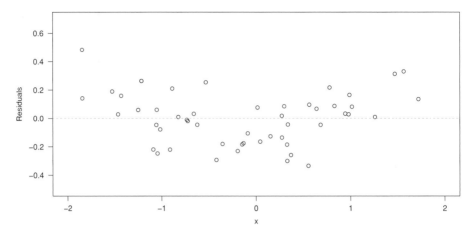

Figure 6.9 Residual plot. Difference between a sample of y_i and the prediction based on a calibrated straight-line model, $\beta_0 + \beta_1 x_i$. The residuals have a distinct structure, with positive values for low and large x_i and negative in between. This structure indicates that a straight-line regression may not be suitable but that a parabola might be more appropriate.

possible candidate models and select the one that optimises the chosen selection criterion. Especially in case of complex models, this procedure might be computationally not feasible.

In principle, two automated procedures – or a combination of both – are possible to reduce the computational burden: in forward selection, one starts with the simplest conceivable model (e.g. an offset without any predictors) and increases the complexity stepwise. In the case of predictor selection, one would in each step try out all candidate predictors and add that predictor to the model that most improves the model fit. In backward elimination, one would start with the most complex model (e.g. including all candidate predictors) and simplify the model stepwise. That is, one would in each step eliminate that predictor that decreases the model fit least.

6.3.5   Model Checking

Model selection criteria inform about the relative performance of models. A model including a specific predictor might be better than a model without that predictor but might still miss crucial effects. Moreover, the selected distribution might be inappropriate. For instance, a normal distribution is in general a good choice to model daily temperatures, but it is likely to fail to describe the lower tail of winter temperatures. Before finally accepting a selected model, one should therefore always check whether the model appropriately describes the data. Such mode checking – including visual inspections – should always complement more detailed model evaluation studies (Chapter 15).

First one should investigate the structure of the residuals (see Figure 6.9). Changes in the spread with changes in the predictor indicate heteroscedasticity, that is, that the assumptions regarding a constant variance are not justified. Systematic deviations of the residuals from the zero line indicate that the model structure is too simple (e.g. the linearity assumption might not hold) or that predictors are missing. This check might also reveal outliers that affect the whole regression relationship. For a nice illustration, see Wilks (2006).

Furthermore, one should investigate the distribution of the residuals by a QQ plot. To this end, one would eliminate the influence of the predictors by removing the predicted mean for each data point and then investigate the QQ plot of the remaining residuals (see Section 6.2.2).

6.3.6   Cross-Validation

Once a model has been selected and initial checks have been carried out, one should test how the model performs in predicting independent data, that is, data which has not been used for calibrating the model. The idea to use independent data is, as in model selection (Section 6.3.4), again to eliminate artificial skill from overfitting.

The basic idea of cross-validation is to leave out a fraction of the data under model calibration and use it only for validation, and to repeat the calibration and validation on different subsets of the data (e.g. Wilks 2006). The model is calibrated to the calibration data and used to predict the withheld validation data. Several different strategies of partitioning the data into calibration and validation data have been developed. Consider a data set of length N. The simplest evaluation on independent data is the hold-out evaluation: the data is split into two non-overlapping subsets. The model is then calibrated on one subset, and used to predict the other subset. A major disadvantage of the hold-out method is that each data point is only used either for calibration or validation, hence sampling errors are relatively high (therefore this method is usually not called cross-valiation).

An approach to optimally make use of the data is the leave-one-out approach: each data point is left out once. Each time, $N - 1$ data points are used for calibration and one data point for validation. Leaving out each data point once ultimately gives N validation points. The leave-one-out cross-validation is of limited use for correlated samples: in such a case, the neighbours of the left-out point carry information about the point itself, thus artificial skill would not be fully removed. A generalisation would be the leave-p-out approach.

As an alternative, the k-fold cross-validation is widely used. Here the time series is divided into k blocks (the folds) of approximately equal length. Each block then once serves as validation data set, and the others are then used for calibration. Again, all N data points are used for validation. The folds can be defined randomly but are often chosen as consecutive blocks.

Predictive skill is evaluated either by averaging performance measures across all folds, or by concatenating all predicted blocks and then calculating one performance measure for the concatenated time series. A wide range of performance measures can be used for evaluation. The most widely used is correlation or root mean squared error, but other measures developed specifically for, for example, forecast verification (Jolliffe and Stephenson 2003) are available for targeted analyses. Evaluation in a climate context has several peculiarities and will be discussed in Chapter 15.

Even though its purpose is different from model selection, cross-validation is mathematically similar. In fact, for linear models, a model selection based on cross-validation and root mean squared error is identical to a model selection based on the AIC (Stone 1977).

6.4   Stochastic Processes

A random process $\{X_i\}$ is a random variable dependent on time steps i (or space or any other domain). In the following, we will only consider processes that are discrete in time, that is, $i = 0, \pm 1, \pm 2, \ldots$. One observed record of a random process is called realisation. The collection of all possible realisations of a random process is called an ensemble.

A random process $\{X_i\}$ is specified by a multivariate probability distribution with marginal density $f_i(x)$ at time i. The expected value and variance are given as

$$\mu_i = E(X_i) = \int_{-\infty}^{\infty} x\, f_i(x) dx \tag{6.74}$$

$$\sigma_i^2 = E((X_i - \mu_i)^2) = \int_{-\infty}^{\infty} (x - \mu_i)^2 f_i(x) dx \tag{6.75}$$

To capture the dependence in time, the joint densities have to be considered, for example, at times $i_1, i_2, i_3, p(X_{i_1} \leq a_1, X_{i_2} \leq a_2, X_{i_3} \leq a_3)$.

6.4.1   Stationary Random Processes

A process is called completely stationary if the joint probability distribution does not change in time. In other words, the joint probability distribution of $\{X_{i_1}, X_{i_2}, \ldots, X_{i_n}\}$ is equal to the joint probability distribution of $\{X_{i_1+k}, X_{i_2+k}, \ldots, X_{i_n+k}\}$ for any time shift k and any times $i_1, i_2, \ldots, i_n$. The concept of complete stationarity is rarely of practical interest as it can never be inferred from a finite data set. Therefore, one alternatively considers stationarity up to a desired order, defined by the statistical moments up to this order. For instance, a random process is called second-order stationary if the expected value $E(X_i) = \mu$ and the variance $E((X_i - \mu)^2) = \sigma^2$ are constant in time and the covariance $E(X_{i_1} X_{i_2})$ depends only on the time lag $k = i_1 - i_2$. In the following, we will only consider second-order stationarity and refer to it simply as stationarity.

For a stationary process $\{X_t\}$, the autocovariance function is defined as

$$R_k = cov(X_i, X_{i+k}) = E((X_i - \mu)(X_{i+k} - \mu)), \tag{6.76}$$

with $R_0 = \sigma^2$, and the autocorrelation function as

$$\rho_k = \frac{R_k}{R_0} = \frac{cov(X_i, X_{i+k})}{\sigma^2}, \tag{6.77}$$

with $i = 0, \pm1, \pm2, \ldots$. The autocorrelation measures the linear dependence between the process at different times separated by a lag k. By construction, $R_0 = \sigma^2$ and $R_k \le R_0$ for all k. Accordingly, $\rho_0 = 1$ and $\rho_k \le 1$ for all k. For real valued processes (i.e. all processes we consider in this book), the autocovariance and autocorrelation functions are even functions, that is, $R_{-k} = R_k$. Equivalently, for two stationary processes $\{X_i\}$ and $\{Y_i\}$, the crosscovariance function is defined as

$$\Gamma_k = cov(X_i, Y_{i+k}) = E((X_i - \mu_X)(Y_{i+k} - \mu_Y)), \tag{6.78}$$

and the crosscorrelation function as

$$\gamma(\tau)_k = \frac{\Gamma_k}{\sigma_X \sigma_Y}. \tag{6.79}$$

6.4.2　Special Random Processes

A random process $\{X_i\}$, $i = 0, \pm1, \pm2, \ldots$, is called white noise (see Figure 6.10) or a purely random process if

$$cov(X_{i_1}, X_{i_2}) = 0$$

for all $i_1 \ne i_2$. Some authors define white noise even more strictly as a sequence of independent random variables.

Markov Processes
In statistical climatology, one often considers linear Markov processes of first order, for example, to model time dependence of temperature or the sequence of wet and dry days. A first-order Markov process is defined such that the probability distribution at time i depends only on the actual realisation of the process at time i_1, but not directly on any realisations at previous time steps:

$$p(X_i \mid X_{i-1}, X_{i-2}, X_{i-3}, \ldots) = p(X_i \mid X_{i-1}). \tag{6.80}$$

A specific variant of a first-order Markov process is a Gaussian autoregressive process of order one (AR[1] process, see Figure 6.10), defined as

$$X_i = aX_{i-1} + \epsilon_i, \tag{6.81}$$

where the noise follows a normal distribution, that is, $\epsilon_t \sim \mathcal{N}(0, \sigma^2)$. If $a < 1$, the process is asymptotically stationary (for sufficiently large i). Then the variance of the AR[1] process converges to $\sigma_X^2 = \sigma_\epsilon^2/(1 - a^2)$, and the autocovariance function becomes $R_k = cov(X_i, X_{i+k}) = \sigma_\epsilon^2 a^k/(1 - a^2)$. The autocorrelation function is given accordingly as

$$\rho_k = R_k/R_0 = a^{|\tau|}, \tau = 0, \pm1, \ldots \tag{6.82}$$

Thus, the autocorrelation function of an AR[1] process decays exponentially; when $a < 0$, it alternates in sign. Realisations of an AR[1] process are created by simulating a sequence of random numbers ϵ_i and then applying Equation 6.81 (the initial condition might be assumed as $x_0 = 0$).

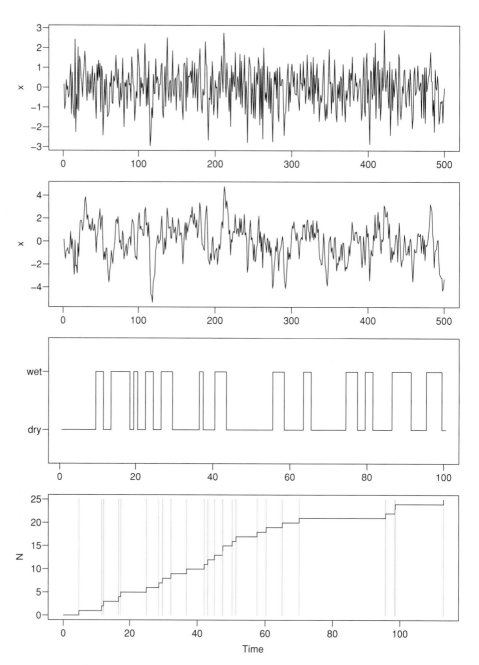

Figure 6.10 Realisations of selected stochastic processes. From top to bottom: Gaussian white noise, AR1 process with Gaussian increments, two-state Markov chain and Poisson process.

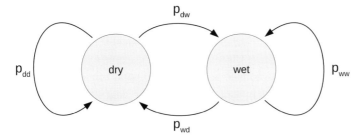

Figure 6.11 Markov chain. The process is defined by two states (here dry and wet) and a matrix of transition probabilities p_{ij}.

Markov chains are a specific type of Markov process with a discrete event space. Let K_i denote a discrete random variable at time i, describing, for instance, a weather state. The transition probabilities from state k to state l are defined as

$$p_{kl}(t) = Pr(K_i = l | K_{i-1} = k). \tag{6.83}$$

Two-state time-discrete first-order Markov chains are often used to simulate precipitation occurrence (see Figures 6.10 and 6.11). The transition probabilities between wet and dry states are given by a matrix of probabilities, describing wet-wet (p_{ww}), wet-dry (p_{wd}), dry-wet (p_{dw}) and dry-dry (p_{dd}) transitions. The transition probabilities for one initial state have to add up to one, such that $p_{ww}+p_{wd} = 1$ and $p_{dd}+p_{dw} = 1$. Realisations of such a process are created by simulating a sequence of uniformly distributed random numbers $u_i \in [0, 1]$. If, at time i, the model is in state k, the model remains in state k if $u_i \leq p_{kk}$ or else transitions to the complementary state l.

If a Markov process does not represent an observable system property, but rather determines a time varying model parameter, it is called a hidden Markov process.

Poisson Process
Some weather generators model precipitation as a continuous process of rain cell arrivals. The model for the arrivals is based on Poisson processes (see Figure 6.10). Such processes are a specific variant of counting processes, which describe the number $N(t)$ of events occurring within a time interval t, $t \leq 0$. A counting process $N(t)$ is a Poisson process if

- $N(0) = 0$
- its increments are independent
- the number of events in any interval t is a Poisson random variable with mean λt.

Within a time interval t, the expected number of events is $E(N(t)) = \lambda t$. The waiting times (or inter-arrival times) between consecutive events are exponentially distributed with rate parameter λ.

6.5   Pattern Methods

In this section, we discuss linear, multivariate statistical techniques that are suitable to characterise climate data given at many locations, namely principal component analysis (PCA), maximum covariance analysis (MCA) and canonical correlation analysis (CCA). While PCA is mostly used to describe one data set, for instance a time- and space-dependent sea-level pressure field, MCA and CCA are designed to capture relationships between two fields, for instance between sea-level pressure and precipitation. The central idea in PCA is to describe a data set that consists of a large number of partially dependent variables, such as the sea-level pressure at different locations, by means of only a few variables, while retaining as much as possible of the variability present in this data set. The purpose of MCA and CCA is to find a small number of variables that capture as much as possible of the link between two fields.

In Section 6.5.1 we introduce a unified terminology for all methods, mainly following Bretherton et al. (1992). PCA is discussed in Section 6.5.2, MCA and CCA in Section 6.5.3. The terminology refers to spatial fields of one scalar physical variable given at different times (i.e. with time as the sample index), which is the typical case in climatology. This is known as S-mode analysis. The formalism can be directly transferred to other cases, for instance to time series of one variable at different locations (space as sample index, T-mode) or to different variables at one location given at different times (O-mode, P-mode).

6.5.1   Expansion of Fields

Consider a space- and time-dependent, real-valued, scalar field $\hat{s}(\mathbf{x}, t)$ where $\mathbf{x} \in R^3$ denotes the space variable and t denotes time. Let $\bar{s}(\mathbf{x})$ denote the temporal mean of this field over the period under consideration. The centred or anomaly field is then defined by

$$s(\mathbf{x}, t) := \hat{s}(\mathbf{x}, t) - \bar{s}(\mathbf{x}) .$$ (6.84)

Throughout Section 6.5 mainly anomaly fields will be used.

In practical applications such fields will be given only at a finite number of locations and time steps. For instance, daily precipitation might be given at the locations of the rain gauges $\mathbf{x}_i$ and the days t_j. For brevity the discrete character of time in practical applications will be suppressed in the notation where not explicitly needed. It is convenient to replace the space- and time-dependent field $s(\mathbf{x}, t)$ by a finite-dimensional, vector-valued time series $\mathbf{s}(t) = (s_1(t), s_2(t), \ldots, s_N(t))^T$ defined by

$$s_i(t) := s(\mathbf{x}_i, t)$$ (6.85)

with $1 \leq i \leq N$, where N is the number of observation sites. The vectors $\mathbf{s}(t)$ are elements of a vector space R^N which will be called state space $\mathcal{E}_{\mathrm{st}}$.

All multivariate statistical methods discussed here use an expansion of the data time series of the form

$$\mathbf{s}(t) \leftarrow \tilde{\mathbf{s}}(t) := \sum_{k=1}^{M} a_k(t)\, \mathbf{p}_k \,. \tag{6.86}$$

In S-mode, the vectors $\mathbf{p}_k$ can be interpreted and displayed as space-dependent scalar fields and thus are called patterns. The time series $a_k(t)$ are called time expansion coefficients. Specific forms for both depend on the analysis method used. For $M < min(N, T)$, where T is the number of time steps in the data set, one might in general not be able to recover the original time series $\mathbf{s}(t)$. This is indicated by using the "$\leftarrow$" symbol.

In all methods discussed in this section, the time expansion coefficients are calculated as linear combinations of the original data

$$a_k(t) = \sum_{i=1}^{N} u_{ik}\, s_i(t) = \mathbf{u}_k^T\, \mathbf{s}(t) \,. \tag{6.87}$$

The column vectors $\mathbf{u}_k$ are called weight vectors, $\mathbf{u}_k^T$ denotes the transposed vectors, and again the specific form depends on the method.

6.5.2 Principal Component Analysis

Principal component analysis is one of the most frequently used multivariate techniques in climate research. It serves a twofold purpose in the investigation of meteorological and climatological data sets. First, it provides information about the spatial organisation of fields by identifying the dominant patterns of variability. Second, it yields optimally compressed data sets, and thus is an ideal filter for subsequent analyses such as studies of temporal variability or of links between two data sets.

Discussions of PCA can be found for instance in Jolliffe (1986), Jolliffe (1990) and Jolliffe (1993), in the outstanding book by Preisendorfer and Mobley (1988), which highlights a number of conceptually deep aspects of this technically simple method and in a more recent review paper by Hannachi et al. (2007). A very concise explanation can also be found in an appendix of Peixoto and Oort (1992). This section draws on elements from all of these sources.

Based on work of Kronecker and Christoffel, the mathematical foundation for PCA was laid by Beltrami (1873) and Jordan (1874), who developed the singular value decomposition of a general square matrix, which is the algebraical core of PCA. A geometrical interpretation of PCA in the way outlined below was given by Pearson (1901) along with an application in a biological context. A first application in psychometry can be found in Spearman (1904), followed by further applications in psychometry and physics in the 1930s (e.g. Hotelling 1933, 1935, Eckart and Young 1936, 1939). According to Preisendorfer and Mobley (1988, page 6),

The first application of PCA in meteorology appears to have been made at the Massachusetts Institute of Technology by G. P. Wadsworth and two colleagues, J. G. Bryan and C. H. Gordon

(1948)....In test calculations over the North Atlantic Wadsworth was faced in 1944 with the daunting task of finding by hand-calculator methods the 91 eigenvalues of a 91 × 91 matrix **S**. Over several weeks, by several people, these were determined out to seven eigenvalues and associated eigenvectors.

PCA was also studied in Russia emphasising theoretical aspects (Obukhov 1947, 1954, 1960, Bagrov 1959). Computer-based calculations were performed by Lorenz (1956) in prediction studies for 500 hPa geopotential height. From the late 50s onwards, PCA had increasingly been used in meteorology, climatology and oceanography, mainly due to the increasing availability of computers (White et al. 1958, Glahn 1962, Kutzbach 1967, Trenberth 1975).

Properties of PCA

PCA in S-mode employs a coordinate transformation in the state space $\mathcal{E}_{st}$ in order to find patterns such that an approximate expansion of the data using only a small number $M \ll N$ of patterns will retain the maximally possible variability. In this sense, it finds the dominant patterns of variability in the order of decreasing importance. Due to this property, the respective time expansion coefficients can be considered as variables that are optimally adapted to a given data set.

The key property of the coordinate transformation can be outlined as follows. The T data vectors $\mathbf{s}(t)$ point from the origin of $\mathcal{E}_{st}$, which represents the mean field or mean data vector $\bar{\mathbf{s}}$, to points in $\mathcal{E}_{st}$. In general, this cloud of points will not be spread isotropically in $\mathcal{E}_{st}$ but will rather show some preferred directions in which the data varies most. PCA aligns the patterns in the expansion (6.86) with these directions. This is equivalent to requiring that the variances of the time-expansion coefficients are subsequently maximised. The patterns are subject to the constraint of mutual orthogonality in $\mathcal{E}_{st}$. The length of the patterns can in principle be freely chosen; a convenient and frequently used choice is to give them unit length.

Let us now consider this approach in more detail. Expressed in the original variables the data vectors $\mathbf{s}(t)$ refer to a standard orthonormal basis (ONB) $\{\mathbf{e}_k\}$ of $\mathcal{E}_{st}$, that is,

$$\mathbf{s}(t) = \sum_{k=1}^{N} s_k(t)\, \mathbf{e}_k \,. \tag{6.88}$$

Thus, every observation location is associated with a particular dimension in the state space. Note that the term ONB requires the definition of a metric $\mathbf{g}$ or equivalently of an inner product on the state space, which in this case is given by

$$\mathbf{e_k} \cdot \mathbf{e_l} := \mathbf{g}(\mathbf{e}_k, \mathbf{e}_l) := \delta_{kl} \tag{6.89}$$

where $\delta_{kl} = 1$ for $k = l$ and $\delta_{kl} = 0$ otherwise.

With respect to this metric, PCA determines a new ONB subject to the criterion that the variance of the new coordinates is subsequently maximised. The basis vectors $\mathbf{eof}_k$ of this alternative ONB are usually called empirical orthogonal functions (EOFs), and the time expansion coefficients or new coordinates $pc_k(t)$ are called principal components

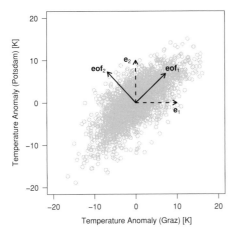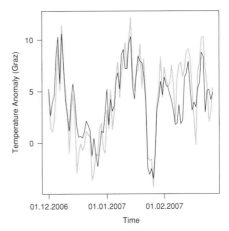

Figure 6.12 Left-hand panel: EOFs for the two-dimesional data set given by daily temperatures in Graz (Austria) and Potsdam (Germany); right-hand panel: observed daily temperature in Graz (grey line) and approximate expansion using the product of the first PC and EOF (black line). Based on ECA-D data (Klein Tank et al. 2002), for the period 1979–2008.

(PCs). Hence the general form of the expansion (6.86) can be replaced by

$$\mathbf{s}(t) = \sum_{k=1}^{N} s_k(t) \, \mathbf{e}_k = \sum_{k=1}^{N} pc_k(t) \, \mathbf{eof}_k \qquad (6.90)$$

and approximate expansions are given by

$$\tilde{\mathbf{s}}(t) = \sum_{k=1}^{M} pc_k(t) \, \mathbf{eof}_k \qquad (6.91)$$

with $M \ll N$.

A two-dimensional example is shown in Figure 6.12. In this figure, the old and new basis vectors are scaled to a length larger than 1 in order to be suitable to be displayed. Note that in the two-dimensional case the direction of $\mathbf{eof}_1$ is in general not identical to any of the two regression lines obtained when regressing the second variable onto the first or vice versa, as it is not the deviations of the data points from the line in any of the two standard coordinate axes that is minimised but the deviation orthogonal to the line.

Like patterns in general, the EOFs are expressed in terms of the old ONB $\{\mathbf{e}_i\}$

$$\mathbf{eof}_k = \sum_{i=1}^{N} eof_{ik} \, \mathbf{e}_i \qquad (6.92)$$

where $\mathbf{E} := (eof_{ik})$ is the transformation matrix from the original to the EOF basis system and the component eof_{ik} is the value of the kth EOF at the location $\mathbf{x}_i$. Following terminology from factor analysis, which is a related but different technique, these components are also often referred to as *EOF* loadings or as PC loadings, while the values of $pc_k(t)$ are also known as *PC* scores.

Before we can derive how the EOFs can be actually calculated, we need to know how to determine the PCs. The weight vectors in Equation 6.87 and thus the PCs follow from the rules of coordinate transformations. Since the EOFs form an ONB, the PCs are obtained by projecting the data anomaly vectors orthogonally onto the EOFs, that is,

$$pc_k(t) = \mathbf{eof}_k \cdot \mathbf{s}(t) . \tag{6.93}$$

The comparison of Equations 6.90 and 6.93 reveals that the patterns and the weight vectors are identical. This holds since $\{\mathbf{eof}_i\}$ are defined as an ONB. In this case, the weight vectors can also be referred to as loadings, which is used by some authors (Jolliffe 1986, 1990, 1993). Note that in some studies the EOFs are orthogonal (which is a necessary condition) but are not scaled to unit length. In these cases, the equality of patterns and weight vectors does not hold (and the use of "loadings" in the literature becomes inconsistent) and Equation 6.93 takes the more general form

$$pc_k(t) = \frac{\mathbf{eof}_k}{\|\mathbf{eof}_k\|^2} \cdot \mathbf{s}(t) \tag{6.94}$$

where $\|\mathbf{eof}_k\|^2 := g(\mathbf{eof}_k, \mathbf{eof}_k)$.

Calculation of EOFs
Having defined PCs and EOFs, we need to know how to find them. We present a deduction for the case of orthonormal EOFs which is essentially taken from Peixoto and Oort (1992). In mathematical terms, we want to maximise the variance of the time-expansion coefficients

$$\frac{1}{T-1} \sum_{i=1}^{T} (pc_k(t_i))^2 = \frac{1}{T-1} \sum_{i=1}^{T} (\mathbf{eof}_k \cdot \mathbf{s}(t_i))^2 \tag{6.95}$$

subsequently for $1 \le k \le N$ subject to the condition that $\{\mathbf{eof}_k\}$ is an ONB. Whether the variance is calculated using the factor $\frac{1}{T-1}$ or $\frac{1}{T}$ is not relevant for the following deduction.

Let $\mathbf{S} = (s_{ij})$ be the data matrix defined by

$$s_{ij} := s_i(t_j)$$

then Equation 6.95 can be rewritten as

$$\frac{1}{T-1} \sum_{i=1}^{T} (\mathbf{eof}_k \cdot \mathbf{s}(t_i))^2 = \frac{1}{T-1} \left(\mathbf{eof}_k^T \mathbf{S} \mathbf{S}^T \mathbf{eof}_k \right) = \mathbf{eof}_k^T \mathbf{C} \, \mathbf{eof}_k$$

where the covariance matrix $\mathbf{C} = (c_{ij})$ is defined by

$$\mathbf{C} := \frac{1}{T-1} \mathbf{S} \mathbf{S}^T$$

The elements of this $N \times N$ real symmetric matrix are given by

$$c_{ij} = \frac{1}{T-1} \sum_{l=1}^{T} s_i(t_l) \, s_j(t_l)$$

The diagonal elements are the variances of the old variables. In S-mode PCA, these are the variances of the station records.

Introducing Lagrange multipliers, maximising $\mathbf{eof}_k^T \, \mathbf{C} \, \mathbf{eof}_k$ subject to the orthonormality constraint on $\{\mathbf{eof}_k\}$ leads to the eigenvalue problem

$$\mathbf{C} \, \mathbf{eof}_k = \lambda_k \, \mathbf{eof}_k.$$

Therefore the EOFs are the normalised eigenvectors of the $N \times N$ covariance matrix. Since covariance matrices are symmetric and positive semi-definite, all eigenvalues are real and non-negative. The eigenvalues λ_k are arranged in decreasing order $\lambda_1 \geq \lambda_2 \geq \cdots \geq \lambda_N \geq 0$.

Note that when the covariance matrices do not have maximal rank the EOFs are still eigenvectors of this matrix. However, the basis of the subspace that is mapped onto $\mathbf{0}$, that is, the basis of the kernel, may be chosen arbitrarily. Non-unique EOFs are also obtained if two or more eigenvalues are identical (Preisendorfer and Mobley 1988). In practical applications the eigenvalues of covariance matrices will rarely be identical but will often be closely spaced. This issue is discussed in Preisendorfer and Mobley (1988), and a summary is given in what follows. North et al. (1982) give a rule of thumb for the uncertainty $\delta \, \mathbf{eof}_i$ of the eigenvector $\mathbf{eof}_i$ in the direction of another eigenvector $\mathbf{eof}_k$

$$\delta \mathbf{eof}_i \sim \frac{\delta \lambda_i}{\lambda_i - \lambda_k} \, \mathbf{eof}_k \qquad (6.96)$$

where $\delta \lambda_i$ is the sampling error or standard deviation of λ_i. According to Girshick (1939), it can be estimated for large sample size T by

$$\delta \lambda_i \sim \sqrt{\frac{2}{T}} \, \lambda_i.$$

If the sampling error of an eigenvalue is comparable to the difference to the next one, then the two eigenvectors are not well distinguishable, and a degeneracy problem similar to that of equal eigenvalues occurs. For a comprehensive discussion of the sampling errors of EOFs, see also von Storch and Hannoschöck (1985).

Most of the statistical or mathematical software packages provide routines to solve this problem. The straightforward approach is to diagonalise the covariance matrix. However, one can also make use of the fact that the singular value decomposition (see also Section 6.5.3) of the data matrix yields the EOFs and PCs simultaneously as left and right singular vectors (e.g. Hannachi et al. 2007).

Variances of the PCs, Selection Rules and Physical Interpretability
Let $\check{\mathbf{C}} = (\check{c}_{ij})$ denote the covariance matrix with respect to the new PC-coordinates

$$\check{c}_{ij} = \frac{1}{T-1} \sum_{k=1}^{T} pc_i(t_k) \, pc_j(t_k).$$

Using the representation (6.93) for $pc_i(t_k)$ yields

$$\check{\mathbf{C}} = \mathbf{E}^T \, \mathbf{C} \, \mathbf{E}.$$

Since the EOFs are the eigenvectors of $\check{\mathbf{C}}$ it follows that in the EOF basis the covariance matrix becomes diagonal with the eigenvalues λ_k as diagonal elements

$$\check{\mathbf{C}} = \begin{pmatrix} \lambda_1 & 0 & \cdots & 0 \\ 0 & \lambda_2 & \cdots & 0 \\ \vdots & \vdots & \ddots & \vdots \\ 0 & 0 & \cdots & \lambda_N \end{pmatrix}.$$

Thus the eigenvalues λ_k are equal to the variances of the PCs. Furthermore, it follows from the diagonality of $\check{\mathbf{C}}$ that the temporal covariances and therefore also the correlations between the PCs are zero. Note that this makes PCs very suitable predictors for multiple linear regression, as no issues related to colinearity of the predictors arise.

Since the diagonalisation of a symmetric matrix is unique up to scaling of the new basis, the PCs and EOFs are the only field expansion according to Equation 6.86, satisfying simultaneously the conditions that the patterns are spatially orthonormal and that the expansion coefficients are uncorrelated in time. This can thus be taken alternatively as the defining property of PCs and EOFs.

A measure for the importance of an individual PC-EOF product in the expansion Equation 6.86 is the fraction of total variance that is explained by the respective term. The total variance is defined by

$$\sigma_{tot} := \frac{1}{T-1} \sum_{i=1}^{N} \sum_{j=1}^{T} s_i^2(t_j).$$

This is equal to the trace of the covariance matrix $\mathrm{tr}(\mathbf{C})$. The trace of an arbitrary matrix is invariant with respect to orthogonal coordinate transformations, and thus the total variance can alternatively be expressed as the sum over the eigenvalues λ_i

$$\sigma_{tot} = \mathrm{tr}(\mathbf{C}) := \frac{1}{T-1} \sum_{i=1}^{N} \sum_{j=1}^{T} pc_i^2(t_j) = \sum_{i=1}^{N} \lambda_i.$$

For the fraction of total variance explained by an individual PC, it follows that

$$\sigma_i := \frac{\frac{1}{T-1} \sum_{j=1}^{T} pc_i^2(t_j)}{\sigma_{tot}} = \frac{\lambda_i}{\sum_{i=1}^{N} \lambda_i}.$$

A measure for the accuracy of an expansion using the leading M EOFs is given by the cumulative variance fraction

$$\sigma_M^{cum} := \frac{\sum_{i=1}^{M} \lambda_i}{\sigma_{tot}}.$$

Closely related to the fraction of variance explained by a certain EOF is the question how many of the EOFs describe a signal, that is, a systematic property of the data. EOFs that are completely determined by noise should be neither interpreted nor used for subsequent analyses. A variety of methods have been developed to distinguish meaningful from noise-dominated EOFs, and overviews are given in Jolliffe (1986) and Preisendor-fer and Mobley (1988). Three groups of methods can be distinguished. Methods in the first group are widely used and are based on the assumption that signals are associated with comparatively large amounts of variation explained by the respective EOFs. An early specification of "comparatively large" is given by Craddock and Flood (1969), who found that "in meteorology the noise eigenvalues are in geometrical progression". Together with the work of Farmer (1971), who further investigated this issue, this led to a selection rule frequently applied in practice: If the logarithms of the eigenvalues are plotted against the EOF indices, the noisy eigenvalues lie on a straight line, whereas the signal eigenvalues show a positive deviation from that line. Based on assumptions on the covariance structure of the noise, Preisendorfer and Barnett (1977) and Preisendor-fer et al. (1981) provided a theoretical understanding of this property, leading to a further refinement of this approach named "rule N", which is often used in applications of PCA. There are other selection rules in this group which are based on different assumptions on the data and which are discussed in Jolliffe (1986) and Preisendorfer and Mobley (1988).

The other two groups are mainly considered by Preisendorfer and Mobley (1988) and comprise selection rules which analyse the temporal behaviour of the PC time series or the similarity of the EOFs to theoretically estimated patterns. PC-oriented selection rules make use of the fact that in meteorological applications (or more general in S-mode PCA) the PC time series associated with a signal should be distinguishable from white noise. For the application of the EOF-oriented methods, some knowledge of the system under consideration is prerequisite.

Even though EOFs and PCs are statistical constructs, individual EOFs may often be linked to specific climatic phenomena. But also for trivial spatially correlated fields (e.g. spatially correlated noise), characteristic patterns emerge as EOFs. The actual patterns depend on the domain shape. For a rectangular area, these would be a leading EOF with uniform sign, followed by a sinusoidal dipole in the direction of the longer dimension, a dipole in the shorter direction and then multipole patterns with an increasing number of nodes (Buell 1975). Although leading EOFs represent large-scale phenomena such as ENSO or NAO the physical interpretability of EOFs is not guaranteed. This issue has been discussed in many publications (e.g. Buell 1979, Legates 1991, Dommenget and Latif 2002), and whether a given EOF has a physical meaning needs to be critically assessed in each case.

The interpretability of EOFs and PCs can also potentially be improved by rotation of EOFs of PCs, which is a coordinate transformation in the subspace of the leading EOFs. The transformation can be used to obtain EOFs or PCs with desirable properties. For instance, rotated EOFs with high positive loadings on some locations and loadings close to zero at others will group locations into areas of regionally coherent variability described by the respective rotated PC (Richman 1986, 1987, Preisendorfer and Mobley 1988). Alternatively the new basis system can be chosen such that the rotated PCs are either high or close to zero in which case individual EOFs can be viewed as typical situations (Preisendorfer and Mobley 1988). So-called orthogonal rotations use orthogonal transformation matrices and criteria for the loadings or scores that aim at obtaining the desired properties. But note that the application of an orthogonal matrix to an orthogonal basis system only leads to an orthogonal new basis if the input vectors are also normalised to the same length. That is, depending on the normalisation of the EOFs an orthogonal rotation might result in non-orthogonal rotated EOFs. Moreover, the original EOFs and PCs are the only basis system for which the patterns are orthogonal and the time expansion coefficients are uncorrelated. As pointed out by Jolliffe (1995), a blind use of rotation routines may thus lead to serious misinterpretation of the results.

6.5.3   Coupled Patterns

While PCA is applied to one field, we now consider two types of space- and time-dependent, real, scalar fields. A typical example in downscaling would be a pressure field and a precipitation field. In this section, we will discuss methods that find linearly coupled patterns in these two fields. Links between two fields found with these methods can aid process understanding and can also be used to estimate one field from the other. The two most frequently employed methods are maximum covariance analysis (MCA) and canonical correlation analysis (CCA).

Using the notation introduced in Section 6.5.1, we represent these fields as time-dependent vector fields

$$\mathbf{s}(t) \in I\!R^{N_s}$$

$$\mathbf{z}(t) \in I\!R^{N_z}.$$

Both fields are assumed to have zero temporal mean with respect to every component. We call $\mathbf{s}(t)$ the left field and $\mathbf{z}(t)$ the right field. Note that the dimensions of the two fields $I\!R^{N_s}$ and $I\!R^{N_z}$ can be different.

Analogously to Equations 6.88 and 6.90, we can represent both fields as linear combinations of patterns $\mathbf{p_k}$ and $\mathbf{q_k}$ with time-expansion coefficients (TECs) $a(t)$ and $b(t)$

$$\mathbf{s}(t) = \sum_{k=1}^{N_s} a_k(t)\, \mathbf{p_k} \tag{6.97}$$

$$\mathbf{z}(t) = \sum_{k=1}^{N_z} b_k(t)\, \mathbf{q_k}. \tag{6.98}$$

The time-expansion coefficients $a(t)$ and $b(t)$ can be calculated by projection of the data onto weight vectors, analogously to equation 6.87, as linear combinations of the data

$$a_k(t) = \sum_{i=1}^{N_s} u_{ik}\, s_i(t) = \mathbf{u}_k^T\, \mathbf{s}(t) \tag{6.99}$$

$$b_k(t) = \sum_{i=1}^{N_z} v_{ik}\, z_i(t) = \mathbf{v}_k^T\, \mathbf{z}(t). \tag{6.100}$$

Note that this notation can be used for cases in which the patterns and the weight vectors are not identical. It will be shown that this is the case for CCA but not for MCA.

A central element of MCA and CCA is the $N_s \times N_z$ cross-covariance matrix $\mathbf{C} = (c_{ij})$ of the two fields, which is defined as

$$c_{ij} = \frac{1}{T-1} \sum_{l=1}^{T} s_i(t_l) z_j(t_l) , \qquad (6.101)$$

where T is the number of time-steps t_l. For the remainder of this section, the time-step index l is suppressed where not explicitly needed, and the abbreviation

$$\langle s_i(t) z_j(t) \rangle := \frac{1}{T-1} \sum_{l=1}^{T} s_i(t_l) z_j(t_l)$$

is used; analogous expressions will be written in the same way.

Maximum Covariance Analysis

The purpose of MCA is to find orthogonal coupled patterns $\mathbf{p}_k$ and $\mathbf{q}_k$ such that the TECs $a_k(t)$ and $b_k(t)$ have maximum covariance. It can be shown that the patterns with this property are the singular vectors of the cross-covariance matrix between the two fields (e.g. Bretherton et al. 1992).

Singular value decomposition is one of the basic matrix decompositions known in linear algebra. Every real $N_s \times N_z$ matrix $\mathbf{A}$ can be factorised by singular value decomposition as

$$\mathbf{A} = \mathbf{L}\hat{\mathbf{S}}\mathbf{R}^T ,$$

where $\mathbf{L} \in \mathbb{R}^{N_s \times N_s}$ and $\mathbf{R} \in \mathbb{R}^{N_z \times N_z}$ are orthogonal matrices. The matrix $\hat{\mathbf{S}} \in \mathbb{R}^{N_s \times N_z}$ consists of a $\min(N_s, N_z) \times \min(N_s, N_z)$ diagonal matrix $\mathbf{S}$ and the zero matrix $\mathbf{0}$ according to

$$\hat{\mathbf{S}} = \begin{pmatrix} \mathbf{S} \\ \mathbf{0} \end{pmatrix} \qquad \text{if } N_s > N_z$$

$$\hat{\mathbf{S}} = \mathbf{S} \qquad \text{if } N_s = N_z$$

$$\hat{\mathbf{S}} = (\mathbf{S}\ \mathbf{0}) \qquad \text{if } N_s < N_z .$$

The diagonal elements $\sigma_i \in \mathbb{R}$ of $\mathbf{S}$ are non-negative and are called singular values. If $R \leq \min(N_z, N_s)$ is the rank of $\mathbf{A}$, there are at most R nonzero singular values. These are ordered by convention

$$\sigma_1 \geq \sigma_2 \geq \cdots \geq \sigma_R \geq \sigma_{R+1} = \cdots = \sigma_{\min(N_s, N_z)} = 0 .$$

The first R column vectors of the matrices $\mathbf{L}$ and $\mathbf{R}$ are unique (up to degeneracy if two or more singular values are identical) and are called left singular vectors and right singular vectors, respectively.

Let $\mathbf{l}_k \in I\!\!R^{N_s}$ and $\mathbf{r}_k \in I\!\!R^{N_z}$ with $1 \leq k \leq R$ denote the left and right singular vectors. Then the singular value decomposition can be alternatively written as

$$\mathbf{A} = \sum_{k=1}^{R} \sigma_k \, \mathbf{l}_k \, \mathbf{r}_k^T .$$

Since $\mathbf{L}$ and $\mathbf{R}$ are orthogonal matrices, the $\mathbf{l}_k$ and $\mathbf{r}_k$ form two sets of orthonormal vectors with respect to the standard metrics in $I\!\!R^{N_s}$ and $I\!\!R^{N_z}$, respectively. The following equations are satisfied

$$\mathbf{A}^T \mathbf{l}_k = \sigma_k \, \mathbf{r}_k \tag{6.102}$$

$$\mathbf{A} \mathbf{r}_k = \sigma_k \, \mathbf{l}_k . \tag{6.103}$$

Note that in case of a square symmetric matrix $\mathbf{A}$, the singular values are the absolute values of the nonzero eigenvalues of $\mathbf{A}$, and $\mathbf{L} = \mathbf{R}$ contains the eigenvectors.

Applying SVD to the cross-covariance matrix (6.101) yields the patterns with the MCA properties. Thus, the link between the patterns and the cross-covariance matrix is given by

$$\mathbf{C} = \sum_{k=1}^{R} \sigma_k \, \mathbf{p}_k \, \mathbf{q}_k^T . \tag{6.104}$$

As the sets of singular vectors $\mathbf{p}_k$ and $\mathbf{q}_k$ are orthonormal within the left and right field respectively, the kth TEC of the pattern is obtained by orthogonal projection of the data at each time step onto the pattern, which means that the weight vectors $\mathbf{u}_k$ and $\mathbf{v}_k$ are identical to the patterns $\mathbf{p}_k$ and $\mathbf{q}_k$, that is,

$$a_k(t) = \sum_{i=1}^{N_s} p_{ik} \, s_i(t) = \mathbf{p}_k^T \, \mathbf{s}(t) \tag{6.105}$$

$$b_k(t) = \sum_{i=1}^{N_s} q_{ik} \, z_i(t) = \mathbf{q}_k^T \, \mathbf{z}(t). \tag{6.106}$$

Left and right TECs for different k are uncorrelated, whereas the TECs within one field can be correlated (Bretherton et al. 1992).

Canonical Correlation Analysis
The purpose of CCA is to find weight vectors such that the first pair of TECs have maximum correlation and the following pairs maximise correlation subject to the constraint that the TECs within one field are mutually uncorrelated (e.g. Bretherton et al. 1992, von Storch and Zwiers 1999). TECs between the fields for different indices are also uncorrelated.

In CCA the weight vectors $\mathbf{u}_k$ and $\mathbf{v}_k$ are also known as adjoint canonical patterns, and the patterns $\mathbf{p}_k$, $\mathbf{q}_k$ are called the canonical patterns. In CCA the general solution for the adjoint canonical pattern is given by the eigenvector equation

$$\mathbf{C}_{ss}^{-1} \, \mathbf{C}_{sz} \, \mathbf{C}_{zz}^{-1} \, \mathbf{C}_{sz}^T \, \mathbf{u}_k = \lambda_k \, \mathbf{u}_k \tag{6.107}$$

where $\mathbf{C}_{ss}$ and $\mathbf{C}_{zz}$ denote the covariance matrices of $\mathbf{s}$ and $\mathbf{z}$ and $\mathbf{C}_{sz}$ the cross-covariance matrix between the two fields. A similar equation holds for $\mathbf{v}$. Pairs of adjoint patterns are related through

$$\mathbf{v}_k = \eta\, \mathbf{C}_{zz}^{-1}\, \mathbf{C}_{sz}^T\, \mathbf{u}_k \qquad (6.108)$$

with a suitable normalization factor η determined by the normalization convention

$$\mathbf{u}^T\,\mathbf{C}_{ss}\mathbf{u} = 1 \qquad\qquad \mathbf{v}^T\,\mathbf{C}_{zz}\mathbf{v} = 1\,. \qquad (6.109)$$

Patterns and adjoint patterns are related through

$$\mathbf{p}_k = \mathbf{C}_{ss}\,\mathbf{u}_k \quad \text{and} \quad \mathbf{q}_k = \mathbf{C}_{zz}\,\mathbf{v}_k\,. \qquad (6.110)$$

The weight vectors are used to define the TECs $a_k(t)$ for the patterns $\mathbf{p}_k$ through

$$a_k(t) = \sum_{i=1}^{N_s} u_{ik}\, s_i(t) = \mathbf{u}_k^T\,\mathbf{s}(t) \qquad (6.111)$$

$$b_k(t) = \sum_{i=1}^{N_z} v_{ik}\, z_i(t) = \mathbf{v}_k^T\,\mathbf{z}(t). \qquad (6.112)$$

As CCA includes matrix inversions its solutions can be very sensitive to sampling effects. The problem is substantially reduced by performing CCA only in the subspace of the leading PCs, that is, by pre-filtering the data sets with PCA prior to CCA. In this case only low-dimensional matrices have to be inverted, and the mathematics also simplifies (e.g. von Storch and Zwiers 1999).

Estimation of One Field from the Other
Both MCA and CCA can be used to estimate one field from the other. The estimates for $\mathbf{z}(t)$ are based on the field expansion Equation 6.98, however with using only a few leading modes M and estimated TECs $\hat{b}_k(t)$ that are calculated using MLR with the TECs $a_i(t)$ as predictors,

$$\mathbf{z}(t) = \sum_{k=1}^{M} \hat{b}_k(t)\,\mathbf{q}_k \qquad (6.113)$$

As MCA, CCA and MLR all yield linear estimates of one field from the other the question arises how they are linked. For the special case when one of the fields is one dimensional Widmann (2005) has shown that the methods are identical if the multidimensional field is estimated from the one-dimensional time series and that when estimating the time series from the multidimensional field MCA is equivalent to using weights given by the regression map, which is different from the CCA and MCA estimates. The general case has been comprehensively investigated by Tippett et al. (2008), who showed that the methods are equivalent if all modes are used, but not in the usual case when only a few modes are employed. Although MLR maximises by construction the explained variance in the predictands and CCA maximises correlations of TECs it is nor clear a priori which method yields the better results when applied to independent data where overfitting problems may arise.

6.6   Further Reading

Standard introductions to statistics in climate research have been published by von Storch and Zwiers (1999) and Wilks (2006). The latter book also contains concise introductions to clustering algorithms and Bayesian inference. An excellent introduction to statistical modelling in general can be found in Davison (2003); a more specific introduction to extreme value theory has been written by Coles (2001). For useful introductions to principal component analysis, refer to Preisendorfer and Mobley (1988) and Hannachi et al. (2007). A comprehensive overview of methods for finding coupled patterns is given in Bretherton et al. (1992), and the links between the different methods are further discussed in Widmann (2005) and Tippett et al. (2008).

7 Reference Observations

Observations are key to any modelling application; they tie the model to reality. In statistical downscaling, observations are needed for both calibration and evaluation. They are either needed as predictor or predictand data.

In the Global Climate Observation System (GCOS) report to the United Nations Framework Convention on Climate Change (UNFCCC) at the fourth conference of the parties (COP-4; GCOS 1998) it is stated that in practice, available observations often have major deficiencies with respect to climate needs. While considerable progress has been made since then (e.g. Karl et al. 2010), fundamental limitations persist. Observational data, both gridded and station data, should not naively be considered true. All observations are a measurement of a physical quantity, often followed by some post-processing. As such they are based on some type of model. This is obvious for gridded observations or reanalyses, which are derived via some interpolation or numerical weather prediction model. But also station measurements are based on some technical device (a thermometer inside a shelter, a bucket), which is assumed to correctly represent a meteorological variable (air temperature, precipitation). Therefore, observations are subject to often substantial random and systematic errors. For any useful application of observational data in a given context, it has to be known how it has been generated in order to understand what this data actually represents and what its limitations are.

7.1 Predictand Data

Statistical downscaling is often desired to downscale to station data (in particular in PP). There is, however, a growing demand to provide full fields and gridded products.

7.1.1 Station Data

Historical climate observations, in particular before the 1970s, are typically based on station-based measurements. These have in turn been used to calibrate locally operating impact models. To generate projections of the impacts of climate change, impact modellers therefore often demand downscaling to these station locations. Most PP methods in fact attempt to provide such results.

In many regions such as Europe and North America, a dense network of weather stations exists, especially of rain gauges (Figure 7.1). In some regions, however, stations are not regularly shared, are sparse or even no station data exists. Most stations are

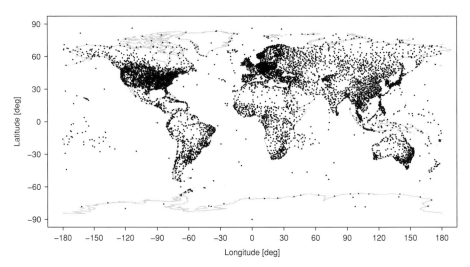

Figure 7.1 Weather stations used for the CRUTEM4 monthly gridded global temperature data set (Jones et al. 2012). Available from http://www.metoffice.gov.uk/hadobs/crutem4/data/download .html.

concentrated in populated areas. In particular in high elevations, few stations are in operation. For instance, in the Alps there are hardly any stations providing daily measurements above 2000m (Frei and Schär 1998, Schwarb 2000). In many countries the network is dense only after the 1960s. Even more of concern, the global observational network has been declining towards the end of the 20th century, in part also because the sharing of data is often not automatic (UNFCCC 1997). While this trend has been stopped, the situation is still bad in many regions of the world (Karl et al. 2010).

Available station data is often affected by substantial systematic errors. For instance, until the late 19th century, thermometers were often mounted on walls, causing a positive temperature bias (Trewin 2010). These errors may depend on season and weather situation and may be stronger for extremes (Brandsma and Van der Meulen 2008). Precipitation observations are also affected by systematic biases (e.g. Groisman and Legates 1994). The catch of precipitation by a gauge decreases with increasing wind speed; this undercatch is particularly strong for precipitation falling as snow and may cause underestimations of up to 50%. Further precipitation biases result, for example, from evaporative losses.

In a climate change context, errors are particularly relevant if they change in time. Indeed, existing station networks have often been designed for weather purposes, not for long-term monitoring of climate change. As such, they have experienced substantial changes that introduced inhomogeneities in the time series such as discontinuities or artificial trends (e.g. Trewin 2010, Venema et al. 2012). Sometimes these changes were deliberate modifications of measurement devices (often to improve measurements) or measurement practice, sometimes they were caused by changes in the local conditions, and sometimes stations were relocated.

For instance, the Stevenson screen is in almost universal use since the 1920s, but before that temperature measurements were not based on standardised devices (Trewin 2010). Similarly, early rain gauges were mounted on rooftops; as it was realised in the mid-19th century that these gauges underestimated precipitation by some 30 to 50% – for snow by up to 70% – they have been relocated to lower elevations of 1 to 2 meters (Auer et al. 2005). Also observing practice has changed: the way daily mean temperature is calculated, the time when observations are read or the units and precision of data recording (Trewin 2010).

Additionally, the regional and local environment of stations has often changed. The most prominent example is the urban heat island effect, which amplified the warming in cities compared to the surrounding countryside (Kukla et al. 1986, Parker 2010). In particular, local changes such as building of houses or car parks has increased temperatures, whereas the start of irrigation cools. Vegetation changes in general modulate local temperatures.

Finally, many sites have been relocated, sometimes only by some meters, often by several kilometres. Depending on the local environment, elevation, topography or distance from coast, such relocations can introduce major inhomogeneities. For instance, many stations have been relocated from city centres to airports – this effect can counteract the urban heat island effect.

If these effects were purely random from station to station, they would approximately cancel out at large scales. But many such inhomogeneities are coherent in space and may introduce large-scale inhomogeneities. For instance, after World War II, many new stations have been set up at airports (and others have been relocated). Böhm et al. (2001) found a bias in temperature trends in the Alps because of the decreasing urbanisation of the station network.

Against this background, the World Meteorological Organisation (WMO) published a list of recommendations (Aguilar et al. 2003): data series should if possible be kept homogeneous. If changes cannot be avoided, all relevant metadata including station history, geographical data, local environment, instrumentation and observing practice should be recorded. Ideally, one should conduct parallel recordings with the old and new conditions for several years to be able to easily quantify inhomogeneities. Unfortunately, metadata are not always recorded, often only in the local language, and they are often not complete. Parallel recordings are rarely done, mostly only at selected stations for research purposes.

Numerous statistical algorithms have been developed to detect and remove inhomogeneities. Most of these methods are based on the relative homogenisation principle (Venema et al. 2012): neighbouring stations are assumed to be exposed to almost the same climate signal, such that the difference between two homogeneous station series would be approximately constant in time. Differences between pairs of stations or to a reference time series can thus be used to detect and remove inhomogeneities. Modern approaches do no longer require a homogeneous reference time series (e.g. Caussinus and Mestre 2004) Homogenisation is mostly carried out on monthly or annual scales to reduce noise, although some approaches have been developed that work with daily data in a sufficiently dense network (e.g. Della-Marta and Wanner 2006, Toreti et al. 2010, Mestre et al. 2011, Trewin 2013).

Some researchers have argued that homogenisation should be avoided because it can introduce physical inconsistencies between different variables or may impose the – potentially not representative – climate signal of a reference series to other series. But it is now widely acknowledged that homogenisation – if possible – improves estimates of long-term temperature variability and trends (e.g. Auer et al. 2005, Venema et al. 2012). Recently, the international surface temperature initiative has been launched to homogenise temperature records and thus to construct an improved estimate of global mean temperature change (Thorne et al. 2011). The EU COST Action ES0601 "advances in homogenisation methods of climate series: an integrated approach" (HOME) has evaluated a wide range of homogenisation methods, based on both real data as well as simulated data with inserted inhomogeneities. Venema et al. (2012) summarise the results and present a list of recommendations. They demonstrate that absolute homogenisation – not making use of neighbouring stations – can make data even more inhomogeneous. Relative homogenisation improved temperature in almost all cases. Homogenisation for precipitation is much more difficult than for temperature, due to the shorter decorrelation length in space. Only the best homogenisation methods improve monthly precipitation sums. The best-performing methods where so-called direct methods that function with inhomogeneous references series.

In any case, users of observational data should have a rough understanding of the quality of the underlying station time series: are they homogenised? And if so, how have they been homogenised?

7.1.2   Gridded Data

Often, observational data sets are required on a regular grid – mainly for two reasons. First, users of climate data often desire full fields of climate information. Second, the output simulated by dynamical climate models typically has to be interpreted as area average values. Hence, a scale gap exists between the model output and station observations. This scale gap maybe bridged by inferring statistical relationships between point and grid scale (e.g. Osborn and Hulme 1997, Director and Bornn 2015, Haerter et al. 2015), but typically, station observations are gridded to the model resolution.

Gridded data sets are mostly based on gridding of station data, but may – especially in recent decades – also originate from remote sensing with, for example, satellites. More recently, radar data has become available. Sometimes post-processed reanalyses are used.

Any gridding of station data uses a more or less sophisticated statistical model that translates point values with irregular spacing into full fields of grid-box area averages (Frei et al. 2003). Topography has a strong imprint on the meteorological fields: temperature decreases with height, and precipitation is controlled by orography (Roe 2005). Station networks are usually not dense enough to infer this fine structure from station data alone (Widmann and Bretherton 2000). Hence covariates describing the local topography have to be built into the model. The imprint is particularly strong on long timescales; the anomaly time series are comparably smooth in space. Most gridding algorithms therefore separate the gridding of the climatological mean field from that of the anomalies (e.g. New et al. 1999, 2000, Isotta et al. 2014). Similarly, some authors separate the gridding of monthly mean values from daily anomalies about these fields (e.g. Haylock et al. 2008). Covariates are then only included into the estimation of the mean fields, not the anomalies.

A widely used model for generating climatological mean fields of precipitation is the Precipitation-elevation Regressions on Independent Slopes Model (PRISM) by Daly et al. (1994). Within the domain of interest facets are defined: single connected domains with a reasonably constant slope orientation. Facets are defined at a suitable spatial scale depending on the available data density. Within a facet, a regression model is build that predicts mean precipitation as a function of elevation. Large-scale data sets with relatively coarse grids typically do not construct facets but simply use elevation as a covariate (e.g. New et al. 1999, Haylock et al. 2008). Anomalies are gridded based on geostatistical approaches such as kriging (Diggle and Ribeiro 2007) or interpolation (Shepard 1984), usually – as discussed earlier – without additional covariates.

A major problem with gridded data sets is the often low density of the underlying station network. In particular at high elevations above 2000m, observations are sparse (Frei and Schär 1998, Schwarb 2000). Before the 1960s national station networks were much sparser than nowadays (e.g. Auer et al. 2005, Maraun et al. 2008). But also over recent years, fewer and fewer stations are in operation in several regions, either because they have been replaced by radar measurements (Isotta et al. 2014) or because the responsible countries do not have the resources to maintain the network (Karl et al. 2010).

Ideally, one would require an average station distance smaller than the decorrelation length of the considered variable at the temporal aggregation of interest. This is, however, hardly the case even for very dense networks. For instance, in the Alps the average rain gauge distance is 61km, which is just about the decorrelation length of daily precipitation at 42km (Auer et al. 2005).

The actual grid resolution should be chosen such that at least one station is available per grid box (Haylock et al. 2008). Most data sets do not fullfill this condition. For example, the first version of the E-OBS data set has an average number of 0.14 stations per 25km × 25km grid box, with much lower densities over Scandinavia, southwest and eastern Europe (Haylock et al. 2008). Over recent years, many additional stations have been included, but still the network is sparse over many regions. Even the recent daily Alpine precipitation data set (5km × 5km; Isotta et al. 2014), has a varying station density from about 0.25 station per grid box down to only 0.08 stations per grid box. As a result, such data sets may still have the statistics of station data rather than area averages.

For the estimation of precipitation climatologies, so-called totalisers (gauges that record over longer periods) are used to complement the daily data. Still, the representation of fine structures is very much dominated by the statistical treatment of orographic influences. The representation of day-to-day variability is especially dependent on the station network density, because here the mismatch between decorrelation length and mean station distance is important. Note that the correlation length depends on the type of event: large-scale frontal precipitation might be much better captured by a sparse network, whereas many localised thunderstorms will not be recorded.

These issues have to be kept in mind when interpreting gridded observational data sets. The sparser the network, the more the results are dominated by the statistical model. Grid box values at a given time may in such cases represent expected values rather than actual observations. Variability due to localised events will be underestimated, and the resulting fields will be too smooth (Hofstra et al. 2010, Kysely and Plavcova 2010, Maraun et al. 2012). This holds in particular for convective precipitation extremes. Thus it is important to complement gridded data sets by prediction intervals (Section 6.3) that give the range within which actual values are expected to be (Daly et al. 1994).

The severity of these problems depends of course on the variable, the considered aspect, the complexity of the terrain and the desired temporal resolution. Temperature interpolations are comparably simple, but in complex topography local phenomena such as inversions might be missed. The decorrelation length of precipitation increases very much with temporal aggregation (Auer et al. 2005). Hence it is useful to carefully balance spatial and temporal resolution. Of course, gridded data sets are affected by inhomogeneities of the underlying station data.

Gridded data sets are increasingly based on remote sensing techniques. In particular, rain radars have been used to create long-term gridded precipitation data sets. Radars transmit pulses of microwave radiation at a wavelength of some centimetres. Rayleigh scattering occurs at raindrops and ice crystals (typical diameters of a few millimeters), the scattering at cloud dropletsand haze is negligible. A receiver records the back-scattered signal. The elevation and azimuthal angles as well as the travelling time of the back-scattered signal define the precipitation location. The attenuation of the originally transmitted signal, that is the reflectivity, is linked to the precipitation intensity (Rogers and Yau 1996). A key advantage of radar data is the full-field coverage at a high spatial-temporal resolution. Uncertainties and systematic biases in intensity estimates result from differences in raindrop-size distributions, enhanced reflectivity by hailstones or melting snow, damped reflectivity by downdrafts and evaporation of rain at low elevations (e.g. Austin 1987). The quality of radar measurements is rather poor in mountainous terrain, especially during winter when precipitation growth takes place at low levels (Haiden et al. 2011). To combine the high spatial temporal resolution of rain radar data with the comparably accurate intensity estimates of rain gauges, combined products are developed (e.g. Haiden et al. 2011).

Most station-based gridded data sets with global coverage, in particular for precipitation, are available only at the monthly scale (e.g. Harris et al. 2014), because the station density in many regions is not high enough to provide reliable information at the daily scale. Yet the desired temporal resolution is often much higher. Station-based gridded data sets has been produced at the daily scale, however, in many regions with a very low station density (e.g. Xie et al. 2010, Schamm et al. 2015). More recently, therefore, reanalysis data (see Section 7.2) has been bias corrected at the monthly scale against gridded observations to serve as daily observations (e.g. Sheffield et al. 2006, Weedon et al. 2011). As these data sets are based on bias correction, they suffer from many of the problems discussed in Chapter 12. Rust et al. (2015) find discontinuities in temperature changes between two calender months in the WATCH forcing data set (Weedon et al. 2011). Sometimes, when no gridded observations are available, reanalysis data is directly interpolated to a higher resolution without any bias correction. For instance, Maurer et al. (2002) interpolate daily 10m wind fields from the NCEP/NCAR reanalysis (Kalnay et al. 1996) to a 1/8-degree grid. Such data sets suffer from the same representativeness problems as discussed in Section 12.7, that is, they likely do not represent the full variability at the target resolution. Over the last years, gridded data sets with a high spatial-temporal resolution have been constructed based on satellite measurements (e.g. Joyce et al. 2004), sometimes calibrated against station data (e.g. Adler et al. 2003, Funk et al. 2015). These data sets typically span rather short periods only but may provide valuable information in data-sparse regions.

To avoid misuse and misinterpretation of gridded data the product has to be understood in terms of the underlying raw data and the statistical model used for interpolation, in particular the treatment of topography. A list of the most commonly used gridded predictand data sets is given in Appendix B.3.

7.2   Predictor Data

PP statistical downscaling models are calibrated on observed predictors (Chapter 11). MOS methods, by contrast, use climate model predictors under calibration (Chapter 12). But still, both approaches require observed predictor data to evaluate their skill (Section 15.3): PP methods can directly be driven with observed large-scale predictors, and MOS approaches can be evaluated with input from reanalysis-driven RCMs or even directly with reanalysis data. Discrepancies of long-term statistics between observed and simulated predictands would then result mainly from errors caused by the (dynamical and statistical) downscaling. To calibrate and evaluate models for representing long-term predictand trends, also the predictor data has to correctly represent long-term trends. Most statistical downscaling methods use daily predictors.

Nowadays most observational predictor data sets are based on reanalyses, which are obtained by assimilating observations into weather forecasting models. Data assimilation combines empirical information with a dynamical model and yields physically consistent state estimates, which cannot be obtained by a merely statistical interpolation of the observations. It is a crucial step in weather forecasting used to find the starting conditions for the forecasts. In contrast to weather forecasting, reanalyses use a fixed version of an atmospheric model and assimilate more input observations. Most weather forecasting models use so-called 4D-variational data assimilation, which assimilates observations during a time window of several hours in one step, older versions used 3D-variational methods, which sequentially process the observations. The mathematical solution of the data assimilation problem requires solving a minimisation problem. In variational data assimilation this problem is solved directly, while so-called Kalman filters provide a suitable approximation in the sequential case. Details on the various methods in meteorology are discussed, for instance, in Kalnay (2003).

The standard reanalyses starting in the mid-20th century are NCEP/NCAR (Kalnay et al. 1996), ERA40 (Uppala et al. 2005), ERA-Interim (Dee et al. 2011b) and JRA-55 (Kobayash et al. 2015). They are all based on variational methods and assimilate surface and upper-air direct observations as well as satellite products. The assimilated meteo-rological variables include temperature, pressure and wind speed but not precipitation, which is calculated using model parameterisations. For a general discussion of reanal-ysis data refer to, for example, Kistler et al. (2001), which provides an overview of the NCEP/NCAR reanalysis. The MERRA (Rienecker et al. 2011) reanalysis, which begins in 1980, assimilates also satellite information related to the composition of the atmo-sphere. Recently, two reanalyses spanning the whole 20th century have been released by the US National Oceanic and Atmospheric Administration (NOAA; Compo et al. 2011) and the European Centre for Medium-Range Weather Forecasts (ECMWF; Poli et al. 2016). They assimilate only surface information, with the former using an Ensem-ble Kalman Filter and the latter a 4D-variational approach. In general the assimilated variables in a reanalysis can be expected to be very close to reality, whereas substan-tial biases may exist for variables that are not assimilated, such as precipitation (e.g. Widmann and Bretherton 2000).

The quality of reanalysis products depends on the assimilation scheme and the under-lying dynamical model, as well as on the assimilated observational data (Bengtsson et al. 2007). In particular over the Southern Hemisphere, their quality is limited by a lack of observations. The ERA-40 reanalysis, for example, appears to under-represent blocking frequency in the Southern Hemisphere in pre-satellite years. Furthermore, biases exist, for example, in surface heat and freshwater fluxes over the oceans; evapotranspiration and precipitation are globally not in balance.

Reanalysis data sets still provide a generally reliable tool for assessing the large-scale state of the climate system. But the available products have considerable limitations for studying changes of climate (Bengtsson et al. 2007, Thorne and Voss 2010); see also the discussion by Dee et al. (2011*a*) and Thorne and Vose (2011). The most important rea-son is changes in the observing system: of course, one aims to assimilate as many obser-vations as possible into the system. This implies, however, that the amount of assimilated data changes with the development of the observational system in time. Trend estimates are particularly unreliable if they reach back before the end of the 1970s, when satel-lite observations have been included. Simmons et al. (2004) have compared surface air temperature variability in reanalysis data with the CRUTEM2v observational data set (Jones and Moberg 2003). They find that short-term variability in general agrees well, but long-term trends in the reanalyses are considerably smaller than observed. From the late 1970s onwards, ERA40 trends were close to those in CRUTEM2v. Bengtsson et al. (2004) find that long-term temperature trends in the lower troposphere are artificially amplified by changes in the observing system but are consistent with satellite-based trend estimates after 1979. Similarly, integrated water vapour trends in ERA-40 are about twice as strong as expected from Clausius-Clapeyron scaling, again an artefact of changes in the observing system. For reanalyses covering only the satellite era, trends of assimilated variables are generally well represented (Simmons et al. 2010): the ERA-Interim temperature trends are in good agreement with CRUTEM3 data (Brohan et al. 2006) when sampled over the same area. Trends in specific humidity agree generally well with HadCRUH data (Willett et al. 2008), although with slightly different trends depending on the considered region. Trends in precipitation, which are not assimilated into ERA-Interim, however, differ rather strongly from the Global Precipitation Climatology Centre (GPCC) data (Schneider et al. 2008), in particular over Africa and South America and even after 1979. These discrepancies can also arise from substantial uncertainties in the observational data sets themselves (e.g. Adler et al. 2003). The 20th-century reanalyses are particularly limited for studying climatic changes. They are not only affected by the introduction of satellites in the late 1970s but additionally by the strong density changes in the global station network in the first half of the 20th century. As a result, for example, Krueger et al. (2013) found artificial trends in storminess in NOAA's 20th-century reanalysis compared to observational data (which have themselves been assimilated into the reanalysis), and Befort et al. (2016) found different long-term trends in storminess in both the ECMWF and NOAA reanalyses.

In a downscaling context, biases and the representation of day-to-day variability in typical predictor fields are also relevant. Brands et al. (2012) analyse the differences between ERA40 and the NCEP-NCAR reanalysis in the distributions in geopotential height, temperature and specific humidity at different pressure levels. For geopotential height and temperature, substantial differences exist in the tropics, which however vanish after a mean adjustment. For specific humidity, substantial differences exist globally, which also remain after a mean adjustment. Correlations between both data sets at the daily scale are surprisingly low over the tropics, in particular for temperature and humidity. These issues very much limit the use of reanalysis data as downscaling predictors in the tropics.

7.3 Further Reading

An overview of inhomogeneities is presented in Trewin (2010). As a prominent example, the urban heat island effect is discussed by Parker (2010). A comprehensive overview of the performance of homogenisation methods is given by Venema et al. (2012). Daly et al. (1994) provide a good overview of gridding problems. Reanalysis problems are discussed by Bengtsson et al. (2007), and Thorne and Voss (2010), Dee et al. (2011a) and Thorne and Vose (2011) have published an in-depth debate of suggestions for making reanalyses suitable for representing long-term changes.

8 Climate Modelling

Simulations with dynamical global climate models are the primary source of information about future climate and provide the input for statistical and dynamical downscaling methods. In this chapter, we give an overview on climate modelling. The level of detail is chosen such that the reader can gain a fundamental understanding of the basic principles; being comprehensive on technical aspects is not the purpose. Introductory texts on climate modelling include McGuffie and Henderson-Sellers (2005), Barry and Chorley (2009), O'Hare et al. (2014). More detail can be found in Kalnay (2003), Neelin (2010), Holton and Hakim (2013), Stocker (2011), Goosse (2015) and comprehensive discussions of many aspects in Trenberth (1992), Washington and Parkinson (2005) and Satoh (2013).

We begin with a short presentation of the different types of climate models and simulation setups. Section 8.2 covers one typical example for the equations that are at the core of climate models. Section 8.3 summarises the most important aspects of numerical integration. Section 8.4 discusses the ways how unresolved processes are included through parameterisations, outlining the overall concept and going into more details for some aspects that are particularly important in the context of downscaling. The chapter is concluded by discussion of model performance, again with a focus on aspects that are most relevant for downscaling (Section 8.5).

8.1 Model and Simulation Types

Models are simplified versions of the real world. They are smaller than reality because the number of locations and variables (i.e. the dimensionality of the state space) is lower and because not all processes are represented. They are simpler because processes are included in a idealised way. Models are also closed, with only a few external forcing factors, whereas reality is open, with a large number of unpredictable, external forcings. There is a spectrum of models. Very simple, conceptual models only represent the most important processes, and their purpose is to aid understanding through simplifying complex systems. At the other end of the spectrum are quasi-realistic models, which are as complex and realistic as the process understanding and practical limitations allow. Global general circulation models (GCMs) and regional climate models (RCMs) are of this type. They can be used to make predictions for the future, as an experimental tool to test hypothesis about processes and to interpolate observations in a physically

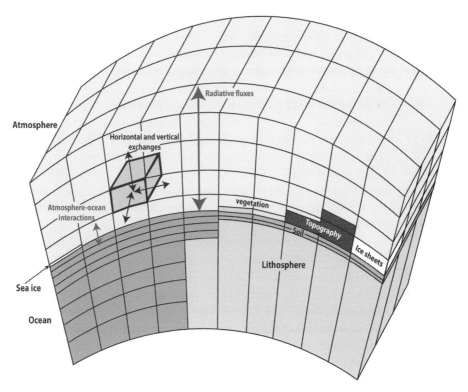

Figure 8.1 Schematic representation of a three-dimensional grid used in GCMs. Adapted from Goosse (2015).

consistent way. Climate models include components for different parts of the real climate system.

8.1.1 Global General Circulation and Earth System Models

Usually GCMs comprise atmosphere, ocean and sea ice components. They are coupled so they can exchange energy, momentum and moisture. Many recent versions also contain model components for dynamic vegetation and the terrestrial and ocean carbon cycle, and some also for atmospheric chemistry and water isotopes. These more comprehensive models are known as Earth system models. In climate models the simulated states are given on three-dimensional spatial grids and at discrete time steps, and the model equations describe how the system evolves in time. For computational efficiency some equations may also be formulated using two-dimensional wave-like basis functions, so-called spectral harmonics. The horizontal grid resolution for the atmosphere component of GCMs in climate applications is usually on the order of 100km, the number of vertical levels between 30 and 100, starting at the surface and including the troposphere and to varying extents the stratosphere. A schematic illustration for a GCM grid is shown in Figure 8.1.

There are four main types of GCM simulations:

- Weather forecasting simulations, in which initial conditions are specified and the goal is to forecast atmospheric and ocean states at individual times (lead times 1–10 days) or the distribution of these states over short periods (seasonal forecasts). This type of simulation exploits the deterministically predictable range of the system; at longer lead times the sensitivity of the outcome to small changes in the initial conditions due to the chaotic nature of the system makes it essentially unpredictable. An exception is slow processes related to the ocean circulation and ice cover, which may lead to predictability for longer lead times and which is explored in decadal predictions. Uncertainty that stems from errors in the initial conditions can be quantified by ensemble simulations based on a range of initial conditions, with ensemble sizes of about 50 members.
- Quasi-equilibrium climate simulations, which are performed for periods of a few hundred years using a constant set of external forcings, that is, Earth orbital parameters that affect incoming solar radiation, solar output, volcanic aerosol concentrations and greenhouse gas concentrations. The purpose is to determine the distribution of states when the system is in equilibrium for these forcings. The change between these distributions for two different forcings, for example, pre-industrial and doubled CO_2 atmospheric concentrations, yields an estimate for the response of the climate system to the forcing change. As the adjustment of the deep ocean and the cryosphere might take hundreds of years and perfect equilibrium may not be reached, the phrase "quasi-equilibrium" is preferred. If forcings for a reference period such as the pre-industrial period are chosen, these simulations are also called "control runs". Although initial conditions have to be chosen to start the simulations, the simulated period is so long that there is no deterministic predictability from them.
- Transient, forced climate simulations, in which time-dependent forcings are used. For simulations for the past, these forcings are observed or reconstructed (Masson-Delmotte et al. 2013). For the future plausible scenarios for greenhouse gas concentrations are required (see Section 9.1.1); currently these are the representative concentration pathways (RCPs; van Vuuren et al. 2011). Transient GCM simulations have been conducted for the last millennium (e.g. Bothe et al. 2015). A larger set of GCMs has been run for 1850 AD through 2100 AD within the CMIP5 (Coupled Model Intercomparison Project) framework (Taylor et al. 2012). The purpose is in principle the same as for quasi-equilibrium simulations, that is, to determine the response of the distribution of states to changing forcings. However, these simulations simulate the delayed, non-equilibrium response of the real world to changing forcings, and the simulated climate change signal is thus expected to be more realistic. Analogously to the quasi-equilibrium simulations the information from the initial conditions can not be used in a deterministic way, but changes in initial conditions will still lead to a spread in the outcome of the simulations. This initial condition uncertainty can be quantified in the same way as for weather forecasts by using initial value ensembles (see Section 9.2). Due to the longer simulation periods and computational constraints the ensemble sizes are limited to only a few members.

- Data assimilation simulations, in which the GCMs are used in combination with atmosphere or ocean observations to obtain optimal and physically consistent estimates for the state of the system (e.g. Kalnay 2003). Data assimilation is used to find the initial conditions for weather forecasts and for atmospheric reanalyses, which provide global, three-dimensional atmospheric states for a comprehensive list of variables for the last few decades with 6h to daily resolution (Kalnay et al. 1996, Kistler et al. 2001, Uppala et al. 2005, Dee et al. 2011b). Reanalyses starting at the beginning of the 20th century have also become available (Compo et al. 2011, Poli et al. 2016). Predictors for fitting downscaling models are often taken from these reanalyses. The approach has recently also been adapted to reconstruct climate for the pre-instrumental period from proxy data (e.g. Goosse et al. 2006, Widmann et al. 2010, Matsikaris et al. 2016, Steiger et al. 2017)

Simulations with atmosphere GCMs can also be performed using observed sea surface temperatures instead of coupling the model to an ocean model. This can be useful, for instance, to identify errors in the atmospheric model. Atmosphere-only GCMs with high resolutions can also be run with prescribed sea surface temperatures. Such simulations are conducted, for example, to study the sensitivity of the atmosphere to different oceanic processes or to resolve the atmosphere at reasonable cost at a very high resolution in climate change simulations.

Another type of global climate model that is computationally less demanding is an Earth system model of intermediate complexity (EMICs). It uses coarser grids and simplified dynamical equations and is used for very long simulations for past and future climate (e.g. Goosse et al. 2006, Goosse 2017). As it lacks regional detail compared to GCMs, it is not used as a starting point for downscaling.

8.1.2 Regional Climate Models

RCMs are used for dynamical downscaling over a limited domain (Rummukainen 2010). They are forced at the lateral domain boundaries and – typically – sea surface by values from GCMs. To evaluate the RCM or to simulate present climate at a high resolution, RCMs can also be forced with boundary conditions from reanalysis data. Regional climate modellers refer to such simulations with "perfect boundaries" as hindcast simulations. The nesting of the RCM into the driving GCM is typically one way, that is, the RCM does not feed back into the GCM (see, e.g., Lorenz and Jacob 2005 for a two-way nesting application). To reduce the gap in resolution between the driving and nested models, also several subsequent nests can be applied.

The GCM values are, obviously, provided at a lower spatial and temporal resolution than required by the RCM (temporal data is typically only available every 6 hours) and need to be interpolated. To avoid numerical instabilities (and to enable a numerical solution in the first place), a so-called sponge zone of typically 10 grid boxes at the boundary of the RCM domain is created. Here, the GCM boundary conditions are gradually blended into the RCM (Rummukainen 2010).

At the boundaries, the RCM is (more or less) consistent with the driving GCM. Within the domain, the RCM simulates weather consistent with its own topography, dynamical core, resolution and parameterisations and develops its own internal variability. This implies that the large-scale atmospheric states within the model domain can be different from the GCM states over the domain. Depending on the context, this "independence" may be desirable or undesirable. In particular in mountainous terrain, deviations from the large-scale GCM may be a key source of added value. But if a single regional-scale event is to be simulated (e.g. with perfect boundaries), regional-scale internal variability should be closely synchronised with the driving model. To imprint such synchrony, spectral nudging has been developed (Kida et al. 1991, Waldron et al. 1996, von Storch et al. 2000): over a range of spatial scales (typically large scales), the numerical solution of the RCM atmospheric field is nudged towards the field of the driving model.

RCMs typically comprise components for the atmosphere and the land surface. Coupled ocean–atmosphere RCMs have been developed, for example, to better simulate the climate of the Mediterranean (e.g. Somot et al. 2008) or Baltic Sea ((e.g. Döscher et al. 2002, Schrum et al. 2003). Such models, in particular if they also represent other components of the climate system such as sea ice (Schrum et al. 2003), are sometimes referred to as regional climate system models (Rockel 2015). State-of-the-art RCMs in climate applications typically have a horizontal resolution of the atmosphere of about 12 to 25 km (e.g. Mearns et al. 2012, Jacob et al. 2014). To better resolve complex topography and small-scale processes, higher-resolution simulations of about 1km resolution are becoming established (Prein et al. 2015). In these so-called convection-permitting simulations, deep convection (in some cases also shallow convection) is no longer parameterised but explicitly resolved. The lower-resolution RCMs are usually run over continental-scale areas and the very-high-resolution versions, for instance, over individual European countries. Recently, a CORDEX flagship pilot study has been launched to create a multimodel ensemble of convection-permitting simulations spanning the Northern Mediterranean and the Alps.

8.2   Dynamical Core of General Circulation Models

8.2.1   Equations of Motion in a Non-Accelerated Reference Frame

The motion of a fluid or gas, such as the air in the Earth's atmosphere, can be completely described by three physical laws. The first one is Newton's second law, which states that forces $\mathbf{F}$ lead to accelerations $\mathbf{a}$, that is to changes in either the value or the direction of velocity $\mathbf{v}$, and that the force needed to achieve a certain acceleration is proportional to the mass m of the object on which the force acts ($\mathbf{F} = m\mathbf{a}$). In addition there is the so-called continuity equation, which states that mass in a fluid or gas is neither destroyed nor generated, and thus the mass in a given volume decreases (increases) if more of the fluid or gas leaves (enters) that volume than enters (leaves) it. The third law describes the fact that energy only changes its form but can not be created or destroyed.

For instance, absorption of sunlight reduces the energy contained in the radiation but increases the thermal energy through an increase in temperature, or mechanical work done through compression leads to an increase in temperature. These laws allow us to calculate from the current state of the fluid its change in time, that is they allow us to predict the temporal evolution.

In order to introduce the basic concepts in this chapter with as few technical complications as possible we do not yet consider the effect of the Earth rotation and describe the motion in a standard, non-accelerated reference frame (known as inertial frame). Furthermore we use standard Cartesian coordinates, that is an orthogonal coordinate frame x,y,z where a change of 1 in either coordinate refers to a length of 1m.

The state of the fluid is fully described by the three velocity components u, v, w in the three coordinate directions x, y, z, by the density ρ and by the temperature T. The pressure p can be calculated from ρ and T through the ideal gas law and is thus strictly speaking not needed to describe the state of the system, but the form of the equations is simpler if it is used too. Note that all the state variables u, v, w, ρ, T, p depend on space and on time t, for instance $T = T(x, y, z, t)$, but for brevity this is not explicitly included in the notation. Using these state variables and coordinates, the equation of motions take the following form:

$$\rho \frac{Du}{Dt} = \frac{\partial p}{\partial x} + F_x \tag{8.1}$$

$$\rho \frac{Dv}{Dt} = \frac{\partial p}{\partial y} + F_y \tag{8.2}$$

$$\rho \frac{Dw}{Dt} = -\frac{\partial p}{\partial z} + \rho g + F_z \tag{8.3}$$

$$\frac{D\rho}{DT} = -\rho \left(\frac{\partial u}{\partial x} + \frac{\partial v}{\partial y} + \frac{\partial w}{\partial z} \right) \tag{8.4}$$

$$\rho c_v \frac{DT}{Dt} = -p \left(\frac{\partial u}{\partial x} + \frac{\partial v}{\partial y} + \frac{\partial w}{\partial z} \right) + Q_{rad} + Q_{cond} + Q_{int} \tag{8.5}$$

The left sides of the first three equations (8.1, 8.2, 8.3) represent for each of the three coordinate directions x, y, z the forces per unit volume that lead to the accelerations $\frac{Du}{Dt}, \frac{Dv}{Dt}, \frac{Dw}{Dt}$ in each of these directions. The right sides state that these forces are given by the pressure gradient forces $\frac{\partial p}{\partial x}, \frac{\partial p}{\partial y}, \frac{\partial p}{\partial z}$ and friction forces F_x, F_y, F_z. The fourth equation (8.4) represents the conservation of mass and states that the density ρ changes in time (LHS) if the velocity field has a divergence or convergence (RHS). The last equation (8.5) represents the conservation of energy. It states that the temperature T changes in time (LHS) if there is divergence or convergence in the flow (compression and expansion lead to temperature changes) or if there is direct heating which can be caused by radiative heating Q_{rad}, conductive heating Q_{cond} or internal heating Q_{int} (condensation, evaporation and friction). The terms in this equation represent energies per unit volume and thus the LHS contains the density ρ and the specific heat constant c_v which together link a given temperature change to the associated change in internal energy.

The time derivative $\frac{D}{Dt}$ is called the "total" or "material" derivative and describes a change following a given flow parcel. This is is known as the Lagrangian perspective and is the direct generalisation of the mechanics and thermodynamics of individual objects to continuous fluids. However, often this perspective is not very practical, as the state variables are usually given as a function of space and time rather than following fluid parcels in the flow. The latter view is called the Eulerian perspective, and in this formulation derivatives can be calculated with respect to time or to the three space coordinates; these derivatives are called partial derivatives and denoted by $\frac{\partial}{\partial t}, \frac{\partial}{\partial x}, \frac{\partial}{\partial y}, \frac{\partial}{\partial z}$. The material and the partial derivatives are linked through

$$\frac{D}{Dt} = \frac{\partial}{\partial t} + u\frac{\partial}{\partial x} + v\frac{\partial}{\partial y} + w\frac{\partial}{\partial z} \tag{8.6}$$

This equation states that a change in time of any variable when following the flow can be caused either by a change in time at the current location, which is given by $\frac{\partial}{\partial t}$, or by moving with the velocity (u, v, w) in a situation where the value of this variable changes with location. The spatial gradient is given by the partial derivatives $\frac{\partial}{\partial x}, \frac{\partial}{\partial y}, \frac{\partial}{\partial z}$, and the way these derivatives are multiplied with the velocity means that a change of the variable only occurs if the movement is across the isolines.

8.2.2   Equations of Motion in a Rotating Reference Frame

Centrifugal and Coriolis Force

We now consider a situation in which motions are described within an accelerated reference frame, specifically a rotating reference frame as given by an observer on the Earth's surface who uses a local, orthogonal reference frame, which rotates with the Earth. The key problem when using accelerated reference frames is that Newton's Second Law can not be directly applied. For instance, in a situation with no forces acting, the motion would be non-accelerated when described in an non-accelerated reference frame, consistent with Newton's Second Law. However, when described from an accelerated reference frame the motion would appear accelerated, yet the forces, which are not affected by the choice of reference frame, would still be zero, and thus the situation would not be correctly described by Newton's Second Law. Conversely, an object that does not change its position in an accelerated reference frame, and thus is not accelerated in it, would be accelerated in a non-accelerated reference frame. In order to cause the acceleration of the object, forces must be present, which still would be measured in the accelerated reference frame. Again having non-zero forces but no accelerations in the accelerated reference frame is not consistent with Newton's Second Law.

In order to still be able to apply Newton's Second Law to calculate accelerations from given forces one needs to introduce additional forces in an accelerated references frame. The forces that are identical in non-accelerated and accelerated reference frames can be viewed as the real forces, whereas the additional forces that need to be introduced in accelerated reference frames are sometimes referred to as apparent forces, although to an observer in the accelerated reference frame they feel as real as the other forces. In rotating reference frames, which are a special case of accelerated reference frames, the apparent forces are the centrifugal and the Coriolis force; both will be explained in what follows.

The centrifugal force can be illustrated by a rotating observer who holds a weight on a string, for instance a rotating hammer-thrower. Before releasing the weight, the position of the weight relative to the rotating hammer-thrower does not change, yet the athlete has to use a substantial horizontal force to hold the weight. From the perspective of non-rotating observers (the fact that they rotate with the Earth can be neglected in this example), the situation is clear: the weight moves on a curved path, and the force applied by the hammer-thrower leads to the necessary acceleration. This force is called the centripetal force. However, from the hammer-thrower's perspective an inwards pull is needed just to keep the weight in its position. For the hammer-thrower this feels as if there was a force pulling the weight horizontally away which has the same strength as the force the athlete applies but points in the opposite direction. This force is formally introduced as the centrifugal force when describing motions in rotating reference frames. The hammer-thrower's description of the situation would then be that he has to counteract the centrifugal force with the force he applies so that the total force is zero, and hence the weight is not accelerated and stays in a fixed position. The magnitude of the centrifugal force (and of the centripetal force) is

$$F_{cent} = m \omega^2 r \tag{8.7}$$

where m is the mass of the object, ω the rotation rate and r the radius of the circular motion.

When describing atmospheric motions from a reference frame that is fixed on the Earth's surface and thus rotates with the Earth the centrifugal force has to be applied with the rotation rate $\Omega = 2\pi/86000s$ given by the Earth's rotation and using the radius of the latitude circle that corresponds to the location of the observer (i.e. $r = a \cos \phi$ with $a = 6371$km and geographical latitude ϕ). However, in practice the centrifugal force does not explicitly appear in the equations of motion because the definition of "horizontal" is such that the centrifugal force does not have a horizontal component and the vertical component is also implicitly included by combining the "gravitation" force which is given by the attraction of masses and points directly towards the centre of mass of the Earth and the centrifugal force to "gravity". The practical expression of the adjustment of "horizontal" (i.e. of equipotential surfaces) is the fact that the Earth's surface is not a sphere but an ellipsoid, with the equator radius about 21 km larger than the radius to the poles.

After the centrifugal force has been introduced Newton's Second Law can be applied to objects that do not move relative to the rotating reference frame. However, when objects move relative to the rotating reference frame an additional apparent force, the Coriolis force, needs to be introduced. Derivations for its magnitude for movements along latitude or longitude circles on the Earth can be found in Holton and Hakim (2013) and Lynch and Cassado (2006). The general case follows from the standard laws of coordinate transformations (e.g. Holton and Hakim 2013). The magnitude of the

Coriolis force is proportional to the the the magnitude $|\mathbf{v}|$ of the velocity and is given by

$$F_{cor} = 2\, m\, \omega\, |\mathbf{v}|\, . \tag{8.8}$$

Note that $\mathbf{v}$ stands for the three-dimensional velocity vector and should not be confused with the velocity component v.

For reference frames moving with the Earth's rotation, the Coriolis force has horizontal and vertical components. In practice only the horizontal component is relevant because its magnitude is similar to that of the horizontal pressure gradient force, whereas in the vertical direction the pressure gradient force and the gravity are a factor of 100 to 1000 larger than the vertical component of the Coriolis force, which therefore can be neglected. The horizontal component for a given $|\mathbf{v}|$ is zero at the equator and maximum at the poles and is given by

$$F_{cor,h} = 2\, m\, \Omega\, |\mathbf{v}|\, \sin\phi \tag{8.9}$$
$$= m\, f\, |\mathbf{v}| \tag{8.10}$$

with $f = 2\,\Omega\, \sin\phi$ being the Coriolis parameter, which is introduced to shorten the notation. The Coriolis force points to the right of the velocity in the Northern Hemisphere and to the left in the Southern Hemisphere. The two horizontal components in x and y direction are given by

$$F_{cor,x} = m\, f\, v \tag{8.11}$$
$$F_{cor,y} = -\, m\, f\, u. \tag{8.12}$$

The Primitive Equations

We will now look at a set of equations that is representative for the equations used in many global climate or weather forecasting models. They are known as the primitive equations and apply to so-called hydrostatic models. High-resolution global or regional models are non-hydrostatic and employ a partially different set of equations. The specific form varies depending on the coordinate systems and state variables used, but the set presented here includes the essential elements. The primitive equations differ from the equations of motions introduced earlier for a non-accelerated reference frame in several aspects:

- they include terms for the Coriolis force
- they include terms that take into account the spherical geometry of the Earth
- in the vertical direction they use the approximation that gravity is balanced by the vertical pressure gradient force, and thus the vertical accelerations are zero. This is known as the hydrostatic balance.
- they use pressure rather than height as a vertical coordinate

Before presenting the primitive equations, we will briefly discuss the hydrostatic balance and pressure coordinates.

The gravity force only acts in the vertical direction and is a factor of 100 to 1000 times stronger than the Coriolis force, friction or the horizontal components of the pressure gradient force. In order to achieve a situation of approximate equilibrium, the gravity force needs to be balanced to a large extent by the vertical pressure gradient force. Otherwise the air would just move towards the surface, creating low pressure and convergence of air at high levels, which in turn would lead to an increase of mass in the total air column and increase the pressure in the lower levels until a sufficiently strong vertical pressure gradient force would be established. Compared to the gravity force and to the vertical pressure gradient force, the other forces, namely the vertical component of the Coriolis force and friction, are very small. Furthermore the sum of all the forces, which causes vertical accelerations, is very small compared to the gravity and vertical pressure gradient force. In other words, the situation is very well approximated by assuming that the gravity force is balanced by the vertical pressure gradient force. This is known as the hydrostatic balance, which, using the standard vertical coordinate z, has the form

$$g = -\frac{1}{\rho}\frac{\partial p}{\partial z}. \tag{8.13}$$

Strong vertical accelerations, for instance those occurring in convection or over steep orography are associated with an imbalance between gravity and the vertical pressure gradient force. Therefore, this approximation cannot be used in high-resolution numerical models, in which these processes become relevant.

Some of the equations of motion get simplified if one uses the pressure p as the vertical coordinate instead of the height z. This is possible because pressure decreases monotonously with height. In pressure coordinates, the vertical velocity $w = \frac{\partial z}{\partial t}$ gets replaced by $\omega = \frac{\partial p}{\partial t}$. Note that following standard nomenclature ω can thus either denote the angular velocity as in Equations 8.7 and 8.8 or the vertical velocity in pressure coordinates with the meaning defined by the context.

In these coordinates pressure cannot be used anymore to describe the state of the system and a useful choice is to replace it by the geopotential Φ. The geopotential at a certain height or pressure level is the potential energy at that location relative to sea level (i.e. to $z = 0$). For the lower part of the atmosphere this energy is approximately proportional to the geometric height z, but because the gravity acceleration g^* (which combines gravitation and centrifugal acceleration) decreases with increasing height and also varies horizontally the proportionality is not exact. Thus surfaces of constant geopotential are not horizontal. The simplest form for geopotential is obtained in z-coordinates as

$$\Phi(x, y, z) = \int_0^z g^*(x, y, z')\, dz' \tag{8.14}$$

while in pressure coordinates it is

$$\Phi(x, y, p, t) = \int_{slp(x,y,t)}^p g^*(x, y, p)\frac{dz}{dp'}\, dp' . \tag{8.15}$$

Like all other state variables the geopotential depends on the coordinates, that is horizontal location and time, and on the vertical location which is now given by pressure. For a given time and pressure level the geopotential can be represented as a map. For a given pressure level this map replaces the pressure maps on heights that correspond to that pressure level, and it has approximately the same structure as the pressure maps.

When using the hydrostatic approximation, pressure as the vertical coordinate, the geopotential as a state variable and geographical latitude ϕ and longitude λ as horizontal coordinates the equations of motion we obtain are known as the primitive equations. They include the apparent forces due to the Earth's rotation and take the following form:

$$\left(\frac{Du}{Dt}\right)_p = \left(f + u\frac{\tan\phi}{a}\right)v + \frac{1}{a\cos\phi}\frac{\partial\Phi}{\partial\lambda} + F_\lambda \tag{8.16}$$

$$\left(\frac{Dv}{Dt}\right)_p = -\left(f + u\frac{\tan\phi}{a}\right)u + \frac{1}{a}\frac{\partial\Phi}{\partial\phi} + F_\phi \tag{8.17}$$

$$\frac{\partial\Phi}{\partial p} = -\frac{RT}{p} \tag{8.18}$$

$$0 = \frac{1}{a\cos\phi}\frac{\partial u}{\partial\lambda} + \frac{1}{a\cos\phi}\frac{\partial(v\cos\phi)}{\partial\phi} + \frac{\partial\omega}{\partial p} \tag{8.19}$$

$$c_p\left(\frac{DT}{Dt}\right)_p = \frac{RT}{p}\omega + Q \tag{8.20}$$

In the (t, ϕ, λ, p) coordinates which we now use, the total derivative is given as

$$\left(\frac{D}{Dt}\right)_p = \frac{\partial}{\partial t} + \frac{u}{a\cos\phi}\left(\frac{\partial}{\partial\lambda}\right)_p + \frac{v}{a}\left(\frac{\partial}{\partial\phi}\right)_p + \omega\frac{\partial}{\partial p}. \tag{8.21}$$

Note that the eastward and northward velocity components u, v (i.e. velocities in the direction of λ and ϕ) are still measured in m/s (rather than in degrees per second) and that the horizontal partial derivatives need to be calculated on a surface of constant pressure, which is indicated by the subscript p.

We now discuss the individual terms in the primitive equations. One contribution to the eastward acceleration $\frac{Du}{Dt}$ is the pressure gradient force which is given as $\frac{1}{a\cos\phi}\frac{\partial\Phi}{\partial\lambda}$. It consists of the derivative of the geopotential in eastward direction and of a factor that accounts for the fact that the horizontal distance that is associated with a given change in longitude depends on the Earth's radius and the latitude. The other contributions are the friction F_λ, the Coriolis term fv and a so-called curvature term. The latter is needed because any horizontal motion on the Earth is following the Earth's curvature and thus requires a centripetal acceleration. If the motion is not along a latitude or longitude circle (u and v are different from zero in this case) this acceleration has an east–west component, which is expressed by $u\frac{\tan\phi}{a}v$. Likewise the acceleration in a northward direction $\frac{Du}{Dt}$ is caused by a a sum of accelerations due to the pressure gradient, friction, the Coriolis force and a curvature term. Having introduced the hydrostatic balance (8.13) means there is no acceleration in the vertical direction anymore, that is one of the five time derivatives in the original set of equations of motion has disappeared. Another time derivative is removed by using pressure as the vertical coordinate and geopotential as a state variable, because the continuity equation (8.19) becomes independent of the geopotential Φ (which is the equivalent of the density ρ in the old version). The new form of the continuity equation essentially states that the flow is divergence free in the new coordinates. Solving the equations of motion numerically gets considerably simpler if only three rather than five time derivatives need to be calculated. The variables with time derivatives (u, v, T) can be updated first (the equations are called prognostic equations), and then the two equations without time derivatives (which are called diagnostic equations) can be used to calculate the remaining variables (Φ, ω).

The new coordinates also lead to a simplification of the thermodynamic equation (8.20). The temperature T changes either because of heating represented by Q or because of a non-zero vertical velocity ω. If $Q = 0$ and the movement is on a surface of constant pressure there is no temperature change, because there is no contraction or expansion of the air.

8.3 Numerical Integration

The equations discussed in the previous section are formulated using continuous space and time. However, in climate and weather forecasting models they are solved on a discrete spatial grid or using an equivalent spectral representation and using finite time steps. The spatial and temporal derivatives in these equations can thus not be calculated anymore in a straightforward way, as the definition of derivatives is based on mathematical limits which require infinitesimally close data points. In climate models this problem is solved by using approximations for derivatives that only use the differences between gridpoint values.

This finite difference approach is illustrated in Figure 8.2, which shows that the slope of the tangent at a gridpoint can be approximated by the slope of straight lines through adjacent gridpoints. Let x_j be the location of the jth gridpoint for which we want to calculate an approximation for the derivative $\frac{\partial \psi}{\partial x}(x_j)$ of a function $\psi(x)$. If only the gridpoints directly left and right are used there are three lines that can be drawn, one between $(x_{j-1}, \psi(x_{j-1}))$ and $(x_j, \psi(x_j))$, one between $(x_j, \psi(x_j))$ and $(x_{j+1}, \psi(x_{j+1}))$ and one between $(x_{j-1}, \psi(x_{j-1}))$ and $(x_{j+1}, \psi(x_{j+1}))$. The slopes of these lines are called the backward, forward and central approximation for the derivative at x_j. The central approximation is thus given for instance by

$$\frac{\partial \psi}{\partial x}(x_j) \approx \frac{\psi(x_{j+1}) - \psi(x_{j-1})}{2\,\Delta x} \tag{8.22}$$

where Δx is the distance between the gridpoints.

In order to illustrate the use of finite differences in solving differential equations, we now consider the one-dimensional advection equation, which describes the advection of a conserved quantity $\psi(x, t)$ with the flow. It results for instance from the thermodynamic equation 8.5 in the case of divergence-free flow, no heating and reduction to one dimension, when a temperature anomaly would just propagate with the flow. For a conserved quantity the total derivative is zero,

$$\frac{D\psi}{Dt} = \frac{\partial \psi}{\partial t} + u\frac{\partial \psi}{\partial x} = 0 \tag{8.23}$$

where u is the flow velocity.

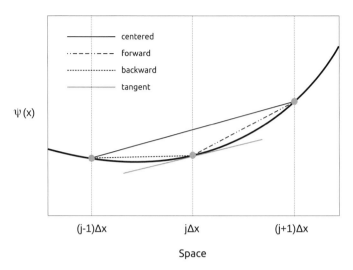

Figure 8.2 Approximation of the derivative of a function by finite differences..

For brevity we introduce the notation $\psi_j^n := \psi(x_j, t_n)$ with t_n denoting the nth timestep. Replacing the partial derivatives in time and space with their central approximations and considering gridpoint j and timestep n yields

$$\frac{\psi_j^{n+1} - \psi_j^{n-1}}{2\,\Delta t} + u\,\frac{\psi_{j+1}^n - \psi_{j-1}^n}{2\,\Delta x} = 0 \qquad (8.24)$$

with timestep Δt. This can be rearranged to

$$\psi_j^{n+1} = \psi_j^{n-1} - \alpha\left(\psi_{j+1}^n - \psi_{j-1}^n\right) \qquad (8.25)$$

with the Courant-Friedrich-Levy (CFL) number

$$\alpha := \frac{u\,\Delta t}{\Delta x}. \qquad (8.26)$$

Equation 8.25 allows us to calculate the value of ψ at gridpoint j and time $t + 1$ from the values of ψ at times n and $n - 1$ and at locations $j - 1, j, j + 1$. It thus allows us to predict ψ for the next timestep from its values at the current and previous timestep. As the value ψ_j^n is not needed in this specific approach, it is known as the leapfrog scheme. Many other schemes have been developed using various combinations of approximations for the derivatives, some based on values that are more than one gridpoint or timestep away from the gridpoint and time of interest (e.g. Kalnay 2003).

The scheme is only numerically stable if the CFL criterion $\alpha \leq 1$ is satisfied. As a consequence a reduction of the grid spacing Δx needs to be accompanied by a reduction of the timestep Δt, which would lead to an eight-fold increase in computing time for a doubling of the horizontal resolution and even more if the number of vertical levels is also increased. The exact form of the stability criterion depends on the numerical scheme, but in all cases a reduction of the grid spacing also requires a reduction in the timestep.

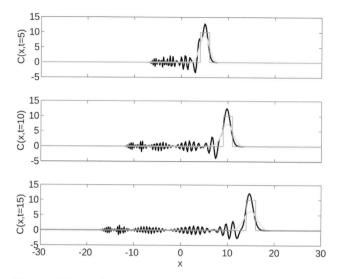

Figure 8.3 Propagation of a rectangular anomaly for the one-dimensional advection equation after 5, 10 and 15 timesteps. Thin grey line: original solution; thick grey line: upstream scheme; black line: leapfrog scheme. Adapted from Stocker (2011).

The discretised equation is different from its continuous form, and as a consequence its solutions are not identical to the solution of the continuous equation at the discrete gridpoints and timesteps. The main errors introduced by the discretisation are numerical diffusion, that is a broadening and flattening of anomalies, and numerical noise, which both contribute to the overall errors in weather forecasting and climate models. Different discretisation schemes lead to different numerical errors, and the aim is to use methods for a given model grid and set of model equations that are as exact and computationally efficient as possible. An example for the numerical errors in the leapfrog scheme and the so-called upstream scheme is given in Figure 8.3.

8.4 Parameterisations

8.4.1 Introduction

Climate simulations usually have a length of a few hundred years, for example 1850 AD to 2100 AD in the simulations conducted as part of the Coupled Model Intercomparison Project Phase 5 (CMIP5). Some simulations are also extended further into the future or conducted for past periods of 1000 to 10,000 years in length in palaeoclimate research. The available computing power limits the spatial resolution of GCMs to typically about 100km and of RCMs to about 10km, with lower resolutions for very long climate simulations and higher resolution for shorter periods and for weather forecasting applications. The operational global weather forecasting model at the European Center for Medium Range Weather Forecasts for instance has currently (2017) a horizontal resolution of 9km, and RCMs are in some cases run with resolutions around 1km. The typical number of vertical levels is about 50 to 100.

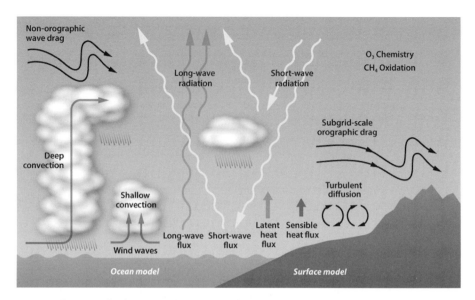

Figure 8.4 Parameterised processes in atmospheric general circulation models. Figure provided by ECMWF https://www.ecmwf.int/en/research/modelling-and-prediction/atmospheric-physics. Copyright given through Creative Commons license version 4.0 (https://creativecommons.org/licenses/by-nc-nd/4.0/legalcode).

There are many process in the real atmosphere and ocean that occur on spatial and temporal scales smaller than the model resolution. In the atmosphere (Figure 8.4) these include

- radiation absorbed, scattered and emitted by molecules, aerosols and cloud droplets
- cloud microphysics
- convection
- boundary layer processes
- drag caused by orography atmosphere-ocean interaction

Sub-grid atmospheric processes can be classified in three types:

- Internal heating that is not caused by the compression or expansion of air (non-adiabatic heating) but by radiation processes and by condensation or evaporation of water. These processes are included in the primitive equations through the heating term Q (Eq. 8.20), but the specification of this term requires a formulation of radiative processes, as well as of clouds and precipitation.
- Motions smaller than the grid scale, for example boundary layer turbulence, shallow and deep convection, orographic drag. All of them involve vertical transport of momentum and in many cases also of heat, moisture and substances such as aerosols or chemical substances. These processes are included in the continuous equations but not in the discretised version.
- Processes that involve variables not included in the basic model, for instance land-surface processes, chemistry, aerosol or carbon cycle.

If these processes were only relevant on the sub-grid scales and if only information at the grid-scale is required they could be ignored. However, the sub-grid processes strongly influence the larger-scale state of the atmosphere. Examples are radiative processes, which affect temperatures on a global scale, cloud formation, which affects the radiation, deep convection in the inter tropical convergence zone, which transports large amounts of heat into the upper atmosphere and affects the entire global circulation, and boundary-layer turbulence, which transports heat and moisture from the surface to the atmosphere on large spatial scales. Sub-grid processes do not only affect the mean atmospheric state but also temporal variability on all timescales. The development of the daily weather depends for instance on the dynamics in low-pressure systems, which are strongly influenced by diabatic processes as well as heat and moisture fluxes from the ocean. Interannual to decadal variability, such as the El Niño – Southern Oscillation, is largely driven by coupled atmosphere-ocean processes (see Section 2.3). In the context of climate change the feedback mechanism linked for instance to changing cloud cover affect the climatic response to the increasing concentrations of greenhouse gases.

It is thus clear that sub-grid processes can not just be ignored in numerical models but need to be included at least in an approximate way. This is the purpose of parameterisations, which attempt to represent the effect of sub-grid processes on the grid scale. It is common to all sub-grid processes that they depend on the resolved variables, and it is this dependency that needs to be approximately captured. Radiation absorption and scattering by molecules depends for example on the number of molecules in a grid cell, hence it is proportional to the density, which can be calculated from the resolved variables. Cloud formation depends on the available moisture and on the temperature, which both are resolved variables. Boundary-layer turbulence depends on the resolved vertical temperature profile as well as on the resolved wind shear, while orographic drag is influenced by the resolved wind speed and vertical temperature profile.

Climate models typically include more than 20 different parameterisations, and it is beyond the scope of this introduction to climate modelling to discuss them comprehensively. Instead we will focus in the next two subsections, on a few examples for boundary layer turbulence, convection and cloud parameterisations, as the latter two strongly influence the simulated precipitation, which is often a target variable in downscaling, and turbulence and convection parameterisations are closely related. More details on parameterisations can be found for instance in Trenberth (1992) and Kalnay (2003); in-depth discussions are provided by Stensrud (2009) and ECMWF (2016).

8.4.2   Reynolds Averaging

In this section we will look more closely at sub-grid motions and show explicitly in an example why they are relevant on the resolved scales. The approach is based on decomposing a given atmospheric equation into resolved and sub-grid components, which is known as Reynolds averaging and was originally applied to separate timescales in turbulence theory. For demonstrating the main principle different atmospheric equations can be used. We will follow Kalnay (2003) and consider the prognostic equation for water vapour; analogous separations hold for heat or momentum. Note that this equation needs to be solved in numerical models in addition to the primitive equations.

The water vapour content per unit volume is given by ρq, where ρ is the density of the air and q the water vapour mixing ratio. Considering the water vapour fluxes into and out of a fixed volume it can be shown that

$$\frac{\partial \rho q}{\partial t} = -\frac{\partial \rho u q}{\partial x} - \frac{\partial \rho v q}{\partial y} - \frac{\partial \rho w q}{\partial z} + \rho E - \rho C \tag{8.27}$$

where we have used Cartesian coordinates x, y, z, the respective velocity components u, v, w and the evaporation and condensation rates E, C. The equation states that changes in the water vapour content in a volume can be caused through water vapour fluxes into the volume and by difference between condensation and evaporation. In numerical models, an approximation of this equation is used in addition to the primitive equations shown earlier for calculating the temporal evolution of water vapour.

The velocity and the water vapour mixing ratio can have substantial sub-grid variability, whereas the density can be assumed to be uniform within a grid cell. When using the Reynolds averaging approach we represent the local values as grid-cell averages denoted by an overbar, and anomalies from this average are denoted by a prime, for example

$$u = \bar{u} + u' \tag{8.28}$$

$$q = \bar{q} + q' \tag{8.29}$$

and analogously for v and w, E and C. If we insert this decomposition into Equation 8.27 and then calculate the grid-cell average we obtain

$$\frac{\overline{\partial \rho (\bar{q} + q')}}{\partial t} = -\frac{\overline{\rho (\bar{u} + u')(\bar{q} + q')}}{\partial x} - \cdots + \overline{\rho (\bar{E} + E')} - \overline{\rho (\bar{C} + C')} \tag{8.30}$$

where "..." represents analogous terms in y and z direction. We now write down the individual terms of the products and use the fact that grid-cell averages of grid-cell averages leave a quantity unchanged and that the grid-cell average of anomalies is zero, for example $\bar{\bar{q}} = \bar{q}$ and $\bar{q'} = 0$. The remaining terms are

$$\frac{\partial \rho \bar{q}}{\partial t} = -\frac{\partial \rho \bar{u} \bar{q}}{\partial x} - \frac{\partial \rho \bar{v} \bar{q}}{\partial y} - \frac{\partial \rho \bar{w} \bar{q}}{\partial z} - \frac{\partial \rho \overline{u'q'}}{\partial x} - \frac{\partial \rho \overline{v'q'}}{\partial y} - \frac{\partial \rho \overline{w'q'}}{\partial z} + \rho \bar{E} - \rho \bar{C}$$

$$\tag{8.31}$$

In addition to the grid-cell average terms there are now terms that include sub-grid variables. This means that the temporal development of the grid-cell average water vapour content represented by $\frac{\partial \rho \bar{q}}{\partial t}$ cannot be calculated simply from the grid-cell averages of ρ, q, u, v, w inserted into Equation 8.27. The new terms are spatial derivatives of grid-cell averages of products of sub-grid anomalies. The term $\overline{w'q'}$ for instance quantifies the covariance of upward velocity w and water vapour mixing ratio q within a grid cell. To better understand the physical meaning, consider a grid cell with an upward sub-grid convection given by a velocity anomaly $w' > 0$ in one half of the cell and a downward velocity anomaly $w' < 0$ of equal magnitude in the other half. In many cases

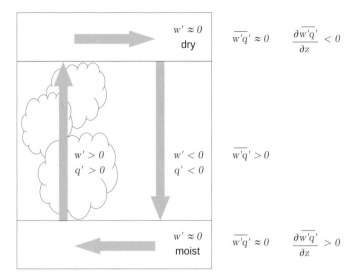

Figure 8.5 Sub-grid variability of vertical velocity, water vapour content and their covariance in a convective situation.

these velocity anomalies will be associated with water vapour anomalies because the upward branch will carry moist air upward and the downward branch will carry dry air downward, that is $q' > 0$ where $w' > 0$ and $q' < 0$ where $w' < 0$ and thus in both halves $w'q' > 0$ (Figure 8.5). If in the cells above and below there is no sub-grid circulation the product is $w'q' = 0$. The overall effect of the convection will be that the cell below the convection cell will get dryer and the cell above more moist, which is captured by the term $\frac{\rho \overline{w'q'}}{\partial z}$ for these cells. Similar arguments can be made for the other directions or for temperature or momentum transport. It is the purpose of convection parameterisations to find approximations for sub-grid covariance terms and their dependency on the resolved variables. A few examples will be given below.

8.4.3 Turbulence and Convection Parameterisations

Among the three covariance terms in Equation 8.31, the vertical term $\frac{\overline{w'q'}}{\partial z}$ is usually much larger than the horizontal terms because the large vertical gradients in water vapour content and in wind speed or potential temperature lead to large covariances $\overline{w'q'}$ (and analogously for temperature and momentum) in case of sub-grid up- and downward motion. The vertical covariance term is relevant in boundary-layer turbulence as well as in shallow and deep convection and is parameterised separately in the two cases.

In the case of turbulence, one might assume that the boundary layer is well mixed within a grid cell and ignore the sub-grid fluxes, that is set

$$\rho \overline{w'q'} = 0 , \tag{8.32}$$

which is known as zeroth-order closure (e.g. Deardorff 1972). If the assumption is made that the turbulent fluxes are proportional to the gradients of the respective quantity, which is known as turbulent diffusion, a first order closure can be used

$$\rho \overline{w'q'} = -K\frac{\partial \overline{q}}{\partial z} \qquad (8.33)$$

In this case the task is to specify a realistic diffusion parameter K, and a potential short-coming is that a constant K may not always be appropriate. Higher-order closures use additional prognostic equations for first-order covariance terms such as $\overline{w'q}$ and parameterise higher-order covariance terms such as $\overline{w'w'q}$ (for more details see Stensrud 2009, Kalnay 2003 and references therein). There are also a number of non-local boundary layer parameterisations, which take into account the structure of the entire boundary layer rather than just the local gradients. A realistic simulation of the boundary layer is not only relevant for the local conditions but the sensible and latent heat fluxes in the boundary layer may also affect the energy available for deep convection and thus influence the atmosphere on large scales.

Turbulence parameterisations account for small-scale, sub-grid motions but not for sub-grid convection, which is happening on larger spatial scales. The need for convection parameterisations was recognised in early GCM simulations. The vertical temperature profile obtained from radiative equilibrium is convectively unstable, and when there is no parameterisation of convection the model develops an unrealistic convective circulation on the resolved scale, known as vertical noodles. The problem was first addressed by Manabe et al. (1965) by simply adjusting the vertical temperature profile to the moist adiabatic lapse rate $\approx 6.5K/km$ when this lapse rate would have been exceeded while conserving the overall energy. Moisture that saturates through the adjustment is rained out. Although this ad-hoc approach has solved the problem of grid-cell convection the circulation obtained was not realistic. A second example for a moist convective adjustment scheme is the Betts-Miller scheme (Betts and Miller 1986), which calculates reference profiles in a more sophisticated way and uses an adjustment over time; the scheme produces a more realistic convection and is still in use in a few models.

Another early convection parameterisation is the Kuo scheme (Kuo 1965). It calculates the convective activity from the large-scale moisture convergence and is an example for a "deep-layer control scheme", in which information over most of the vertical extent of the troposphere rather than only local conditions is used as input for the parameterisation. It states that precipitation PR is given by

$$PR = (1 - b)M \qquad (8.34)$$

where M is the moisture convergence which has contributions from the convergence of the atmospheric flow as well as from moisture transport from the surface. The heating due to convection contains a contribution from the condensation and one from the vertical covariance terms for potential temperature (for details see also Stensrud 2009). The heating rates are calculated by assuming that the effect of the convection will be a relaxation of the temperature profile to the moist adiabatic lapse rate and that the speed at which this occurs increases with the moisture convergence. A key assumption of the Kuo scheme is thus that the strength of convection is limited by large-scale moisture convergence. It has been argued (Fritsch et al. 1976, Raymond and Emanuel 1993) that this is not in agreement with observations and fundamentally violates causality.

The majority of convection parameterisation schemes developed later are formulated closer to actual physical processes. These dynamic schemes may take some time to get the atmosphere towards a stable temperature profile rather than adjusting it instantaneously as in the moist adiabatic adjustment scheme or after a prescribed time as in the Kuo scheme. They are formulated as mass flux schemes. The earliest example is the Arakawa-Schubert scheme (Arakawa and Schubert 1974). A more recent version which is still in use in many models is the Tiedke scheme (Tiedtke 1989). We will now outline the mass flux approach using convective moisture transport as an example and following the discussion in Bechtold (2015).

The goal is to represent the vertical covariance term for water vapour flux $\rho\overline{w'q'}$ in Equation 8.31 in a grid cell that includes several cumulus elements in a way that is linked to properties of the convective elements. We denote the area covered by cumulus elements as a, the total grid cell area as A and the fractional area covered by cumulus elements as $\sigma = a/A$. As before, we assume that the density ρ is uniform but that there is sub-grid variability in w and q. We now express this sub-grid variability as the mean values over the "environment" area outside the cumulus elements w^e, q^e and the mean values within the cumulus elements w^c, q^c. The grid-cell averages are then given by

$$\overline{w} = \sigma w^c + (1-\sigma)w^e \tag{8.35}$$

$$\overline{q} = \sigma q^c + (1-\sigma)q^e \tag{8.36}$$

$$\overline{wq} = \sigma\overline{wq}^c + (1-\sigma)\overline{wq}^e \tag{8.37}$$

As in Equation 8.31 we use

$$\overline{w'q'} = \overline{wq} - \overline{w}\,\overline{q} \tag{8.38}$$

The second term on the RHS is due to the mean vertical circulation and is given by

$$\overline{w}\,\overline{q} = (\sigma w^c + (1-\sigma)w^e)(\sigma q^c + (1-\sigma)q^e). \tag{8.39}$$

The first term on the RHS is decomposed according to Equation 8.37. For the two parts accounting for the cumulus and environment areas we now apply Reynolds averaging separately and neglect correlations within these two regions, which yields

$$\overline{wq}^c = w^c q^c, \quad \overline{wq}^e = w^e q^e. \tag{8.40}$$

Inserting these into Equation 8.37 and in turn into Equation 8.38 we obtain

$$\overline{w'q'} = \sigma(1-\sigma)(w^c - w^e)(q^c - q^e). \tag{8.41}$$

Using the small-area approximations

$$1 - \sigma \approx 1, \quad w^c \gg w^e \tag{8.42}$$

and multiplying with the density to represent a flux we obtain

$$\rho\overline{w'q'} = \rho\sigma w^c(q^c - q^e). \tag{8.43}$$

By introducing the convective mass flux $M^c = \rho \sigma w^c$ this can be rewritten as

$$\rho \overline{w'q'} = M^c(q^c - q^e). \tag{8.44}$$

The mass flux scheme links the sub-grid covariance $\overline{w'q'}$ to actual processes, represented by the convective mass flux M_c. The question is now how to calculate the mass flux. In reality a grid-cell area would be populated by several individual cumulus elements that all have their individual mass fluxes M_i^c. The vertical profile of these mass fluxes depends on the air entering and leaving the convection area at each vertical level, which are known as entrainment and detrainment rates (ϵ_i and δ_i). A frequently used approach is bulk mass flux schemes, in which the individual cumulus elements are replaced with a bulk convection, for which the mass flux is given by $M^c = \sum_i M_i$ and the entrainment and detrainment rates by $\epsilon = \sum_i \epsilon_i$ and $\delta = \sum_i \delta_i$. The temporal development and vertical profiles of the mass flux $M^c(z, t)$ and the moisture $q(z, t)$ are then governed by budget equations for mass and moisture within the bulk convection, which are

$$\frac{\partial M^c(z, t)}{\partial t} + \frac{\partial M^c(z, t)}{\partial z} - \epsilon(z, t) + \delta(z, t) = 0 \tag{8.45}$$

and

$$\frac{\partial \left((q^c(z, t) M^c(z, t))\right)}{\partial t} + \frac{\partial \left((q^c(z, t) M^c(z, t))\right)}{\partial z} \tag{8.46}$$
$$-q^e(z, t)\epsilon(z, t) + q^c(z, t)\delta(z, t) + C(z, t) = 0.$$

As the convective elements typically cover only a small fraction of the grid area, the moisture content outside the cloud can be replaced to a good approximation by the grid-cell average (i.e. $q^e \approx \bar{q}$).

In order to find a solution for the budget equations, one needs to specify the vertical profiles of the entrainment, detrainment and condensation rates and constraints for the temporal development (i.e. the time derivatives) as well as boundary conditions at the surface and at the top of the atmosphere. The entrainment and detrainment rates are only roughly constrained by observations and may vary considerably between different convective situations. Their specification is still an area of ongoing research. Constraints for the temporal development are based on prescribing how long it typically takes to reach equilibrium from a convectively unstable situation; usually shallow and deep convection are treated separately. The condensation rates $C(z, t)$ and formation of precipitation are calculated explicitly using cloud microphysics schemes. Thus the heating, moistening and precipitation generation in convective clouds is simulated by a combination of explicit calculations and implicit representation through the mass flux schemes (for more details see e.g. Stensrud 2009). The boundary conditions at the surface are derived from the convergence of the resolved flow.

Recently the increasing computing power has made it possible to run RCMs with horizontal resolutions of around 1km (e.g. Prein et al. 2015). At these high resolutions convection is partly explicitly resolved and thus the models are called convection permitting. However a substantial part of real-world convection is still occurring on smaller scales, and this spatial scale is known as the grey zone in climate modelling.

At these resolutions, the convection parameterisation used in coarser-resolution models needs either to be modified or switched off. Although in the latter case the real-world processes are not fully resolved, the kilometre-scale convection might be sufficient to produce realistic convective heat and moisture transports. The resolutions needed for an almost fully convection-resolving model are on the order of 25m and will still not be computationally feasible for a few decades for long climate simulations.

8.5   Performance of Climate Models

As stated in Section 4.3, the input to downscaling methods needs to fulfil certain requirements. PP methods rely on large-scale predictors that are bias free simulated; MOS methods use grid-scale predictors. Even if MOS by construction adjusts biases, it cannot overcome errors resulting from a substantial misrepresentation of relevant processes. In fact, for both approaches the predictors need to be realistically simulated for present climate and credibly represent future climate. Yet all models are simplifications of reality, and therefore some aspects of the model output will be substantially different from their real-world counterpart, whereas others will be realistic and provide valuable information. This fact is expressed in the famous quote by statistician George E. P. Box, "All models are wrong, but some are useful" (Box and Draper 1987).

In Chapter 2, we discussed the large- to local-scale processes governing regional climate. These processes determine the variability and change of downscaling predictors. For PP methods, the large-scale processes will play the major role; for MOS methods the whole chain of processes from the large-to local scale will be important. We now will discuss the performance of climate models in simulating these processes and their response to climate change. A general overview of climate model performance, in particular of GCMs, is covered in Flato et al. (2013). Sillmann et al. (2013) evaluate the skill of the CMIP5 GCMs in representing different types of extreme events. Region-specific RCM evaluation results can be found, for example, in Kotlarski et al. (2014) for Europe, Mearns et al. (2012) and Solman et al. (2013) for North and South America, respectively, and Kim et al. (2014) for Africa.

Model errors are mainly caused by the finite – horizontal and vertical – resolution of climate models. The primitive equations are solved numerically by finite differencing schemes (Section 8.3), sub-grid processes are simplified by parameterisations (Section 8.4) and sub-grid details of the orography are not resolved. In many models, the vertical resolution in the stratosphere will be coarse. These simplifications may cause model errors at small and regional scales. Interactions between small and large scales make these errors typically grow into large-scale errors (e.g. Slingo et al. 2003, Volosciuk et al. 2015). See also Section 8.4.1. Model biases may also have remote influences (Wang et al. 2014).

Crucially, errors in the representation of present climate may affect the response to external forcing and will therefore cause model uncertainties of future projections (see Section 9.1.2). For instance, uncertainties in global climate sensitivity mainly result from the representation of cloud feedbacks (Bony et al. 2006, Flato et al. 2013, Bony et al. 2015). Model errors thus limit the credibility of climate change simulations. In the following, we will present a non-comprehensive overview of model errors at large and regional to small scales.

To begin with, however, it is useful to recall the distinction between dynamically and thermodynamically controlled changes (Shepherd 2014, Section 2.3). The former is related to changes in the atmospheric circulation, whereas the latter describes changes independent of circulation changes. Most of our robust knowledge about climate change stems from thermodynamic changes, whereas changes in atmospheric circulation are highly uncertain, due to both model errors and the strong influence of internal climate variability.

8.5.1   Large-Scale Errors

Large-scale errors affect mainly GCM simulations, but they are inherited by downscaling (see also the discussion in Section 17.1). RCMs may introduce further large-scale errors (Becker et al. 2015), but these are typically small compared to GCM errors. Thus, the following discussion mainly applies to GCMs. In a downscaling context, GCMs have to correctly represent the large-scale temperature fields, circulation of the atmosphere and oceans and the hydrological cycle, as well as their response to climate change (see Section 2.1). As discussed earlier, this requires also a realistic representation of the effect of small-scale processes (such as convection, clouds, sea surface fluxes and ocean eddies) on large-scale processes.

Model errors have a characteristic signature in biases of surface climate at all scales (Figure 8.6). Probably the most evident biases in the annual mean temperature field of the CMIP5 ensemble are the warm biases in the upwelling regions of the eastern tropical oceans (Richter 2015). Also the equatorial Atlantic cold tongue is too warm (Richter and S.-P. Xie 2008), whereas the Pacific cold tongue is too strong (Li and Xie 2014). The inter-tropical convergence zone (ITCZ) is generally misplaced, or even split into a double ITCZ such as in the eastern Pacific. Tropical sea surface temperature (SST) biases are not fully understood but have various causes such as a misrepresentation of heat transport by ocean eddies, ocean–atmosphere feedbacks, low stratospheric clouds or orography and in turn of winds and ocean upwelling (e.g. Richter and S.-P. Xie 2008, Li and Xie 2014, Richter 2015). These biases in turn affect longer-term variability and the climate change signal in the tropics (Richter et al. 2014, Ding et al. 2015, Zhou and Xie 2015). Closely linked are major problems in the representation of El Niño Southern Oscillation (ENSO; Bellenger et al. 2014) and the Monsoon systems (Roehrig et al. 2013, Sperber et al. 2013). In case of ENSO, even crucial components such as the Bjerknes feedback between SST and Walker circulation are misrepresented by many models. The resulting uncertainties are so severe that climate model projections disagree even on the sign of ENSO amplitude changes under global warming (Chen et al. 2015). The West African monsoon rainfall is displaced towards the south; future changes are still very uncertain, including drying as well as strong increases (Roehrig et al. 2013).

The representation of mid-latitude climate suffers from strong model biases as well. A common and persistent problem of many GCMs is the North Atlantic cold bias

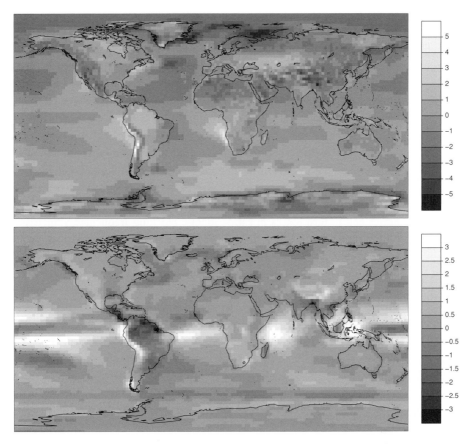

Figure 8.6 CMIP5 multimodel mean bias of annual mean surface air temperature [K] (left) and precipitation [mm/day] (right) for 1980–2005. Reference observations: ERA-Interim reanalysis (temperature) and GPCP data (precipitation), adapted from Figure 9.2 (b) and Figure 9.4 (b), Stocker et al. (2013). Data courtesy of Peter Gleckler and Hongmei Li.

(Flato et al. 2013), owing to unresolved eddies in the Gulf Stream (Eden et al. 2004). Similarly, sea surface temperature fronts in the oceans are not well resolved by ocean models. Such biases have, together with other effects such as the representation of orography, knock-on effects on the mid-latitude storm tracks and jet stream position (Nakamura et al. 2008, Woollings 2010, Keeley et al. 2012). In fact, the North Atlantic storm track in most state-of-the-art GCMs is southward displaced and too zonal in winter, simulating too few and weak cyclones over the Norwegian Sea and too many cyclones in central Europe (Woollings 2010, Zappa et al. 2013). Northern Hemisphere blocking frequency is too low, in particular in boreal winter (Masato et al. 2013). A misrepresentation of circulation regimes may result in qualitatively wrong response to global warming (Dawson et al. 2012). Uncertainties in the large-scale circulation response to global warming are still high (Collins et al. 2013). Uncertainties in the response of local surface climate depend strongly on the considered variables and climatic aspects and the scale and region of interest. European precipitation projections on large scales, for example, are dominated by thermodynamic effects and therefore rather robust (Kendon et al. 2010). In transition regions between projected wetting and drying, however, dynamical changes are relevant and cause high uncertainties in projected precipitation changes. For instance, the RCMs from the CORDEX and the ENSEMBLES projects broadly agree in the projected wetting of Northern Europe and drying of Southern Europe. But in many of the transition regions in Central Europe, the ENSEMBLES and CORDEX ensemble projections simulate qualitatively contradicting changes (Maraun 2013*b*, Jacob et al. 2014).

There is hope that, because of the earlier-mentioned interaction between small and large scales, the increasing resolution of GCMs will substantially reduce large-scale circulation errors. For instance, Scaife et al. (2011) have shown that an increase in horizontal resolution in the ocean model substantially reduces North Atlantic SST biases and in turn biases in Euro-Atlantic blocking frequency. Similarly, Davini et al. (2017) found that a 16km global GCM forced with observed sea surface temperatures can properly capture wintertime European blocking frequency, and Woollings (2010) pointed out that high-resolution GCMs have much more realistic jet stream positions. However, coupled GCMs of sufficiently high resolution to remove major SST and atmospheric circulation biases are currently too expensive for transient ensemble simulations of climate change. This argument is also relevant for PP downscaling of current-generation GCMs: to realistically simulate the required large-scale predictors, the GCM resolution has to be sufficiently high to well represent synoptic-scale weather and climate.

8.5.2   Regional- and Local-Scale Errors

Regional and local model errors are mostly relevant for MOS approaches: their intention is to make use of the full process chain down to the grid scale, but therefore they also rely on the representation of the process chain. Crucially, since MOS may use not only RCM but also GCM output as predictors, the following discussion applies to both dynamical model types. In fact, the finer resolution of RCMs may substantially improve the representation of MOS predictors (see the discussion in Section 17.1). As laid out in Section 2.2, regional processes that are relevant for model performance at the grid scale are the influence of topography, land-atmosphere feedbacks, as well as small-scale dynamical, thermodynamical and microphysical processes.

GCMs still have a rather coarse resolution (see Figure 1.1 in Section 1.1). Thus, GCMs typically do not well represent local climate in regions of complex topography such as coastal regions and mountain ranges (Flato et al. 2013). Here, RCMs may add considerable value (Deser et al. 2011, Hall 2014). But even RCMs of rather high resolution may misrepresent orographic precipitation (Caldwell et al. 2009) or may systematically shift meso-scale atmospheric flow (Maraun and Widmann 2015). The representation of orography affects the simulation not only of present climate but also of climate change. For instance, elevation-dependent warming caused by the snow-albedo feedback is misrepresented in GCMs (Walton et al. 2015), and also changes in Alpine convective precipitation intensity are better represented in RCMs (Giorgi et al. 2016).

Christensen et al. (2008) have shown that temperature and precipitation biases depend nonlinearly on the mean state. Bellprat et al. (2013) trace biases of high temperatures to soil moisture scarcity, and Maraun (2012) demonstrates that much of the RCM spread in temperature projections can be explained by different responses of soil moisture. Similarly, Hall et al. (2008) demonstrate that the uncertainties in temperature and soil moisture projections for the US can partly be explained by uncertainties in simulating the snow-albedo feedback.

Convective precipitation is parameterised in GCMs and standard RCMs (see Section 8.4). These parameterisations have difficulties representing crucial features of convective precipitation such as the diurnal cycle (Yang and Slingo 2001, Prein et al. 2013) and influence the representation of long-term precipitation means (Bukovsky and Karoly 2009). But furthermore, there is growing evidence that changes in summertime extreme precipitation are not well represented by convection parameterisations (e.g. Kendon et al. 2014, Meredith et al. 2015a). In fact, Meredith et al. (2015a) demonstrated that, for example, convective downdrafts, and in turn the response of heavy precipitation to low-level warming and moistening, are not credibly simulated by standard RCMs.

As discussed in the opening paragraph of this section, a suitable resolution is thus crucial for simulating useful MOS predictors. These issues and their consequences will be discussed in detail in Sections 12.6, 12.7 and 12.9.

In any case, the discussions about model performance highlight that a thorough evaluation of the chosen dynamical models is imperative prior to any statistical post-processing. In particular, even though bias correction may adjust model biases, it will not be able to correct fundamental model errors. Model evaluation in a down-scaling context, including dynamical models, will therefore be discussed in depth in Chapter 15.

8.5.3   Representation of Observed Trends

Several studies have been conducted to assess how well coupled climate models reproduce the observed spatial patterns of temperature and precipitation trends. Bhend and Whetton (2013) analysed the CMIP3 and CMIP5 multimodel ensembles and find that observed and modelled temperature trends are consistent over many regions of the globe, but marked differences exist: both ensembles underestimate temperature trends over central Asia and the Indian Ocean in winter; over the Arctic, Europe and Middle East in summer; and over the maritime continent in both seasons. Regions of overestimation are more scattered but coherent, for example, over parts of the Southern Ocean and the Altiplano in South America. Van Oldenborgh et al. (2013) show that the CMIP5 ensemble is overconfident; the spatial variability of warming patterns is too small. In Europe, climate models underestimate observed precipitation trends, mainly because these models do not simulate the observed trends in the atmospheric circulation towards a positive NAO in the second half of the 20th century (Barkhordarian et al. 2013, van Haren et al. 2013a, van Haren et al. 2013b). Whether the discrepancies between simulated and observed trends result from missing forcings, from a wrong response to forcing or from an under-representation of internal variability is not fully understood.

8.6   Further Reading

A good coverage of the physical processes that govern the climate system can be found in Neelin (2010) and with some emphasis on longer timescale in Goosse (2015). Both books also contain concise overviews of climate modelling. More technical details on climate modelling and numerical weather forecasting, for instance with respect to numerical methods and data assimilation, are given in a chapter in Holton and Hakim (2013) and in a comprehensive yet still accessible way in Kalnay (2003). Atmospheric dynamics is comprehensively covered in Holton and Hakim (2013), and clear overviews are also given in Lynch and Cassado (2006) and Randall (2015). An excellent introduction to atmospheric physics and dynamics is Wallace and Hobbs (2006). A general overview of climate model performance can be found in, for example, Flato et al. (2013). An easily accessible though insightful discussion of difficulties in simulating the processes controlling European climate has been given by Woollings (2010), and a review on biases in the tropical oceans has been published by Richter et al. (2016).

9 Uncertainties

Simulations of future climate are affected by inherent uncertainties – an iconic illustration of at least part of these uncertainties is the ensemble projections of global mean temperature (Collins et al. 2013, see Figure 9.1). Understanding, quantifying and attributing these uncertainties is of genuine scientific interest as they reflect both our limited understanding of the system and fundamental limits of predictability. But it is furthermore indispensable if credible and salient climate information is required for decision making (see Chapter 5). In this chapter we will discuss different types of uncertainties in climate projections (Section 9.1) and how they can be assessed by ensembles of climate models (Section 9.1). The actual interpretation of uncertainties will be discussed in Chapter 18.

9.1 Types of Uncertainties in Climate Projections

Uncertainties are often categorised into epistemic and aleatory (O'Hagan 2004). Epistemic uncertainty refers to our limited knowledge – it is thus in principle reducible. Aleatory uncertainty stems from unknowable knowledge and describes the limited ability to predict the behaviour of a system because of randomness intrinsic to the system itself – it is therefore often called ontic or ontologic uncertainty (van Asselt et al. 2002, Foley 2010). As it is a fundamental property of the considered system, aleatory uncertainty is inherently irreducible.

In practice the characterisation of uncertainty as either epistemic or aleatory is not clear cut but often reflects limitations in our current knowledge as well as practical constraints. For instance, the high uncertainty in state-of-the-art climate predictions from seasons to decades ahead may not be fully attributable to intrinsic stochasticity of the climate system but may be partly caused by our insufficient knowledge of initial conditions as well as model biases. Thus, part of the uncertainty characterised as aleatory might turn out to be epistemic and therefore in principle reducible (Hawkins and Sutton 2009).

Before discussing different sources of uncertainty in more detail, it might be helpful to recall how simulations of future climate – at the global and regional scale – are generated. As discussed in Chapter 8, climate models are numerical representations of our theories about atmosphere-ocean dynamics and thermodynamics as well as interactions with other components of the climate system. Uncertainties in climate models

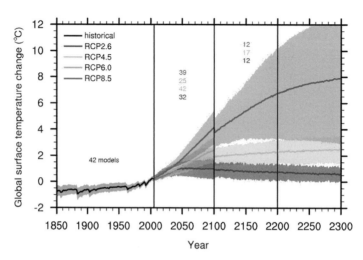

Figure 9.1 Projections of global mean surface temperature based on the CMIP5 ensemble of GCMs, forced with different representative concentration pathways. Adapted from Figure 12.5, Stocker et al. (2013).

arise from our limited understanding of the climate system and from the actual representation of physical laws in numerical models: continuous equations are discretised and often simplified, sub-grid processes are approximated by semi-empirical parameterisation schemes and model parameters are not fully constrained, partly because of observational uncertainties. The climate simulations themselves start from a set of initial conditions and are subject to scenarios of future radiative forcing. Different types of epistemic and aleatory uncertainties obviously affect these simulations. In principle, these types can be summarised into forcing uncertainty, model imperfections and internal climate variability (Stainforth et al. 2007).

9.1.1   Forcing Uncertainty

Forcing uncertainty stems from sources outside the climate system.[1] These sources might be natural such as volcanic and solar radiative forcing or anthropogenic owing to, for example, greenhouse gas emissions and land use change. Forcings can be included either as (time-varying) model parameters (such as the CO_2 concentration) or boundary conditions (such as land use).

Some natural forcings are highly predictable, such as changes in the earth's orbit according to the Milankovich cycles (Berger 1988), but these are irrelevant on centennial timescales. The occurrence of major volcanic eruptions, which considerably affect climate, is unpredictable on climatic timescales. Anthropogenic forcing is in principle

---

[1] Sometimes, factors that are influenced by climate and feed back into the climate system are prescribed as boundary conditions to reduce model complexity. Such factors, which are not strictly forcings, will be discussed in the next section.

unpredictable at scales beyond a few decades. The amount of greenhouse gas emissions depends on population and economic growth, political decisions to mitigate climate change and technological development. The most relevant limitation of predictability, however, is human reflexive uncertainty (Dessai and Hulme 2004): future mitigation decisions will react upon – then past – experiences of climate change and its impacts. These factors make it impossible to even attempt a probabilistic characterisation of forcing uncertainty.

Therefore, different scenarios have been defined to characterise a range of plausible future evolutions of greenhouse gas emissions and other forcings. In the IPCC's fourth assessment report, the scenarios from the IPCC's special report on emissions scenarios (Nakicenovic and Swart 2000) have been used. These scenarios were based on different plausible socio-economic future evolutions. In the IPCC's fifths assessment report, these scenarios have been replaced by so-called representative concentration pathways (van Vuuren et al. 2011). These scenarios prescribe a range of different plausible future evolutions of radiative forcings without assuming any specific socio-economic futures. Similarly, one may prescribe land-use change scenarios to assess the influences of active land-use management on regional climate. Natural forcing is commonly assumed to be constant at average present-day levels.

Any probabilistic assessments of future climate change are conditional on one or more of these emission scenarios. The resulting uncertainty is therefore often called scenario uncertainty. Such conditional climate predictions are called climate projections.

9.1.2   Model Imperfections

Model imperfections comprise two types of uncertainty. First, climate models are sophisticated but still highly simplified representations of climate dynamics. Not all relevant climatic processes are fully understood, and the complexity and accuracy of climate models is limited by computational and observational constraints. Thus, even the most complex state-of-the-art climate models do not realistically represent many relevant aspects of real-world climate, and several potentially relevant processes may not be simulated at all (see Section 8.5). Such model inadequacies (Kennedy and O'Hagan 2001, Stainforth et al. 2007) limit our potential to simulate future climates: whereas our confidence is high that climate models credibly simulate global mean temperature changes, our understanding of future changes in regional climate, in particular extreme events, is much weaker.

Second, even credibly simulated climate processes might be represented differently in different climate models (Smith 2002): different dynamical cores might be chosen, different resolutions, different parameterisations and different model parameters. Such uncertainties, resulting from the individual model choice and specification, have been called model uncertainty. The dynamical core determines how and on which grid the prognostic equations are numerically approximated. Model resolution controls which processes are directly simulated and which processes are parameterised. Furthermore, it influences the representation of resolved processes due to interactions between small and large scales (Slingo et al. 2003, Volosciuk et al. 2015). For a particular sub-grid process several parameterisations may exist (Section 8.4). For instance, many different convection parameterisations exist with their own specific advantages and disadvantages, all producing slightly different precipitation responses to the same forcings. In fact, the largest uncertainty in equilibrium climate sensitivity stems from the uncertainty in how cloud radiative feedbacks are parameterised in climate models (Bony et al. 2006, Dufresne and Bony 2008, Myhre et al. 2013, Bony et al. 2015). Finally, parameter uncertainty arises partly from the fact that finite observational data sets cannot fully confine model parameter values, but also because bulk parameters are in general not directly observable. Furthermore, the actual choice of parameter values is often a compromise to balance the applicability of a parameterisation across different climates. Note that climate model parameters are typically not universal constants; no true parameter values exist but only optimal values in the sense that they optimise the representation of selected processes.

The distinction between model inadequacy and model uncertainty is somewhat arbitrary. For instance, a model with too low a resolution may produce inadequate simulations of extreme precipitation. Nevertheless, the distinction is useful and informative: one could consider model uncertainties as that fraction of uncertainty that can be explored by multimodel and perturbed physics ensembles. From this point of view, the CMIP or CORDEX ensembles sample model uncertainty (which is consistent with the usage by, e.g., Hawkins and Sutton 2009). Model inadequacies then summarise all non-sampled uncertainties. The separation into both types of model imperfection thus depends on the chosen model ensemble. The more processes are represented and the higher the resolution, the more model inadequacy is transferred into model uncertainty. Model inadequacies of an ensemble of the most complex state-of-the-art models cannot be quantified (see Section 9.2).

9.1.3 Internal Climate Variability

Internal climate variability refers to the intrinsically chaotic and, at sufficiently long timescales, effectively stochastic behaviour of the climate system. If started from slightly perturbed initial conditions, the climate system would evolve independently after sufficiently long lead times. Therefore, the resulting uncertainties are often called initial conditions uncertainty (Stainforth et al. 2007). Different components of the system have different characteristic timescales, which are in turn linked to the predictability of the corresponding processes. For instance, the atmosphere has basically lost its memory after a couple of weeks, the ocean mixed layer and sea ice after a few years, the overturning circulation in the ocean after several centuries.

Atmospheric internal variability fundamentally limits the skill of weather forecasts, but basically has no relevance for decadal climate predictions or centennial climate projections: the climate distribution over one or several decades will not be affected by different atmospheric initial conditions (Stainforth et al. 2007 refers to this uncertainty as

---

> **Box 9.1** Model imperfections in statistical downscaling
>
> Statistical downscaling models are simplified statistical representations of meso- and small-scale atmospheric dynamics (see Part II). As discussed in Chapter 4, these models translate, based on a chosen model structure, a set of predictor variables into a set of predictand variables. In this context, model uncertainty is simply sampling uncertainty: parameter estimates are uncertain because the observational time series are finite. Model inadequacies are more relevant: the chosen model structure represents complex physical processes, in PP downscaling (Chapter 11) a cascade from synoptic to local scales, in bias correction on local scales. Any statistical model almost certainly does not adequately represent all aspects of these processes: often linear models are assumed to link predictors to predictands. But in reality, the relationship might be nonlinear, interactions between different predictors might exist and the relationship beyond the observed range might be different from the model extrapolation. Thus climate projections are almost certainly – at least to some extent – inadequate. Furthermore, in PP downscaling, predictors have to be included that capture local present-day variability and long-term climate change. Again, it is likely that not all relevant predictors have been included. The consequences of model inadequacies for building a statistical downscaling model are discussed in Part II of the book. The actual evaluation of downscaling model imperfections is discussed in Part III.

microscopic initial condition uncertainty). The longer the process timescales, however, the more relevant the associated internal variability will be in a climate context (macroscopic initial conditions uncertainty according to Stainforth et al. 2007): changing the initial conditions of the ocean model might result in very different climate distributions, even several decades into the future (Hawkins et al. 2016). On short lead times of a couple of years, uncertainties due to internal climate variability might be reduced by improved observations of oceanic initial conditions and improved forecasting systems (Hawkins and Sutton 2009, Meehl et al. 2009). On longer lead times, the internal climate variability becomes unpredictable but is still relevant, as it might strongly influence long-term climate distributions. Key in this context is the signal-to-noise ratio between the expected, that is, forced climate change signal and the effectively random internal variability. For a high signal-to-noise ratio, the initial conditions uncertainty might be negligible. But in particular at regional scales, internal climate variability has a strong influence on long-term climate distributions (Hawkins and Sutton 2009, Deser et al. 2012, Hawkins and Sutton 2012, Maraun 2013b). Thus, the common understanding that climate projections are a boundary value problem is in general wrong in the sense that initial condition uncertainty is large at the regional scale. It is, however, correct in the sense that for long lead times, the initial conditions are forgotten such that internal variability is essentially unpredictable.

9.2   Assessment of Uncertainties

Assessing uncertainties has two – of course related – aspects: first, model evaluation and, second, quantifying and understanding the uncertainties of future projections. Model imperfections can be evaluated by comparing climate model simulations against observations and by assessing the credibility of simulated future trends. In this context, internal climate variability is relevant for two reasons: on the one hand, it is a property of the climate system and may be imperfectly represented. On the other hand, internal climate variability is difficult to separate from forced trends and thus limits our ability to identify model imperfections. The evaluation aspect will be discussed – from a downscaling perspective – in more detail in Part III.

Uncertainties in climate projections can at least partially be explored by ensembles of climate model simulations (see also Section 8.1):

- Ensembles with different forcings can be employed to assess the sensitivity to greenhouse gas emissions, land use change and other forcing factors.
- multimodel ensembles can be employed to assess uncertainties due to the choice of different model parameterisations, dynamical cores and different resolutions.
- perturbed physics ensembles can be employed to quantify parameter uncertainties.
- initial condition ensembles can be employed to explore the influence of internal variability on long timescales. In particular, the ocean state should be varied to assess the variability of long-term modes of variability (macroscopic initial conditions uncertainty).

The individual contributions of scenario uncertainty, model uncertainty and internal variability for a particular variable, region, season and lead time can be approximately separated by ANOVA type methods (Déqué et al. 2007, Hawkins and Sutton 2009, Yip et al. 2011).

A useful measure to quantify the importance of different sources of uncertainty for a given lead time is the fractional uncertainty (Hawkins and Sutton 2009, see Figure 9.2). For any lead time, it separates the ensemble spread into spread resulting from different considered emission scenarios, the choice of the model and internal climate variability, and relates this spread to the ensemble mean signal at that lead time.

Obviously, fractional scenario (forcing) uncertainty is low for short lead times and grows towards the end of the century. Fractional model uncertainty decreases slightly with lead time; the actual model uncertainty of course grows, but it gets smaller compared to the projected mean change. Fractional uncertainty from internal variability decreases with lead time: for short time horizons, the climate change signal is weak, and almost all changes stem from internal variability. On longer lead times, internal variability stays roughly constant but gets weaker compared to the emerging climate change signal. This time-dependent distribution of uncertainty across different sources depends on the considered spatial scale and variable but also on the considered aspect. On the global scale, internal variability is much less relevant (Hawkins and Sutton 2009); for

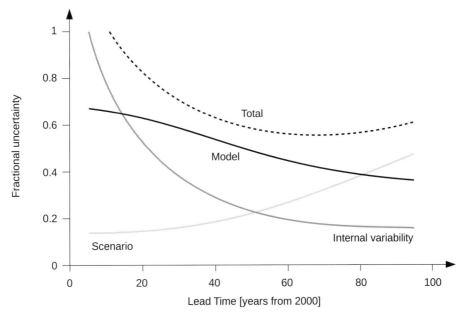

Figure 9.2 Fractional uncertainty of regional decadal mean temperature projections (schematic for an arbitrary region). Adapted from Hawkins and Sutton (2009). ©American Meteorological Society. Used with permission.

precipitation, it is more important than for temperature (Hawkins and Sutton 2011). Model uncertainty will in general be higher for extremes than for mean aspects.

This separation of course has some caveats: first, one cannot attach probabilities to forcing scenarios (see Section 9.1.1), thus the measured spread is rather arbitrary and depends on the available scenarios. Second, the separation is not perfect because (1) internal variability is not well sampled for individual climate models and can thus not be fully separated from model imperfections, and (2) model imperfections are not fully sampled (see discussion in the following section).

9.2.1   Deep Uncertainties

Climate is a complex system and shows emergent behaviour: the system develops macroscopic properties that cannot be reduced to microscopic physical rules. Emergent behaviour is reflected in climatic phenomena at many different scales such as the organisation of convection or the pathways of teleconnections between, say, ENSO and European climate. The relevance of these phenomena varies with space and timescales and may be different in a changed climate. As a consequence, climate projections are subject to deep uncertainties, that is conditions in which we do not know "(1) the appropriate models to describe interactions among a system's variables, (2) the probability distributions to represent uncertainty about key parameters in the models, and/or (3) how to value the desirability of alternative outcomes" (Lempert et al. 2003; the third issue is of course only relevant in a decision making context).

Here it is insightful to recall the distinction between model uncertainties and model inadequacies (Section 9.1.2). The former may be interpreted as the uncertainty sampled by an ensemble, the latter as the remaining, non-sampled uncertainty. In other words: even the most complex model ensemble can only explore part of the uncertainty range – relevant processes, which are not included or not well represented (because of our poor understanding or because of computational constraints) likely cause remaining model inadequacies. Even more, in practice, multimodel ensembles are in general ensembles of opportunity (Tebaldi and Knutti 2007): they consist of a historically evolved set of in general not independent climate models and parameterisations that have not been designed for the purpose of systematically exploring uncertainties. The real uncertainty ranges are thus likely larger than the simulated ranges. Better models will help to transfer non-sampled into sampled uncertainties.

In particular for extreme events, there is evidence that our uncertainty ranges might be too narrow (or centred on the wrong multimodel signal). For instance, Kendon et al. (2014) and Meredith et al. (2015*a*) demonstrated that the warming response of extreme summertime convection might be qualitatively different in models explicitly simulating deep convection from models parameterising deep convection.

9.2.2   Probabilistic Climate Change Projections

Several approaches exist to formally attach probability distributions to climate change projections, for example based on Bayesian statistics (Tebaldi et al. 2005, Buser et al. 2009) or, more simply, on Kernel dressing (Schölzel and Hense 2011). But even conditional on a given emission scenario, such projections should – in the presence of deep uncertainties – not be interpreted as reliable forecasts. A forecast is reliable if the uncertainty range is neither too wide (under-confident) nor too narrow (over-confident; Jolliffe and Stephenson 2003, Wilks 2006). Forecast verification assesses the skill and reliability of a forecast. But we cannot assess the reliability of a projection, and deep uncertainties almost certainly cause projections to be unreliable: if the latest, statistically post-processed, CMIP ensemble, projects a multimodel mean warming of 4°C for the end of the century in a certain region under a certain emission scenario, it is likely that the true forced warming will be different – it might be lower or higher (but if climate in that region is credibly simulated, it is likely not completely different). Thus, to avoid misunderstandings and misleading overconfidence, climate projections should better be interpreted as a (at best) credible range of future climates.

Box 9.2 Does downscaling add uncertainties?

Different modelling steps result in a "cascade of uncertainties" (Schneider 2001): forcing uncertainty feeds into climate model uncertainty feeds into impact model uncertainty. Often it is argued that downscaling adds another "layer of uncertainty" – and often it is implied that one should therefore not apply downscaling.

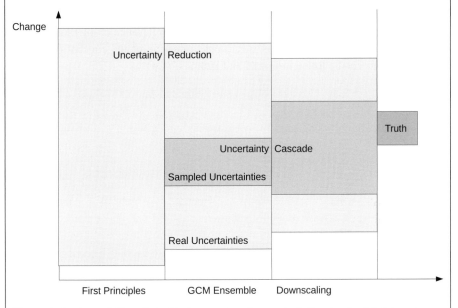

Uncertainties in modelling chain. Dark grey: true change, including irreducible internal variability. Medium grey: cascade of sampled uncertainty. Light grey: (unquantified) real uncertainty about the final output.

In the context of deep uncertainties and scale gaps between model and target resolution, this argument is not necessarily true. Of course, downscaling models will – as any model – suffer from model imperfections, which might increase the spread of the GCM ensemble (see Figure 9.3). But the GCM ensemble might not represent the regional change: local orography and feedbacks might modify the large-scale climate change signal, all with their associated uncertainties. Thus, the sampled GCM uncertainty might be too small or even in the wrong position. In fact, the concept of sampled and unsampled – real – uncertainties might help: based on first principles one might infer a wide range of plausible futures (light grey). An ensemble of GCM simulations will partly quantify uncertainties (medium grey) and at the same time help rule out some futures as implausible. But the GCM does represent large-scale climate change and may (in the hypothetical example) not even capture the (in reality of course unknown) true changes (dark grey). Downscaling, carried out with credible and properly evaluated models, might still increase the sampled uncertainties but might rule out even more futures as implausible and might even shift the uncertainty range closer to the true signal. In some situations, of course, no downscaling might be necessary. Finally, improperly done downscaling might also shift the projections away from the true signal.

9.3   Further Reading

An excellent overview of climate model uncertainties is given by Stainforth et al. (2007). The concept of fractional uncertainty has been introduced by Hawkins and Sutton (2009). Deser et al. (2012) highlight the relevance of internal climate variability for multidecadal projections. A critical discussion of the concept of probabilistic projections can be found in Dessai and Hulme (2004).

Part II

Statistical Downscaling Concepts and Methods

10 Structure of Statistical Downscaling Methods

When developing or selecting a statistical downscaling method, various choices have to be made, implying a variety of assumptions. Depending on the planned application, some of these assumptions might be fulfilled, others might be violated, some of the choices might be useful, others inappropriate. To gain insight into the implications of these choices, to understand differences or similarities of different methods, it is therefore helpful to consider the general ingredients of statistical downscaling methods. Most statistical downscaling methods establish an empirical link between a set of predictors $\mathbf{x}$ and predictands $\mathbf{y}$ by a statistical model $F(.)$,

$$\mathbf{y} \sim F(\mathbf{x}). \tag{10.1}$$

The predictands generally describe local or regional-scale surface weather or climate, whereas the predictors are generally of larger scale – therefore the term "downscaling". For calibrating the statistical model, the predictands are taken from observational data $\mathbf{y}_{obs}$. For the simulation of local climate $\mathbf{y}_{sim}$ based on numerical model output, the derived downscaling function $F(.)$ is then applied to a simulated set of predictors $\mathbf{x}_{sim}$.

Any downscaling method can therefore be characterised according to the different ways of treating and formulating the downscaling function $F(.)$, the predictors $\mathbf{x}$ and the predictands $\mathbf{y}$ under calibration and simulation. Under calibration, predictors can be taken from either observed or numerical model data, resulting in either perfect prognosis (PP) or model output statistics (MOS) methods (Section 10.1). The simulation of downscaled time series can be carried out either deterministically or stochastically (Section 10.2). Methods furthermore differ in their treatment of the local marginal, temporal and multivariate structure (Section 10.3). Finally, of course, methods differ in the choice of the function $F(.)$: for instance, regression models, weather typing approaches, analog methods or a simple mapping of distribution parameters. These transfer functions will be discussed in detail in Chapters 11, 12, 13 and 14.

10.1 Perfect Prognosis versus Model Output Statistics

Whether the predictors under calibration are chosen from either observations or model data defines the two most fundamentally different types of statistical downscaling: perfect prognosis (PP) or model output statistics (MOS; see Figure 10.1).

Box 10.1 A Babylonian Confusion

There is a tendency of downscaling model developers to invent method names which serve more to flatter the developer's vanity rather than to inform users about the characteristics of a method. We are convinced that the widespread Babylonian confusion of method names is an important reason why the statistical downscaling community is so scattered and why the understanding of which method is applicable in which context is so limited. Often, new methods are only marginally different from existing methods, but still completely new names are created. Often names describe rather technical details but do not at all link to the family tree of a method. Often new names are invented for already existing algorithms. Often, names are chosen that ring the wrong bell.

   We urge researchers to be humble in inventing method names. A name should be chosen to guide users, and to give information about the key elements of a method. It should link to existing methods and should not put the focus on marginal technical improvements. In any case, a method should be accompanied by useful metadata that precisely explain the position of a method in the space of possible implementations.

Traditionally, most statistical downscaling methods applied in climate research have been of the PP type: the calibration of the statistical model is carried out completely in the real world, hence both predictor and predictand data are taken from observations (or observation proxies such as reanalyses). The calibrated model is then applied to

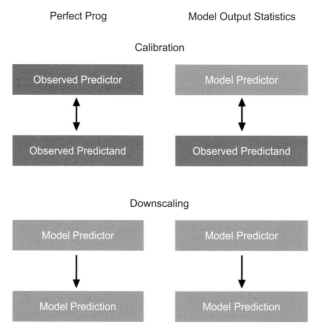

Figure 10.1 Perfect prognosis versus model output statistics. The top shows the origin of predictor and predictand data during calibration, the bottom the same for the actual downscaling.

> **Box 10.2** Calibration for climate change applications
>
> Statistical models are usually developed, calibrated and checked following the standard statistical procedures laid out in Section 6.3. In a climate change application, this procedure is not sufficient. In both PP and MOS, the model calibration will be dominated by day-to-day variability – it is by no means ensured that such models capture climate change trends. Related challenges will be discussed in Sections 11.6.1, 12.9 and 13.4.

predictors derived from climate model data. For this approach to be sensible, the predictors need to be "perfectly" simulated – thus the name. As a direct consequence of the calibration setup, the PP approach cannot correct model biases (see discussion in Chapter 11). Any bias in the simulated predictors will in general cause a biased downscaling.

As PP predictors and predictands are both taken from observations under calibration, they are in temporal correspondence: the predictor on 23 February 2012 corresponds directly to the predictand at the same date. A temporal sequence of pairs of predictors and predictands (x_i, y_i) exists, and a calibration is possible that explicitly makes use of the pairwise correspondence. For instance, a regression model can be built. We call this type of calibration pairwise calibration. An important feature is that pairwise calibrated downscaling models can be inhomogeneous: predictors and predictands can be different physical variables. For instance, one might use geopotential height and humidity fields to downscale precipitation.

In MOS–type downscaling methods, the calibration links predictors from a numerical model to observed predictands. Potential model biases already enter the calibration procedure and are thus taken into account by construction. If a model is systematically, say, a degree too warm, the calibrated MOS will automatically subtract this bias. In climate modelling, MOS is typically not calibrated in a pairwise setting: because free-running climate simulations (from a GCM, potentially downscaled with an RCM) are not in synchrony with observations, no predictor/predictand pairs exist and any calibration can only be based on long-term distribution. We therefore call this type distributionwise calibration. For instance, the simulated long-term mean can be adjusted to match the observed mean. Distributionwise calibration is only practically meaningful when predictors and predictands are of the same physical dimension, that is, when the method is homogeneous. MOS in climate sense is thus almost always a pure bias correction: simulated precipitation is corrected against observed precipitation.

10.2 Deterministic and Stochastic Downscaling

Different types of downscaling methods provide different types of simulated time series. First, consider a typical bias correction setting: an RCM simulates, say, daily temperature at a certain grid box (the predictors), yet the simulation is biased compared to observational data (the predictands). In a post-processing step, this bias is then subtracted from each simulated daily temperature value. Thus, for any predictor value, a unique

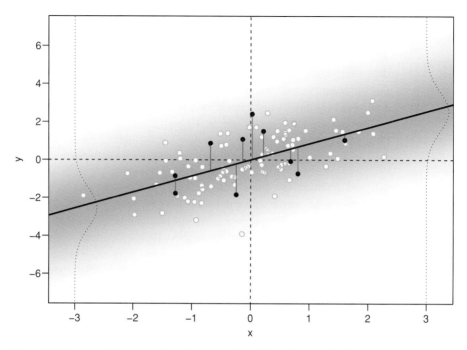

Figure 10.2 Deterministic and stochastic downscaling. The white dots indicate the data on which the model is calibrated. The bold black line depicts the predicted mean of the downscaling model, the grey shading and dotted lines the predicted distribution around the mean. Black dots indicate simulations from the predicted distribution.

corrected simulation is obtained by applying the bias correction. If the bias is 1 K, then a simulation of 14°C will always result in a corrected simulation of 13°C. This bias correction method may thus be termed deterministic. Now consider a typical weather generator that simulates random time series of precipitation, conditioned on the large-scale circulation. We call such methods stochastic.

In this context, deterministic regression models are a special case. Figure 10.2 shows pairs of observations $(x_i, y_i$; white dots), and a fitted straight line model. Often, a predictor time series x_i would be downscaled deterministically as $\mu_i = \beta_0 + \beta_1 x_i$ (the black regression line). But μ_i is actually the expected value of a predicted distribution (shading and dotted lines), given a predictor x_i – the output of a regression model is strictly speaking probabilistic. One could, for example, straight forwardly predict quantiles of the distribution such as the median or the 90th percentile. The downscaled μ_i should therefore be interpreted as a typical value that is expected under specific large-scale conditions. For temperature, it may be justified to treat μ as actual values (the distribution would be rather narrow); for daily precipitation it is not (also because of the skewed marginal distribution, which is not captured by the predictors; see discussion in the next section). These methods are commonly treated as deterministic rather than probabilistic; we will therefore refer to them as deterministic throughout the book. In principle, of course, one can obtain time series of actual values by simulating random numbers (black dots) from the predicted distribution (also called randomisation; von Storch et al. 2000).

Box 10.3 Inflated Regression versus Randomisation

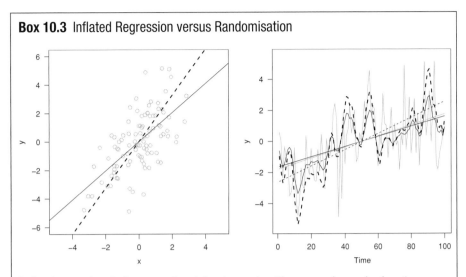

Inflated regression. Left: scatterplot; right: time series. Shown are observed values (grey dots, time series) and trend (grey line); mean predicted by linear regression, including trend (black lines) and inflated regression, including trend (dashed lines).

Often, the predicted mean of regression models has been interpreted as actual simulated values. This perspective has lead to the presumption that statistical downscaling methods are "smoothing", because they would not represent the full local variability. The less variability is explained by the predictors, the stronger is the "smoothing". Monthly temperature values might therefore be sufficiently explained by a conditional mean, whereas daily precipitation intensities are not well represented. Inflated regression has been proposed to better represent the local variance. The predicted variance is rescaled to the observed variance, that is every predicted anomaly is multiplied by a constant factor (see Figure 10.3). Von Storch et al. (2000) and Maraun (2014) demonstrated that this concept is flawed: the reason for the mis-represented variance is that local variability is not fully determined by large-scale variability. Yet inflation does not add unexplained variability; the resulting time series are still fully determined by the predictors. They are too smooth, and trends might be artificially amplified. Thus, variability unexplained by the predictors has to be stochastically simulated ("randomisation") as discussed in this section. Maraun (2013a) showed that inflation may also occur when quantile mapping is used for downscaling (see Section 12.7.1). The effect of inflation will be discussed in Chapter 16.

10.3   Marginal, Temporal and Multivariate Structure

Different downscaling approaches differ very much in their simulation of the statistical climate aspects defined in Section 4.2: the marginal distribution, residual temporal and spatial dependence, and the dependence between different variables. At the one end of the range of possible implementations are deterministic regression models and simple bias correction methods: these simply inherit all the mentioned characteristics from their predictors. For instance, an additive bias correction fully preserves the spatial, temporal and multivariable correlation structure and leaves the marginal distribution unchanged apart from an adjustment of the mean. A PP deterministic regression model represents the local variable as a linear combination of large-scale predictors and thus determines the local marginal distribution and spatial-temporal dependence.

At the other end are pure weather generators: these do not rely on predictors and therefore have to explicitly model all characteristics. This approach of course gives freedom to choose appropriate distributions and dependence models but is ultimately limited by the required balance between the sophistication of the model and the availability of observational data for calibration. For instance, many weather generators are single-site weather generators: they do not simulate spatial dependence at all and would produce uncorrelated weather time series even at close-by sites.

In between, a wide range of methods exists: for instance, conditional weather generators are stochastic PP methods that combine the use of predictors with explicit statistical dependence models, and quantile mapping adjusts marginal distributions. Which approach is sensible depends heavily on the considered variables, statistical aspects and timescales, as well as on the driving climate model: if one is interested in typical daily temperatures, a simple regression model may give a good spatial and temporal representation. If one is interested in daily precipitation intensities, such a model will produce an inadequate marginal distribution and far too smooth a spatial-temporal variability. Here a sophisticated noise model would be required. For the temperature example, bias correction of a coarse GCM might work reasonably well, whereas one would need a high-resolution RCM as input for a useful spatial-temporal bias correction of daily precipitation.

10.4   Further Reading

The classical paper discussing the difference between the PP and MOS approaches is Klein and Glahn (1974). Issues related to inflation have already been recognised by Glahn and Allen (1965). An in-depth discussion on inflation and its adverse effects can be found in von Storch et al. (2000), Maraun (2013a), Maraun (2014) and Maraun (2016).

11 Perfect Prognosis

State-of-the-art GCMs do not simulate regional-scale climate processes and also have a limited ability to represent resolved meso-scale processes. But GCMs often do have reasonable skill simulating synoptic-scale variability, such as the passage of cyclones and anticyclones. Perfect prognosis (PP) statistical downscaling exploits this skill to simulate regional climate. The approach builds upon synoptic meteorology and climatology, which analyses the relationship between synoptic scale and regional weather (Hewitson and Crane 1996).

DEFINITION 11.1 In PP statistical downscaling, a statistical model that links large-scale predictors to local-scale predictands is calibrated to observed data. The statistical model is then applied to predictors simulated by climate models.

Comprehensive reviews about PP statistical downscaling methods for climate change studies have been published by Hewitson and Crane (1996), Wilby and Wigley (1997), Zorita and von Storch (1997), Fowler et al. (2007), Maraun et al. (2010a) and Schoof (2013); region-specific reviews are, for example, Hanssen-Bauer et al. (2005) for Scandinavia and Jacobeit et al. (2014) for the Mediterranean.

In the following, we will first introduce the assumptions underlying the PP approach (Section 11.1). The most widely used PP methods will be presented in Section 11.2. The skill to apply a PP method in a specific context is very much determined by the structure of the method. These issues will be discussed in depth in Section 11.3. Issues related to the use of different predictand variables, in particular impact-relevant predictands, will be summarised in Section 11.4. Crucial for any PP method to work sensibly under climate is a careful predictor choice. Relevant assumptions and potential predictors for temperature and precipitation will be discussed in detail in Section 11.5. Section 11.6 lays out the statistical construction of a PP method, with a focus on climate change applications. A cookbook is finally given in Section 11.7. Several advanced PP methods condition weather generators on large-scale atmospheric predictors. Given the peculiarities of these method types, we will discuss them individually in Chapter 13. The following discussions on assumptions and predictor requirements, however, apply equally to these methods.

11.1   Assumptions

In Section 4.3, we have briefly introduced the general assumptions a sensible downscaling is based upon. These assumptions can be further specified for the PP approach:

- A crucial assumption is already implied by the approach's name: perfect prognosis means that the predictors have to be realistically and bias free simulated in present climate. Additionally, in a climate change context, the response of the predictors to external forcing needs to be credibly represented.
- Informative predictors need to be selected that explain a large fraction of local variability on all timescales of interest. In a climate change context, all relevant predictors need to be included that determine future changes in the predictand variable.
- The statistical model needs to have a suitable structure. That is, the influence of the predictors on the predictand (possibly including interactions between the predictors) needs to be reasonably well incorporated. In a different climate, predictors and predictand will likely assume climatic states that are outside the observed range. Therefore, for the downscaling model to be sensible, its structure needs to sensibly allow for at least moderate extrapolations. Additionally, all statistical climate aspects of interest that vary substantially with climate change have to be conditioned on suitable predictors.

This list is a concise revision of previous discussions about PP assumptions (e.g. von Storch et al. 1993, Hewitson and Crane 1996, Wilby et al. 2004, Cavazos and Hewitson 2005). The two latter assumptions incorporate what is often called the time-invariance or stationarity assumption: if all predictors relevant for climate change are included, and their influence on the predictand is sensibly modelled, also beyond observed states, the model is valid in a future climate. In practice, of course, statistical models are only simple approximations of the mesoscale processes they are intended to bridge, such that time invariance is at best approximately given. Often it is assumed that predictors should not lay outside the range of observed predictors, but this assumption has been shown to be invalid for key predictors (e.g. Wilby et al. 2004). Therefore, in many practical applications, one instead has to assume that the model is suitable at least for moderate extrapolations.

In the following sections, we will discuss how these assumptions may or may not be met by different PP methods and what the implications of these assumptions are for predictor selection.

11.2   Methods

In the following, we give a concise presentation of methods used for PP downscaling. The discussion of methods is far from being exhaustive – the idea is to discuss the major concepts in detail and then present a brief overview of variants and applications. In this section, we do not discuss structural limitations in detail – these discussions can be found in Section 11.3.

11.2.1 Regression Models

Linear Models
The most obvious choice to statistically link predictor to predictand data is to formulate a regression model. In the simplest case, one might use a linear model according to Equation 6.58 in Section 6.3.1. A set of M predictor time series $x_{i1}, \ldots, x_{iM}$ is linked to a predictand time series y_i by

$$E(Y_i) = \mu_i = \beta_0 + \beta_1 x_{i1} + \cdots + \beta_M x_{iM}. \tag{11.1}$$

We deliberately write this equation in the less familiar form of a conditional expected value μ_i to emphasize that a linear regression model predicts an expected value, not actual values, as discussed in Section 10.2. Because the local weather is in general not fully determined by large-scale conditions, residuals η_i between predicted mean μ_i and predictand y_i arise. In case of a linear regression model, these are assumed to follow a normal distribution, $\eta_i \sim \mathcal{N}(0, \sigma^2)$, with zero mean and variance σ^2. This noise is often not explicitly simulated, hence most downscaled time series only represent "typical" local-scale values – such time series are then too smooth. Some authors have suggested "inflating" the downscaled time series to match the observed local variance (e.g. Karl et al. 1990). von Storch (1999) has demonstrated the logical flaw inherent to this approach; for a detailed discussion see Box 10.3 in Chapter 10. Linear regression has been applied mainly to downscale temperature (e.g. Wilby et al. 1999, Benestad 2002, Huth 2002, Benestad 2005, Huth 2005).

Note that linear models can – to some extent – account for nonlinear predictor–predictand relationships by transforming predictors prior to including them into the model. Such transformations include simple functions (e.g. quadratic or square root, exponential or logarithmic, sine or cosine), interactive terms (e.g. products between predictors) or more complex physically based transformations (Section 11.5).

Generalised Linear Models
In many situations the predictand y_i cannot be sensibly modelled by a normal distribution. Consider daily precipitation as an example. The distribution is highly skewed, with a high fraction of zeroes, many moderate events and only few high-intensity events. Often, therefore, daily precipitation models are separated into precipitation occurrence and the intensities on wet days. In such a setting, a generalised linear model (GLM, Section 6.3.2) according to Equation 6.67 might be suitable:

$$g(E(Y_i)) = g(\mu_i) = \beta_0 + \beta_1 x_{i1} + \cdots + \beta_M x_{iM}. \tag{11.2}$$

The conditional mean or expected value μ_i of the chosen distribution is still modelled as a linear function of a set of predictors but may be transformed by a link function $g(.)$.

Different link functions may be chosen for different purposes; for instance, precipitation occurrence can be modelled by a logistic regression (see Section 6.3.2). But also a nonlinear influence of the predictors can be modelled. An exponential link function, for example, allows us to incorporate multiplicative effects of the predictors on the predictands and ensures a positive expected value. The distribution of Y_i, conditional on x_i, is assumed to be from the exponential family. Often, precipitation intensity is assumed to follow a gamma distribution (Katz 1977, Chandler and Wheater 2002). GLMs have widely been used to model daily precipitation (Chandler and Wheater 2002, Yang et al. 2005, San-Martín et al. 2017), often as weather generators with an explicit model of temporal dependence (see Chapter 13).

An extension of GLMs is generalised additive models (GAMS; Hastie and Tibshirani 1990), where the linear combination in x_{ij} and β_j is replaced by a sum of nonparametric smooth functions of x_{ij}. Vrac et al. (2007a) have used GAMs to downscale monthly temperature and precipitation in simulations of the last glacial maximum.

GLMs can describe a wide range of predictor/predictand relationships. In some situations, however, one might want to model the predictor influence on several distribution parameters independently. For instance, predictors might affect mean and variance of the predictor in different ways. Here, the more flexible class of vector-generalised linear models (VGLMs; see Section 6.3.3) may be required. Instead of the conditional mean of a distribution only, a vector of Q distribution parameters $\Theta = (\theta_1, \ldots, \theta_Q)^T$ (consisting, e.g., of mean and standard deviation) of a distribution is predicted:

$$g_q(\theta_q) = \beta_{0,q} + \beta_{1,q}x_{i1} + \cdots + \beta_{M,q}x_{iM}. \tag{11.3}$$

Nonlinear Regression Models

As stated, already linear and generalised linear models offer a wide range of possibilities to represent nonlinear predictor-predictand relationships. So-called artificial neural networks (ANNs) are an alternative, semi-parametric regression-like approach (Hewitson and Crane 1994). ANNs mimic networks of neurons; the neurons are nonlinear functions of a linear combination of input variables, either from predictors or a previous layer of neurons. Different calibration procedures exist that are similar to parameter estimation in regression models. ANNs have, for instance, been used to downscale precipitation over South Africa (Hewitson and Crane 1996), Japan (Olsson et al. 2001) and the UK (Haylock et al. 2006).

Quantile Regression

Quantile regression might be a simple alternative to GLMs or VGLMs (Koenker 2005). Instead of predicting a full distribution, one only predicts a chosen quantile of a distribution. Key to the approach is, therefore, that no distributional assumptions are necessary. To avoid negative values, for example, in the case of precipitation, a censored approach can be used. A disadvantage of the distribution-free approach is that different but close quantiles may not be predicted in the correct order. Quantile regression has been applied, for example, to downscale precipitation (Friederichs and Hense 2007, Cannon 2011, Tareghian and Rasmussen 2013).

Multisite Regression Models

In principle, regression models can simultaneously downscale to multiple sites. The most widely used approach is to simply calibrate independent models to each station separately without modelling residual noise (Huth 2002). Alternatively MCA and CCA, which use linearly coupled patterns between the predictor and predictand fields, can be used (Section 6.5.3). The use of CCA in downscaling has been pioneered by Zorita et al. (1992), and the approach has been frequently used (e.g. Busuioc et al. 1999, Benestad 2001, Benestad 2002, Huth 2002). Downscaling based on MCA has also been used in several studies (e.g. Widmann et al. 2003, García-Morales and Dubus 2007, Chu et al. 2008, Paul et al. 2008, Chu and Yu 2010). The pattern-based methods have the potential advantage of having fewer regression parameters. Moreover, the coupled patterns might be helpful to understand the processes that link the large and small scales, and they constrain the links between the predictands (see also Section 11.3.3). These links can also be constrained by using PCs across target stations as predictands rather than the individual local records (Benestad et al. 2015). If also predictor PCs are used this approach becomes mathematically very similar to the CCA and MCA methods, as all three approaches find linear links between the subspaces of the leading PCs in the predictor and predictand fields, with the only difference being the exact formulation of the link.

Phenomena such as daily precipitation exhibit substantial unexplained small-scale variability. Here, one has to explicitly model the residual inter-site dependence, which is not explained by the predictors. Copulas (Section 6.1.7) are a useful framework for these models (e.g. Bevacqua et al. 2017). Such models, often used in conjunction with weather generators (e.g. Yang et al. 2005, Bárdossy and Pegram 2009), are presented in Chapter 13.

11.2.2   Weather Type Methods

The use of weather type methods has its origins in synoptic meteorology (Hewitson and Crane 1996). Weather types (see Figure 11.1) have long been used to characterise the spatial-temporal evolution of synoptic-scale weather (e.g. Lamb 1972, Hess and Brezowsky 1977, Huth et al. 2008, Philipp et al. 2010). A more specific discussion of different definitions will be given in Section 11.5. As discussed in Chapter 2, regional weather, in particular in mid-latitude winter, is strongly controlled by the synoptic-scale weather. The relationship between the large-scale flow and local scale weather is often rather complex and not well modelled by directly linking predictor fields and predictands by a linear model (for alternatives to better capture predictor–predictand relationships, see Section 11.5). The idea of weather-type-based downscaling is therefore to condition local weather on a set of K categorical large-scale weather types X_k, $k = 1 \cdots K$, which incorporate the most relevant synoptic circulation patterns.

The simplest method would just consider the mean μ of local weather y for a given weather type. At a given day i

$$\mu_i = \mu(X_{k(i)}), \tag{11.4}$$

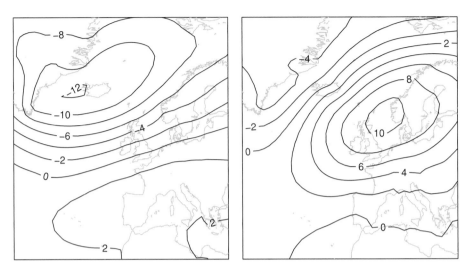

Figure 11.1 Sea-level pressure anomalies [hPa] for two weather types characterising North Atlantic and European weather. Left: positive NAO phase (1st rescaled EOF of the SLP field), right: blocking (2nd rescaled EOF of the SLP field). Based on ERA-Interim data (Dee et al. 2011*b*), 1979–2012.

where $k(i)$ denotes the number of the weather type at day i. The actual $\mu(X_k)$ is estimated by averaging observed weather across all days on which the weather type X_k occurs. This method, however, is rather crude as it ignores all within-type variability of surface weather. It has, therefore, only been applied to derive monthly or annual precipitation sums from integrating over mean precipitation in the sequence of modelled weather types (e.g. Saunders and Byrne 1999).

More advanced downscaling approaches therefore predict a full distribution $G(\theta)$ conditional on a weather type, that is

$$y_i \sim G(\theta_i) \text{ with } \Theta_i = \Theta(X_{k(i)}), \tag{11.5}$$

where $\Theta(X_k)$ is a parameter vector of the distribution, estimated separately for all days where the weather type X_k occurs. This model has, for example, been applied to downscale daily temperatures (Gutiérrez et al. 2013) or precipitation (San-Martín et al. 2017). In most cases, however, these models include a specific Markov component to account for short-term memory. Such weather generators will be discussed separately in Chapter 13.

11.2.3   Analog and Resampling Methods

Analogs have originally been used to assess atmospheric predictability (Lorenz 1969). Two solutions of the equations of motion, starting from similar initial conditions, should evolve similarly up to a certain lead time, with a growing difference. Thus, for two similar observed atmospheric states – the analogs – the temporal evolution should initially be similar. This principle has been applied for empirical weather forecasts (Lorenz 1969,

Kruizinga and Murphy 1983) or short-term climate predictions (Barnett and Preisendor-fer 1978): instead of numerically integrating the equations of motions, one employs an observed weather sequence from the past, starting from an analogue, as prediction.

In downscaling, a similar aspect of analogs is used: two similar large-scale atmo-spheric states may result in similar local surface weather states (Zorita and von Storch 1999). Local weather at a day i, given some state of the large-scale circulation X_i, is therefore downscaled as follows: one first searches for an analog of the large-scale cir-culation $X_j = \text{analog}(X_i)$ at a day i when observations y_j of local weather are available. These observations are then used as prediction of local weather at day i:

$$y_i = y_j \text{ with } j : X_j = \text{analog}(X_i). \tag{11.6}$$

Analogs are defined on a chosen predictor set X, which is often filtered to eliminate small-scale noise, by minimising a cost function such as the Euclidean distance:

$$\text{analog}(X_i) = X_j | (X_i - X_j)^2 \overset{!}{=} \min . \tag{11.7}$$

The best choice of distance metric depends on the target variable, but the Euclidean norm performs satisfactorily in most cases (Matulla et al. 2008). To avoid sampling identical local states for the same (after filtering) large-scale weather situation, that is to introduce a random component, it has been suggested to randomly select between the k most similar atmospheric states (Lall and Sharma 1996). In principle, one could introduce dependence between the sampling of subsequent analogues (see discussion in Buishand and Brandsma 2001).

If the same analog is selected for a set of stations, the analog method by construction preserves spatial and inter-variable dependence. Therefore, the analog method has also been used in conjunction with regression models to simulate spatially coherent random fields. For instance, Wilby et al. (2003) simulated a random sequence of area-average precipitation values, based on a regression model using large-scale predictors. For each day, the observed area average closest to the simulated precipitation value was selected as an analog. Station data were then simply simulated by the observed station data from the analog. Such methods are also called conditional resampling.

A key issue for the success of the analog method is – trivially – the existence of analogs (i.e. states where $(X_i - X_j)^2$ is not just minimised but indeed small). Van den Dool (1994) pointed out that, given the large number of degrees of freedom of the atmosphere, at a global scale essentially no analogs exist. Yet Zorita and von Storch (1999) argue that the analog method is typically applied over a limited domain only and that much of the complexity of a given state can be regarded as noise and filtered out. This line of argument is in particular justified as no prediction in time is sought, and therefore no very precise analog is required, but only a zero lag prediction of local weather. However, the problem indeed exists for rare events, which may be sampled poorly. Very extreme events may not have been observed at all. Therefore, Young (1994) proposed randomly perturbing observed values.

To relax the problem of missing analogs and to allow for extrapolations, Hidalgo et al. (2008) proposed to combine a set of analogs using a weighted average. The approach is homogeneous and uses, for example, large-scale precipitation fields to define analogs for downscaling precipitation. To derive analogs, a high-resolution observational data set is first aggregated to the climate model resolution. For each day in the simulated time series, a set of coarse-resolution analogs is selected from the observations, and a regression model is calibrated between these coarse-resolution analogs and the simulated field. This regression model is then applied to the set of corresponding is high-resolution observational fields yielding a "constructed analog", which serves as a downscaled version of the simulated large-scale field. Simulated large-scale precipitation fields, however, likely do not fulfill the PP condition. To address some of the shortcomings of the constructed analog approach, several modifications have been implemented, including a combination with bias correction (Maurer et al. 2010, Pierce et al. 2014).[1] These methods will be discussed in Chapter 14.

11.3   Structural Skill and Limitations of PP Methods

In Section 4.2, we introduced a set of statistical climate aspects a downscaling method may be required to represent. These are marginal, temporal, spatial and multivariable aspects of the climate distribution. In the following, we will discuss how different types of PP methods by construction represent these aspects. Later, in Part III, we will discuss how specific implementations actually perform.

Here we discuss only skill linked to the structure of downscaling methods. It is obvious that if the predictors are not suitable, a downscaling method will not correctly represent local climate and its changes. Predictor requirements will be discussed separately in Section 11.5.

A weakness common to all PP methods is the inability to convincingly represent local feedbacks. For instance, local warming will enhance evaporation – once the soil has dried out, sensible heating of the lower atmosphere will increase. PP projects the large-scale state of the free atmosphere onto local weather. That is, if the feedback acts on a scale that does not alter the large-scale circulation as resolved in the driving climate model, PP will miss this effect. Some weather generators attempt to capture such feedbacks by simulating temperature dependent on the spell lengths (Section 13.2.4).

11.3.1   Marginal Aspects

Given the fact that PP downscaling models are calibrated statistical methods, they should trivially represent mean climate at a given location. Whether higher moments such as the variance or extremal properties such as specific high quantiles are reliably represented depends very much on the implementation of the method.

Deterministic regression models do not explicitly simulate unexplained variability. These methods then do not represent the full distribution, in particular not extreme

---

[1] When local-scale processes and feedbacks are important, for example in complex terrain or when considering heavy precipitation, the local climate change signal may differ from the large-scale signal. Such problems are relevant for constructed analogs and will not be overcome by standard bias correction; see the discussions in Section 12.7.

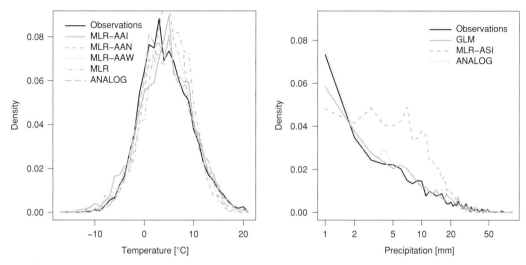

Figure 11.2 Histograms of observed and downscaled winter temperature (left) and summer precipitation (right) at Graz, Austria. The different PP downscaling methods have taken their predictors from ERA-Interim and are cross-validated. Local observations from ECA-D data (Klein Tank et al. 2002), for the period 1979–2008. For details of the methods, see Appendix A.

events. Some authors suggested "inflating" the regression (e.g. Karl et al. 1990), but as discussed in Box 10.3, this approach is not valid: it assumes that local climate is completely determined by large-scale weather and, as a result, does not accurately simulate local weather (see also von Storch et al. 2000, Maraun 2013a, 2014). An alternative is a stochastic approach, that is to explicitly simulate random noise, conditional on the large-scale predictors (e.g. in a VGLM the variance can explicitly depend on the predictors). As discussed, regression models can also be interpreted as probabilistic predictions of local climate: based on the chosen distribution and its predicted parameters, for each day a probability density function is predicted, and quantiles or other quantities of interest can be derived. Quantile regression by construction predicts quantiles only.

Weather type methods in general behave similarly to regression models in terms of marginal properties: they are typically deterministic methods, that is only the expected weather given a certain weather type is predicted, and no realisation of actual weather is generated. Also here, stochastic weather generator approaches have been developed, which provide information about higher moments and extreme events (see Chapter 13). Analog methods can essentially represent all marginal aspects for which a sufficient density of analogs exists. Hence, analogs will have difficulties representing the variability of extreme events (note, however, that in the evaluation presented in Section 16.5 an analog method performed well even for 20-season return levels).

Figure 11.2 illustrates the representation of marginal aspects for winter temperature (left) and summer precipitation (right) at Graz, Austria. The overall marginal distribution of temperature is well captured by all methods, with or without randomisation. This indicates that much of the variability is explained by the predictors. At closer look, however, it becomes obvious that all but the analog method (ANALOG) are biased towards higher temperatures. The deterministic, uninflated methods underestimate the upper tail of the distribution, while the inflated (MLR-AAI) and randomised (MLR-AAW) regression overestimate the lower tail. Differences between different method types become much more obvious for precipitation. The simple deterministic linear regression, even with inflation (MLR-ASI), essentially preserves the histogram of the predictors, that is a Gaussian-like distribution. Both the stochastic generalised linear model (GLM) and analog (ANALOG) methods simulate a realistic histogram. See also Figure 11.3 for some direct comparison of time series.

11.3.2   Temporal Aspects: Persistence and Systematic Variations

Many implementations of PP methods do not explicitly model temporal dependence – it is inherited from the predictors. Deterministic regression and weather type models thus simulate too smooth a temporal sequence. This holds in particular for precipitation at daily scales, which is not well determined by the large-scale circulation, especially during summer. Local temperature variability is typically stronger controlled by large-scale predictors. Also aggregation to longer timescales increases the fraction of explained variability.

Stochastic methods add random noise. If this noise were white, that is without any temporal dependence, it would likely deteriorate the representation of temporal structure. Advanced stochastic PP methods therefore employ weather generators to explicitly model residual temporal dependence (see Chapter 13).

The annual cycle is most often explicitly modelled, for example by calibrating the model individually to different seasons or months or by adding harmonic components. Some models, however, attempt to implicitly include seasonality via the chosen predictors. The former approach is straightforward, but it is in general questionable whether the annual cycle can be assumed unchanged in a future climate. The latter approach is more process oriented. If suitable predictors are found that describe the processes underlying seasonality, changes in the annual cycle might successfully be represented.

Figure 11.3 shows observed and simulated time series for Summer 2003 in Graz, Austria. For temperature (top), basically all methods reproduce the overall temporal variability. Only the white noise randomisation (MLR-AAW) destroys too much of the temporal dependence. For precipitation (bottom), the deviations are much larger. None of the presented methods explicitly models temporal dependence; it is purely taken from the predictors. The inflated deterministic regression model captures the overall evolution but simulates far too smooth a time series with too many wet days. The analog method does not capture the dry–wet sequence and misses many important events. The stochastic GLM more or less reproduces at least the major temporal features.

11.3.3   Spatial Aspects

The behaviour of statistical downscaling methods in representing spatial aspects is similar to that in representing temporal aspects: deterministic regression and weather type models are single-site methods in the sense that they do not explicitly model spatial dependence. It is inherited from the predictors. For instance, a deterministic weather

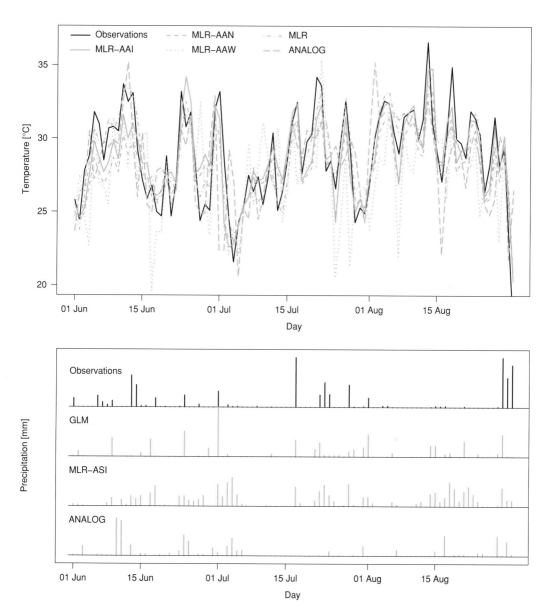

Figure 11.3 Observed and downscaled temperature (top) and precipitation (bottom) during the heat wave of summer 2003 at Graz, Austria. The different PP downscaling methods have taken their predictors from ERA-Interim and are cross-validated. Local observations from ECA-D data (Klein Tank et al. 2002), for the period 1979–2008. For details of the methods, see Appendix A.

typing method would simulate the expected precipitation field given a specific weather type. Such a field is no realisation of a typical precipitation field and thus far too smooth in space. Again, for smooth fields such as for temperature, and at long aggregation times, it might be justified to interpret typical fields as individual realisations.

In contrast to regression for individual predictands, where the link between them is only induced through the predictors, the pattern-based MCA and CCA methods, as well as regression with predictand PCs, directly impose a link between the predictands. The predictand patterns, which are added up with amplitudes estimated from the predictors (Section 6.5.3), contain information about the spatial correlation structure of the predictands. For CCA and regression with predictand PCs, this is due to the explicit prefiltering with the leading EOFs, while for MCA the covariance maximisation condition implies that the patterns are also in the subspace of the leading EOFs. But note, again, that pattern methods – such as all deterministic regression methods – do not explain residual variability which is not imprinted by the predictors. If the explained variance is low, day-to-day variations will be too smooth, amplitudes too low.

Stochastic methods are often single-site methods and therefore do not model the residual spatial dependence that is not explained by the predictors. In particular for daily precipitation, the spatial coherence imprinted by the predictors is typically weak and will be destroyed by the noise. Such methods could simulate a dry, sunny day at one location and an overcast drizzle day at a neighbouring location. Thus, for highly variable fields at short timescales, explicit spatial models are in general required to realistically represent spatial dependence. Skill and limitations of such models will be discussed in Chapter 13.

The analog method resamples observed weather fields including their observed spatial dependence structure. As such, analog methods are an ideal candidate for spatial downscaling. In practice, however, analog methods are often applied to individual stations separately. Stochastic implementations may then select different analogs for different stations and thereby to some extent destroy spatial dependence. For extreme events, the analog method might find only a few (if any suitable) analogs, thus that the spatial variability of extreme events will not be sampled but only a few patterns (which by chance occurred in the observations) will be repeated.

Figure 11.4 shows examples of how different types of PP methods represent residual spatial dependence for 8 December 1993. Observed precipitation clusters mainly along the line Scotland-Netherlands-Germany-Switzerland. All methods rather well simulate the occurrence patterns. The stochastic single-site weather type method (WT-WG) simulates no coherent pattern, the single-site generalised linear model (GLM) simulates a rather broad range of intensities, and the analog method simulates high intensities mainly along the Northern Alps.

11.3.4   Multivariable Aspects

Deterministic regression and weather typing methods, again, represent the expected values for different variables. The conditioning on large-scale predictors prescribes covarying behaviour which is, again, in general too strongly correlated. Stochastic methods simulating each variable individually will destroy this covariance. Multivariate statistical models may thus be required to realistically model the dependence between different variables, in particular at short timescales. The analog method by construction produces a realistic multivariable dependence.

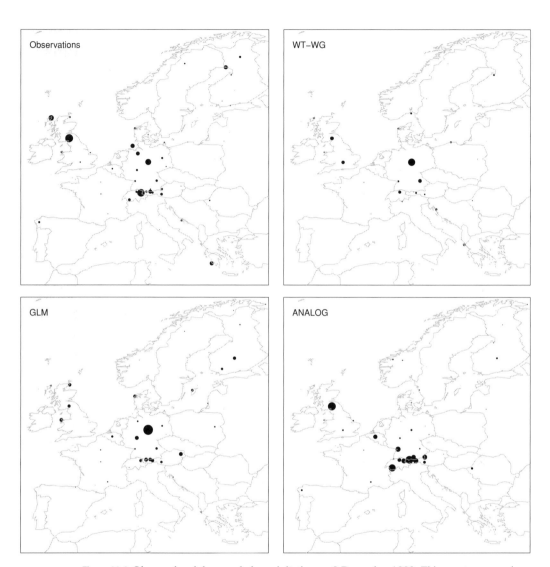

Figure 11.4 Observed and downscaled precipitation on 8 December 1993. This event was one in a row which lead to the 1993 Christmas flood along the lower Rhine. The different PP downscaling methods have taken their predictors from ERA-Interim and are cross-validated. Local observations from ECA-D data (Klein Tank et al. 2002), for the period 1979–2008. For details of the methods, see Appendix A.

Figure 11.5 illustrates the representation of multivariable aspects with summer temperature and precipitation at the airport of Madrid, Spain. Both variables are essentially independent for temperature between 20°C and 33°C, but for higher temperatures, no precipitation occurs. Neither the regression (MLR for temperature, GLM for precipitation occurrence, inflated regression for amounts) nor the stochastic weather type method (which simulates temperature and precipitation separately) reproduce this aspect. Only the analogue method by construction produces a similar relationship.

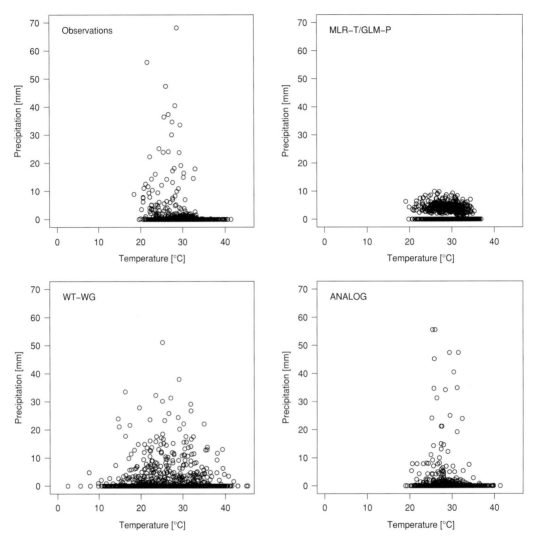

Figure 11.5 Observed and downscaled precipitation during summer at Madrid-Barajas, Spain. The different PP downscaling methods have taken their predictors from ERA-Interim and are cross–validated. Local observations from ECA-D data (Klein Tank et al. 2002), for the period 1979–2008. For details of the methods, see Appendix A.

11.3.5   Simulating Climate Change

All regression type methods are in principle capable of capturing climate change – as long as their model structure allows for moderate extrapolations and predictors informative of climate change have been included. These issues will be discussed in Sections 11.5 and 11.6 in detail.

Care has to be taken when using weather type and analog methods for climate change simulations. Typically, weather type definitions (Section 11.5) only account for pressure fields (or geopotential height fields at upper pressure levels); temperature and atmospheric moisture content are in general not included. For weather forecasting in present climate this choice may be justified, because a specific weather type also controls the temperature and humidity fields. In a future climate with its different radiative forcing, however, the thermodynamic properties of a weather type are likely to change. In such a situation, the climate change signal of the predictands will not fully be captured. This issue will be discussed in more detail in Section 11.5. Similarly, also the dynamic properties of a circulation type might change. The mean flow might get stronger or weaker or may shift. Thus, even the definition of the weather type might not be transferable to a future climate. The limitations of the analog method are similar to those of weather typing methods (Beersma and Buishand 2003, Yates et al. 2003): in a warmer climate, a specific circulation state is likely to be associated with surface weather different from that in present climate. For instance, temperatures will typically be higher, and if moisture is available, precipitation will be more intense. In other words: no real analogs for future weather situations might be available in observations of present climate. As a result, Gutiérrez et al. (2013) find that pure weather type and analog methods heavily underestimate projected temperature changes over the Iberian Peninsula. A way out might be to combine such method types with regression models (e.g. Gutiérrez et al. 2013).

11.4 Predictands

Predictand variables can in principle be any observed climatic variable. Typically, meteorological (near-) surface variables such as temperature, precipitation or wind have been target variables in downscaling. In many impact studies, rather complex indices are required, which are a combination of several meteorological variables over longer timescales, and thus require a reasonable representation of marginal, temporal and multivariable aspects. In such situations, it is conceivable to directly predict desired indices such as the maximum consecutive dry days in a season (Haylock et al. 2006) or indices specifically tailored for impact studies such as the fire weather index or the physiological equivalent temperature (Casanueva et al. 2014). By aggregating several meteorological variables, potentially over time, into one variable, these indices circumvent the development of sophisticated daily, multivariate statistical models. However, the stronger the aggregation, the more questionable it is whether the predictor–predictand relationship is time invariant.

11.5 Predictors

At the beginning of the chapter, we discussed the assumptions which have to be satisfied for a sensible PP downscaling model. Two of these refer explicitly to the predictors:

- The PP assumption requires the predictors to be realistically simulated in present climate and to credibly respond to global climate change.

Table 11.1 Requirements for sensible predictors of a PP model.

Perfect prognosis assumption	
predictors have to be realistically and bias-free simulated in present climate	– should represent synoptic processes without biases – should be from the free atmosphere – has to be satisfied not only for mean fields (or their anomalies) but also for variability
predictors have to be credibly simulated in future climate	– key processes and their response to climate change have to be credibly simulated for each predictor – requires sufficient understanding of driving model
Informativeness assumption	
predictors have to explain a large fraction of variance on the timescales of interest	– high correlation between predictor and predictand – ideally also on longer timescales – physical explanation for the relationship
predictors have to explain the response to climate change	– stable time-invariant relationship – representation of observed trends – pseudo-reality evaluation

- Predictors need to be informative, that is they need to explain a large fraction of local variability on all timescales of interest, including the response of the predictand to climate change.

In the following, we will discuss these assumptions in further detail and what their consequences for predictor selection are. Table 11.1 summarises the relevant details.

11.5.1   The Perfect Prognosis Assumption

The key rationale of PP methods is to identify predictive relationships between observed large-scale and local-scale weather and then to apply these relationships to simulated large-scale weather in climate change simulations. One therefore has to ensure that the driving dynamical model, typically a GCM, realistically simulates the relevant predictors in present climate and furthermore that the response of these predictors to climate change is credibly simulated. Any bias in the simulated present-climate predictors will cause a bias in the downscaled predictand. Any implausible response to climate change will result in an implausible downscaling of the predictors in future climate and thus noncredible regional projections. Importantly, these biases and implausibilities do not only pertain to marginal aspects such as mean or variances but also other, in particular temporal aspects such as blocking frequency, persistence or the onset of the rainy season (Charles et al. 2007).

The PP assumption has been linked to the minimum skillful scale of climate models (Grotch and MacCracken 1991, von Storch et al. 1993): GCMs may have a nominal resolution of the order of 100km, but processes acting on these scales will typically be rather poorly represented. It has been assumed that the minimum skillful scale – the scale at which processes can be considered to be realistically simulated – is several times the nominal resolution, that is at synoptic scales. Moreover, GCMs have a poor representation of topography and as a result of surface climate. Therefore, predictors for PP downscaling should represent synoptic-scale processes in the free atmosphere.

The link to the skillful scale as discussed, however, ignores a major issue for climate projections: the PP assumption has to hold for future climate simulations as well. Testing this requirement not only involves an assessment of mean fields and temporal variability in present climate but ultimately calls for a sufficient understanding of how credibly the driving GCM simulates future climate in the region of interest. It may turn out that in some regions even synoptic-scale predictors do not fulfil the PP assumption. Vice versa, it could be that for smoothly varying predictor variables, even grid-box values may suffice.

Closely linked to this discussion is the "un-biasing" of predictors: it is often argued that predictors should be defined as anomaly fields to eliminate biases in the predictor mean fields and in turn in the predictands (e.g. Karl et al. 1990, Wilby et al. 2002). This technique might indeed be useful to reduce predictand biases, but it does not guarantee that the PP assumption is better fulfilled: as in bias correction (see Chapter 12), a removal of the bias trivially results in an unbiased outcome for the calibrated aspects – here the mean climate of the predictand. Whether all other aspects are realistically simulated and whether the climate change signal of the predictor is credibly simulated has to be established separately.

Assessing whether the PP assumption is fulfilled is a highly non-trivial task – in fact, as it involves statements about the simulation of future climate, it can only be evaluated for present climate, for the future simulation only credibility can be established. We will discuss possible evaluation approaches in Chapter 15.

11.5.2 The Informativeness Assumption

A second key assumption is that the predictors carry information about the predictand's variability at all relevant scales, including long-term trends. For regional climates, which are only weakly forced by large-scale flow, PP will thus have lower skill than for climates dominated by the synoptic-scale circulation (Schoof 2013). To credibly represent the effects of global warming, informative predictors have to carry all relevant information about the predictand response to external forcing. Missing predictors will result in unrealistic changes and apparent time invariances of the other predictor–predictand relationships (Schmith 2008). Informativeness is often measured in terms of correlations between predictor and predictand, but a high correlation does not guarantee informative predictors. First, the correlation is dominated by the timescales with highest variability, that is typically by day-to-day variability. For weather forecasting, high informativeness for these scales might be sufficient. But in climate science, it might be relevant to realistically represent interannual to decadal variability, and it is crucial to credibly represent long-term trends. Informativeness for these scales is not assessed by a correlation between, say, daily or monthly resolved predictor and predictand time series. Second, correlations do not necessarily imply causality, that is a physically sensible influence of the predictor on the predictand.

Consider a situation in which a direct physical influence of a predictor on a predictand is given. For instance, precipitation occurs when the air is saturated, condensation sets in and droplets can grow to the required size to precipitate. Thus, relative humidity may be a sensible predictor for precipitation occurrence (in combination with other predictors describing vertical velocities). In present conditions, for typical temperature values, relative humidity strongly correlates with specific humidity. In a warmer climate, however, the saturation vapour pressure will be higher, and cloud formation will occur at higher values of specific humidity. Using specific humidity as occurrence predictor may thus work in present climate conditions but would lead to an overestimation of the number of wet days when transferred to future climate conditions.

Similarly, gradients rather than mean fields might carry the information – in present climate gradients and means might be highly correlated, but they might change differently under climate change. For instance, zonal wind can be approximated by the north-south gradient of geopotential height at a given level. In present climate, absolute geopotential height might correlate strongly with this gradient, but in a warmer climate, geopotential heights will generally be higher than at present, causing artificial trends, which are not related to circulation changes.

Finally, two processes might be correlated on short timescales because they have a common driver on these scales, but they might respond to different forcings acting on longer timescales. One of these predictors might be informative for climate change, whereas the other process might not. Such a situation might occur if one attempts to predict the annual cycle of the predictand by means of meteorological predictors. A physically motivated model of the annual cycle might be very useful, because it might capture changes in seasonality in a natural way. However, many meteorological variables have a strong annual cycle and will trivially correlate with each other (as long as the cycles are in phase) – whether a physical dependence is given has to be tested separately (e.g. by testing correlations after removing seasonality).

How to select informative predictors is the major topic of the following subsections.

11.5.3   Predictors for Temperature and Precipitation

In the light of the previous discussion, the choice of suitable predictors – that are informative and fulfil the PP assumption – is a crucial step in the development of a PP statistical downscaling model. Here we will illustrate the relevant considerations for temperature and precipitation.

First guidance on suitable predictors may be found in corresponding PP and MOS applications in numerical weather forecasting (e.g. Klein and Glahn 1974, Hall et al. 1999). In a climate downscaling context, several studies have searched for the best selection of predictors in a given context (Huth 1999, Wilby and Wigley 2000, Huth 2002, Cavazos and Hewitson 2005). With the archiving of large climate model ensembles such as CMIP (Meehl et al. 2007a, Taylor et al. 2012), the availability of simulated predictor data has greatly improved.

Table 11.2 Predictors for temperature.

Physical aspect	Examples of predictors
temperature of the free atmosphere	temperature at 850hPa thickness, e.g, of 1000hPa to 500hPa layer
dynamical state	sea-level pressure geopotential height at 850hPa wind direction

Depending on the variable, the selection might be more or less straightforward. Generally speaking the choice of predictors depends on the region, the season, the statistical climate aspect to be downscaled and the considered temporal resolution. In the following, we will only consider predictors for the daily resolution.

Key to a successful choice of informative predictors is a thorough understanding of the mechanisms controlling the variability and long-term changes of the predictands. Given the discussion already, it is essential to choose predictors which have a direct physical influence on the predictand. This is particularly relevant for precipitation predictors in a climate change context, as will be discussed below. Unfortunately, a thorough physical motivation of the predictor selection is only given in a few studies (e.g. Karl et al. 1990, Charles et al. 1999). Physical reasoning should be backed up by a thorough evaluation of downscaled trends, potentially in a pseudo-reality (Section 15.4.2). In this context, it is also relevant to consider an appropriate spatial scale of the predictors. For instance, vorticity evaluated over a large region might be useful to represent a mid-latitude cyclone. Vorticity at the grid-box scale might not well capture the large-scale rotating flow but may (given a high-enough model resolution) be useful to detect the sharp bending of the isobars along a front.

Temperature
The selection of suitable predictors for local surface temperature is relatively straightforward, see Table 11.2 for an overview. The most obvious choice would be to directly choose temperature just above the boundary layer, such as at 850hPa (Huth 1999). The hypsometric equation (Wallace and Hobbs 2006, chap. 3.2) states that the thickness of a layer of air between two pressure levels is proportional to the mean virtual temperature of the layer. Thus, another sensible predictor of temperature is the difference in geopotential height between two pressure levels, for example between 1000hPa and 500hPa (Huth 1999). These predictors represent temperature in the free atmosphere, not surface temperature. The height of the 850hPa layer as well as the actual temperature profile depend strongly on the given weather situation. Therefore, additional dynamical information should be included, either in terms of sea-level pressure patterns or better (because directly simulated by a GCM) geopotential height, for example at 850hPa (Huth 1999). Dynamical predictors alone may have some skill in present climate but will miss the climate change signal (Huth 2002): for instance, geopotential height fields may characterise a specific weather type with a typical temperature pattern. Under future radiative forcing, however, this pattern will become warmer.

Table 11.3 Predictors for precipitation.

Physical aspect	Examples of predictors
	Precipitation occurrence
relative humidity at cloud base	relative humidity at 850hPa dew point depression
vertical velocities	SLP, geopotential height at 850hPa divergence, vorticity tendencies stability indices
	Precipitation amounts
vertical velocities	as above
available moisture	specific humidity moisture advection/convergence

The temperature predictors (temperature above the boundary layer, layer thickness) are directly motivated from first principles, so their link with surface temperature should be valid under climate change. Whether the link between dynamical predictors and surface temperature remains valid cannot be easily corroborated a priori.

Precipitation

For precipitation at the daily scale, predictors for both precipitation occurrence and intensities are required. An overview is given in Table 11.3. In particular, summer precipitation is strongly controlled by local processes such as convective instability – a key question is thus whether informative predictors can be found that can be defined at a scale large enough to fulfil the PP condition.

As discussed in Section 2.2, precipitation occurs, once a layer of moist air is lifted and becomes super-saturated, cloud droplets form and, sustained by updrafts, can grow to a sufficiently large size to fall out of the cloud and to reach the ground without evaporating (Rogers and Yau 1996, Pruppacher et al. 1998). A sensible predictor of precipitation occurrence thus has to represent relative humidity at the cloud base as well as the vertical velocities. If available for the chosen GCM, relative humidity can be directly used as predictor (Wilby et al. 2002). Otherwise, a sensible measure is the dew point depression, that is the difference between the dew point temperature[2] and actual temperature.[3] Of course, cloud bases vary strongly in height, and for a given region and season a typical height has to be chosen. Often, the dew point depression is taken at 850hPa (e.g. Charles et al. 1999). A further option would be to use both temperature and specific humidity as predictors (Wilby and Wigley 2000), but it is questionable whether their effect – relative humidity is a nonlinear function of temperature and specific humidity – can be captured by a linear model.

---

[2] The dew-point temperature is defined as the temperature at which condensation sets in if an air parcel is cooled isobarically.

[3] Using only dew point temperature – a measure of specific humidity – instead of the depression would lead to implausible climate change signals (Charles et al. 1999).

The link between large-scale weather and local vertical velocities depends strongly on the type of precipitation and is connected with different types of instability. In synoptic-scale cyclones, precipitation is controlled by baroclinic instability. Lifting happens across large scales and is directly linked to the synoptic circulation. Often, pressure-related predictors such as sea-level pressure or geopotential height are used. The underlying idea is that vertical lifting is related to horizontal divergence, which typically occurs in cyclonic circulation. In orographic precipitation, vertical velocities are caused by the flow across a mountain and thus again linked to the synoptic circulation. In localised summer convection, convective instability and vertical motion are caused by local diabatic heating of the lower atmosphere and amplified by the latent-heat release through condensation. Here the link with large-scale circulation is only indirect. For instance, a high-pressure system will be accompanied by a rather stable stratification, which suppresses deeper vertical motion. Organised convection in tropical cyclones, monsoonal systems or along squall lines is of course linked to the large-scale atmospheric circulation. The latter and associated large-scale vertical motion can be represented by various dynamical properties based on pressure or geopotential height fields such as winds, divergence or vorticity (Karl et al. 1990, Charles et al. 1999, Wilby and Wigley 2000, Cavazos and Hewitson 2005). The passage of precipitation-bearing weather systems is connected with strong temporal changes in pressure and temperature; thus also such tendencies might serve as predictors (Karl et al. 1990). Finally, measures of regional-scale atmospheric stability such as the K-index (Karl et al. 1990), convective available potential energy (CAPE, Perica and Foufoula-Georgiou 1996) or the total-totals index (TTI, Dayon et al. 2015) measure the stability of the atmosphere and have been used as predictors for convective precipitation – the question of course is whether a GCM would fulfill the PP condition for such predictors.

Precipitation amounts are controlled by vertical velocities and available moisture, where latent heat release by condensating moisture additionally invigorates vertical velocities. Thus, additional predictors describing specific humidity, its advection or convergence are required.

The interpretation of precipitation predictors is often not straight forward, as the chosen variables often only indirectly represent the underlying physical processes. For instance, specific humidity has often been used as a predictor for precipitation occurrence. As discussed, the actually relevant process is saturation, which is directly related to relative humidity. For a given temperature, of course, these two are directly proportional. But a specific humidity that may result in cloud formation in present climate may correspond to an undersaturated atmosphere in a warmer climate. Considering temperature as an additional predictor in a linear model may help in a certain temperature range. In general, however, temperature and relative humidity are linked via the Clausius-Clapeyron relation, which is highly nonlinear (Wallace and Hobbs 2006). Similarly, the meridional wind component has been used as a precipitation predictor. The underlying mechanism is likely that these winds carry moist and warm air northward – that is the actual physical relevant property is moisture advection, not the wind itself. Again, if the moisture carried by these winds increases in the future, the predictor–predictand relationship will change.

11.5.4   Predictor Domain

The size, location and, in general, also shape of the predictor domain have to be chosen such that the PP assumption is satisfied and the predictors are informative.

For the PP assumption to be satisfied, a size considerably larger than a model grid box will typically be required (von Storch et al. 1993). But to be as informative as possible, the domain should be small sufficiently small (Wilby et al. 2004). As discussed, the predictor-predictand relationship should be close to physical processes to avoid time-invariances in the relationship. For continental-scale teleconnection patterns, this requirement might not be fulfilled (Hertig et al. 2015).

The location of the domain is typically centred on the region of interest. Deviations may, however, be reasonable to better represent relevant physical processes (Wilby et al. 2014). For instance, it might be sensible to move or extend the domain towards the origin of the main flow. Over northern Europe in winter, for example, a domain extending to the Atlantic may help to resolve the influence of the westerly flow on local climate. Similarly, it might also be useful to extend the domain to better capture more rare weather types, which might cause important extreme events. For instance, over the south-eastern Alps it might be sensible to extend the domain towards the northern Mediterranean, where Genoa lows are formed that may cause heavy rainfall on their passage to the north-east. Finally, the location (and size) should take the skill of the driving model into account (Wilby et al. 2004). For instance, in complex terrain, the simulated regional flow might be systematically distorted compared to observations, such that the domain should extend into a region where the PP assumption can be assumed to be valid.

A standard approach to select a suitable predictor domain is to estimate correlations between the predictand and a set of candidate predictors across a large enough area (Wilby and Wigley 2000, Wilby et al. 2004). More sophisticated approaches additionally attempt to optimise the domain shape. For instance, Sauter and Venema (2011) select predictor variables, domain size, location and shape based on self-organising maps (Kohonen 1998). A similar approach by Radanovics et al. (2013) employs the growing rectangular domain algorithm. In any instance, it should be assessed whether the resulting predictor fields are informative and satisfy the PP assumption.

Some researchers use variables in single grid box as predictors. Apart from the question whether the PP assumption is fulfilled at such a small scale, it should be clarified whether a grid-box predictor carries the required information. For instance, as discussed in Section 11.5.3, a major reason to include sea-level pressure as predictor is strong link between synoptic-scale horizontal circulation and vertical motion. Yet sea-level pressure at a single grid box does not carry that information. Thus, depending on the chosen predictors, a domain should also not be too small.

11.5.5   Predictor Pre-Processing and Transformations

Predictor fields are typically highly correlated in space but additionally superimposed by considerable noise which does not carry information about the predictand variability. Without a dimension reduction and filtering, one would easily run into the danger of fitting noise to noise, that is overfitting. Thus, predictor fields are typically pre-processed by principal component analysis (PCA; see Section 6.5.2; Huth 1999).[4] A potentially useful side effect is that principal components, the resulting predictor time series, are uncorrelated. Note, however, that PCA does not necessarily construct physically meaningful predictors, in particular if the analysis is applied to multivariable predictor fields.

In reality, the PP assumption is never fully fulfilled such that the empirical orthogonal functions (EOFs) derived from climate model data typically differ from those derived from the large-scale observational data. One solution is to project the climate model simulation onto the EOFs derived from the observations. Alternatively, it has been proposed to derive EOFs that are common to model and observations by applying PCA to the concatenated time series of observed and modelled predictor fields (Benestad 2001, 2011).

Predictor fields may further be transformed to construct more process-oriented predictors (see also Section 11.5.3). Such transformations typically involve spatial-temporal derivatives. For instance, one may calculate meteorological properties such as divergence, vorticity (Wilby and Wigley 2000) or tendencies (Karl et al. 1990), that is temporal changes in the predictor field. Furthermore, one may calculate specific indices representing, for example, atmospheric stability, such as the K-index (Karl et al. 1990), CAPE (Perica and Foufoula-Georgiou 1996) or the TTI (Dayon et al. 2015). Again, it has to be ensured that these are evaluated on a sufficiently large-scale for the PP assumption to hold.

Similarly, weather types (also called circulation patterns or regimes; see Figure 11.1) can be considered a process-oriented predictor transformation. Instead of generating continuous predictors, the large-scale atmospheric circulation is mapped onto a discrete set of characteristic patterns (Lamb 1972, Hess and Brezowsky 1977, Philipp et al. 2007). In downscaling, these are assumed to discriminate between regional weather situations. Weather types typically represent major atmospheric flow regimes and thereby can account for nonlinearities in the predictor–predictand relationship, at the expense of potentially loosing information due to the discretisation. This is particularly relevant for climate change applications, as weather types by construction do neither allow for slow changes in a variable nor for extrapolation. As such, weather types are typically employed to characterise the large-scale dynamical state of the atmosphere, not temperature or humidity fields. In line with this argument, weather types are usually defined based on sea-level pressure or geopotential heights (Lamb 1972, Hess and Brezowsky 1977, Vautard 1990, Plaut and Simonnet 2001, Philipp et al. 2007), or wind fields (Moron et al. 2008). The number of types differs according to the specific situation and may range from only four (Vautard 1990, Plaut and Simonnet 2001) up to almost 30 as in the Grosswetterlagen classification (Hess and Brezowsky 1977).

Early weather typing schemes have subjectively categorised the flow (e.g. Lamb 1972, Hess and Brezowsky 1977). State-of-the-art methods usually employ an objective classification algorithm such as k-means clustering (MacQueen 1967, Plaut and Simonnet 2001), hierarchical clustering (Ward 1963, Casola and Wallace 2007), fuzzy rules (Bardossy et al. 2005), self-organizing maps (SOMs, Kohonen 1998, Wehrens and Buydens

---

[4] In fact, PCA itself is a mere transformation of the data set. Filtering and dimensionality reduction is obtained by simply omitting PCs explaining a low fraction of predictand variance; see Section 11.6.

2007, Leloup et al. 2008) or variants of PCA (Jacobeit et al. 2003). Recently, clustering methods based on statistical models have been developed. These methods model the probability density function of the atmospheric state by a superposition of distributions, for example by Gaussian mixture models (Haines and Hannachi 1995, Hannachi 1997, Fraley and Raftery 2002, Rust et al. 2010, 2013). In such a setting, the number of distributions (i.e. clusters) can be objectively chosen, for example by means of information criteria (Fraley and Raftery 2007; see also Section 6.3.4).

A comparison of different weather types has been carried out by the European Cooperation in Science and Technology (COST) action 733 (http://www.cost733.org, Huth et al. 2008, Philipp et al. 2010, 2014).

11.6   Model and Predictor Selection

Even after a careful choice of model type and candidate predictors that fulfill the PP condition and that are informative, a further selection step is necessary, for mainly two reasons. First, overfitting needs to be avoided, that is the number of predictors needs to be sufficiently low to ensure a stable parameter estimation. Second, a suitable model structure within the given model type needs to be selected. For instance, a variable may enter linearly or quadratically, or a product of variables may enter instead of a sum. Model selection is always a tradeoff between variance and bias: including all relevant predictors within a complex model structure may in principle enable unbiased predictions but may in practice result in high prediction uncertainty when only a limited amout of data is available.

In statistics and statistical weather forecasting, procedures for model selection have been developed, which balance model complexity against model fit (Section 6.3.4). In principle, all feasible candidate models have to be compared, but this procedure is impracticable for a relatively high number of parameters. Therefore, forward selection or backward elimination is typically used. As selection criterion, one may use likelihood ratio tests or information criteria. In the atmospheric sciences, cross-validation has been used. Note that an inclusion of all predictors that are significantly correlated with the predictor will typically lead to overfitting. For details, see Section 6.3.6.

11.6.1   Selecting Model Structure and Predictors for Climate Change

As discussed in Section 11.1, a statistical downscaling model that is applied to climate change simulations requires a model structure suitable for at least moderate extrapolation and needs to include predictors capturing the response to climate change. Developing such a model is far from trivial (Winkler et al. 1997, Huth 2004) and may in many cases be impossible without referring to physical understanding.

Climate towards the end of the 21st century is likely to assume a state which has not been observed during the time during which instrumental data is available. In particular thermodynamic changes, resulting directly from the warming of the atmosphere, will have emerged from internal climate variability, especially for high-emission scenarios (Collins et al. 2013). Thus, the values of temperature and humidity predictors are

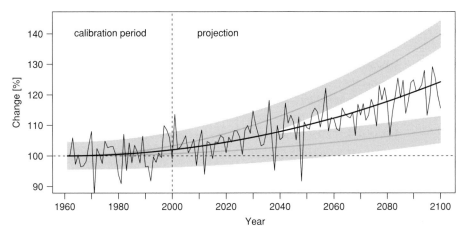

Figure 11.6 Influence of weakly constrained predictor–predictand relationships on the climate change signal. Downscaled projections (grey lines and shading) may diverge substantially from the (hypothetical) true signal (bold black line) merely due to uncertainties in the predictor–predictand relationship. For detail see text.

likely to exceed the range of present-day observations (Wilby et al. 2004). In practice this means that statistical downscaling models have to be valid under at least moderate extrapolations. By definition, only few data points are available in the tail to constrain a model chosen for extrapolation: it might be linear, polynomial or assume a more complex shape. The farther the extrapolation, the stronger the differences and thus structural uncertainties will be. A sensible extrapolation thus relies on a priori knowledge, that is process understanding.

A similar problem arises when selecting predictors that capture long-term changes. Statistical selection criteria will by construction favour predictors that explain a high fraction of the observed short-term (mostly daily) variability. But these predictors may not carry information about long-term climate change. Consider precipitation intensities, which may be predicted by circulation and humidity predictors. In present-day climate, these two might be strongly correlated: a given state of the atmospheric circulation will strongly control the moisture content of the atmosphere. For instance, over Europe a westerly flow will carry moisture from the Atlantic, while an easterly flow will typically advect dry continental air. A statistical model selection will likely select the circulation predictor and – to avoid overfitting – omit the humidity predictor. In a stationary context, this is a perfectly feasible model selection. Long-term warming, however, might slowly increase the available moisture, an effect which will crucially not be represented by the statistical model. At first sight, it may sound reasonable to additionally include the humidity predictor. The fundamental problem, however, is that because the predictor has little explanatory power for present climate and is strongly correlated with other predictors, the predictor–predictand link is not well constrained by observations; hence uncertainties in the associated regression parameters are high. In other words: projections of the predictand are highly uncertain.

Figure 11.6 illustrates the problem of unidentifiable parameters that are of minor importance for present climate but dominate long-term trends. The data is synthetic but could represent, say, winter precipitation. Assume also that we know the present and future state of the atmosphere, such that no climate model is needed. The black time series depicts local winter averages y. A statistical downscaling model is calibrated over the years 1960 to 2000, based on two – uncorrelated – large-scale predictors x_1 and x_2. During the calibration period, x_1 explains about 50% of the variability of y, x_2 about 8%. The latter predictor, however, carries the future signal. Physically, x_1 could represent the atmospheric circulation, x_2 specific humidity. Because of the relatively low correlation between y and x_2, the estimate of the regression coefficient is quite uncertain. The effect of this uncertainty is explored by Monte Carlo simulations. The bold black line represents the projected mean, based on the mean estimated parameter – it is an unbiased estimate and perfectly well represents the future change. The bold grey lines, however, represent the projections based upon the upper and lower end of the 90% confidence interval of the parameter uncertainty. The additional grey bands represent the range of interannual variability. Thus, the initial uncertainty in the parameter estimation may result in projections that range from about +9% to +40%, whereas the true change is about +24%. This uncertainty would be even larger if x_1 and x_2 were correlated and contributes to other sources of uncertainty.

If the predictand shows a strong trend already during the observed record, a cross-validation might help to constrain the model structure and to select relevant predictors. In principle, however, this problem is unresolved. A first important step is to refer to process understanding and to select candidate predictors that have a direct physical influence on the predictand which may be assumed to hold in a future climate. The physical relationship should then also prescribe the statistical model structure. To best constrain the model parameters, it might be useful to filter out short-term variability to separately calibrate the model components which explain long-term changes. Research is needed to explore optimal strategies.

Finally, all statistical climate aspects of interest that may substantially change under climate change have to be conditioned on predictors. For instance, a model might well describe the behaviour of extreme temperatures in present climate. Assume that the model conditions only temperature means on meteorological predictors but leaves the temperature variance unchanged. If the variance would change in the future, this model would fail to plausibly represent extreme future temperatures. Similar issues may arise for changes in the temporal, spatial or multivariable structure.

In any case, this issue highlights the importance of pseudo-reality studies, as will be discussed in Chapter 15. This approach will help to eliminate implausible statistical models, either because their model structure is not suitable for extrapolation or because their predictors are not informative of climate change.

11.7   A Cookbook for Developing a PP Model

In the following, we will summarise the most important steps in developing a PP statistical downscaling model. Several authors have presented similar lists before (e.g. Hewitson and Crane 1996, Wilby et al. 2002). We have extended these by several essential evaluation steps and credibility checks. Figure 11.7 gives a concise overview. The procedure is typically iterative and several steps may have to be repeated. The order of the different steps is to some extent arbitrary.

11.7.1   Initial Considerations and Checks

Prior to the development of any statistical downscaling model, one should acquire an overview of the relevant regional-scale statistical aspects as well as the governing large-scale climatic phenomena. Examples of relevant regional-scale aspects might be seasonal mean temperature, precipitation over a river catchment or specific extreme events such as drought. These phenomena depend strongly on the given user problem, as discussed in detail in Chapter 5. In turn, these phenomena determine characteristic space and timescales and the relevant statistical climate aspects. A method type needs to be chosen that is in principle capable of capturing the desired aspects (see Section 11.3).

Furthermore, it is important to understand the dominant large-scale climatic phenomena. Knowledge about the climate itself and how it is represented by models is important for assessing the PP and informativeness assumptions. Based on the discussion in Section 11.5, one should choose a wide range of physically meaningful predictors that are likely to be informative for present and future climate. To reduce the number of candidate predictors, a screening can be carried out, for example by considering only those predictors which correlate with the predictands. This screening eliminates all predictors which are not informative of present climate (see Section 11.5.2). One should, however, avoid eliminating predictors that are informative of slow changes (see discussion in Section 11.6.1); the screening should thus be informed by process understanding.

At some point during the process, the chosen climate model needs to be checked: does it realistically simulate the predictors in present climate (PP assumption for present climate)? Does it simulate a credible climate change signal for the predictors, or is there evidence for an implausible signal (PP assumption for future climate)? If one of these checks is not passed, other predictors need to be chosen that capture the relevant processes but might be better simulated by the climate model. In some cases, if major large-scale climate phenomena are fundamentally misrepresented, no climate-model-based regional climate projections might be possible at all (see Sections 8.5 and 18.5.2).

11.7.2   Model Development

The actual model development comprises the domain choice (Section 11.5.4), predictor pre-processing (Section 11.5.5) and the selection of the specific model structure and predictors (Section 11.6). Here it is important to recall that a classical model selection that would be successful in weather forecasting might result in a model which well simulates day-to-day variability but fails to represent long-term changes. For possible ways out, see Section 11.6.1.

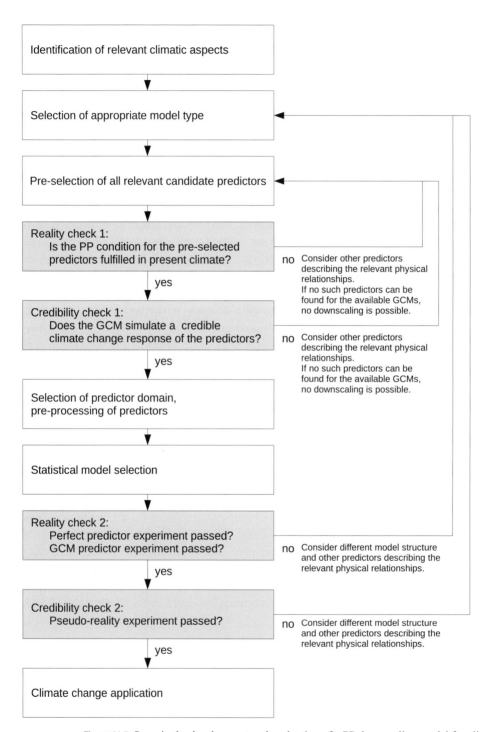

Figure 11.7 Steps in the development and evaluation of a PP downscaling model for climate change studies. The order of the individual steps is to some extent arbitrary.

11.7.3 Model Evaluation: Reality and Credibility Checks

A proper model evaluation for regional climate change simulations has to encompass, in addition to the credibility check for the predictors, a range of experiments to assess the skill of the downscaling method in present climate and to evaluate its likely performance in a different future climate. We will discuss model evaluation in detail in Chapter 15.

11.8 Further Reading

A classical paper on PP statistical downscaling in weather forecasting is Klein et al. (1959). Useful recent reviews are given by Fowler et al. (2007) and Maraun et al. (2010*b*). We have not covered downscaling for variables other than temperature and precipitation in this chapter. Useful insight for wind can be found, for example, in Pryor et al. (2005) and Salameh et al. (2009) and for humidity in Huth (2005). Nonlinear regression models as well as self-organising maps have only briefly been covered. More detail can be found in Hewitson and Crane (1994) and Kohonen (1998), respectively.

12 Model Output Statistics

As discussed in Chaper 8, both GCMs and RCMs are substantially biased compared to real-world climate (Flato et al. 2013, Kotlarski et al. 2014). Impact models such as hydrological models are typically calibrated with observed input data to produce realistic output; if driven with biased climate model output, these models in turn produce biased outcome themselves (Wilby et al. 2000). In addition, as discussed in the previous chapter, climate models – both GCMs and RCMs – often do not provide the resolution desired by impact modellers. For instance, the statistics of grid-box precipitation and station precipitation may be very different. Thus, impact modellers often demand some kind of bias correction and downscaling of climate model data. In an attempt to remove systematic biases and to bridge the scale gap between dynamical model output and the desired resolution, a wide variety of statistical bias correction methods has been developed. For reviews of bias correction methods, refer to, for example, Maraun et al. (2010*b*), Teutschbein and Seibert (2012) and Maraun (2016).

The idea of all these methods is to estimate some distributional properties of the climate system (e.g. long-term means, variances, quantiles) both in climate models and observations, and then infer a transfer function that maps the model property onto the observed property.

With the availability of large climate model ensembles such as those provided by CMIP (Meehl et al. 2007*a*, Taylor et al. 2012) or CORDEX (Giorgi et al. 2009, Giorgi and Gutowski 2016), bias correction has become very popular. Over recent years, the use of bias correction became quite standard in the production of national and global climate change projections (e.g. Maurer 2007, Li et al. 2010, Hagemann et al. 2011, Dosio et al. 2012, Stoner et al. 2013, Girvetz et al. 2013, Hempel et al. 2013, Maurer et al. 2014, CORDEX 2016). These studies have in turn served as input for impact studies (Gangopadhyay et al. 2011, Girvetz et al. 2013, Hagemann et al. 2013, Warszawski et al. 2014) and assessment reports (Cayan et al. 2013, World Bank 2013, Georgakakos et al. 2014). More and more bias corrected data has been made available through online portals by climate service providers and coordinated modelling experiments (Worldbank n.d., prepdata n.d., CORDEX 2016).

Over recent years, a debate about the usefulness and sensibility of bias correction has arisen. Some authors even question the very basis of bias correction (Ehret et al. 2012). At the IPCC Workshop on Regional Climate Projections and their Use in Impacts and Risk Analysis Studies (São José dos Campos, 15–18 Sep 2015), it was concluded that bias correction is prone to misuse, and a code of best practice still needs to be established

(Stocker et al. 2015). We will thus critically discuss the assumptions underlying bias correction and in which situations they might not be fulfilled.

Bias correction has a long tradition in weather forecasting. We will therefore first discuss similarities and differences with numerical weather prediction in the following section. Then definitions and assumptions will be presented in Sections 12.2 and 12.3. After an overview of widely used bias correction methods in Section 12.4, we will discuss the structural skill and limitations of bias correction (Section 12.5) and requirements for MOS predictors (Section 12.6). Being key issues, downscaling in a MOS context, bias correction in presence of location biases and the modification of the climate change signal by bias correction will be discussed separately in Sections 12.7, 12.8 and 12.9. In the final section, a cookbook for the development of a successful bias correction will be given which refers to the relevant sections.

The whole presentation will be limited to existing standard methods such as additive corrections or quantile mapping. Hybrid approaches, which combine bias correction with a further downscaling, will be discussed in Chapter 14.

12.1   Terminology and Relationship with Numerical Weather Prediction

The term "bias correction" is used differently by different communities. For instance, in perfect prognosis (PP) downscaling (see previous chapter), methods using standardised predictors (i.e. where long-term means have been subtracted) are often called "bias correcting".

The concept of bias correction as discussed in this chapter is essentially a special case of model output statistics (Glahn and Lowry 1972, Gneiting et al. 2005) which has been used successfully to post-process numerical weather forecasts for more than four decades. As the term MOS is (rather[1]) precisely defined as the counterpart of PP, and even more provides a broader bird's-eye view on bias correction, we will use MOS as the generic term. We thereby follow the terminology used by Widmann et al. (2003), Rummukainen (1997) and Maraun et al. (2010b).

DEFINITION 12.1 As MOS, we define any method that establishes a statistical transfer function between model and observed data and then applies this transfer function to post-process model data.

In numerical weather prediction, a forecast model is initialised based on observed data and previous forecasts. Thus the prediction, on lead times of a few days, typically does not diverge too strongly from real-world weather. Considering a time series of consecutive weather forecasts, the simulated and observed weather sequences are thus in close synchrony. The synchrony allows for a pairwise calibration (Chapter 10), that is to calibrate a regression model. Depending on the chosen predictors and predictands, such a regression model can be used in different settings: (1) to simply bias correct, that is predictor and predictand would be the same variable (homogenous, Chapter 10). (2) It

---

[1] Some authors restrict MOS to regression models, while others use the broader definition we adopt.

can use a range of different predictors to improve the prediction, or to predict variables which have not been simulated by the numerical model (non-homogenous).

In climate change modelling, climate model and observations are typically not in synchrony, and the calibration can only be distributionwise (Chapter 10). This setup has some important consequences:

- In a distributionwise setting it is only possible to sensibly map between identical variables, hence MOS methods in climate research are typically homogeneous.
- Without synchrony between simulation and observation, performance measures as known from forecast verification (such as correlations, root mean squared errors and other skill scores) are not applicable. That is, the evaluation of skill is far from trivial for MOS methods in climate research.
- Free-running climate models are not forced to stay close to observations, as in numerical weather prediction. Therefore, after a certain spin-off time, climate models drift into their own attractor and are biased on all spatial and temporal scales.

Following this discussion, we use the term "bias correction" as a synonym for homogeneous MOS. In most cases, a bias correction will be calibrated distributionwise.

12.2   Bias Definition

As discussed in Section 4.2, observed and simulated regional climate can be thought of as a sample of two multivariate probability distributions with marginal, temporal, spatial and multivariable aspects. If the simulated and real-world distributions are different, the model is biased. Following the statistical bias definition (see Section 6.2), we define a bias at time t as the systematic difference between a simulated model property $\theta_{mod}(t)$ and an observed property $\theta_{obs}(t)$ as

$$\text{Bias}_\theta(t) = \theta_{mod}(t) - \theta_{obs}(t). \tag{12.1}$$

This definition is in line with that of the World Meteorological Organisation (WMO), which defines a bias as the correspondence between a mean forecast and mean observation averaged over a certain domain and time. We just extend it to properties other than the mean.

From a given finite observational time series (or field of time series), one can only derive estimates of θ_{obs}. The same holds in practice also for θ_{mod}, which could at least in principle be better estimated by an initial-conditions ensemble or long control simulations. Internal climate variability is in fact a severely limiting factor for a reliable bias estimation. Even on 30-year timescales, many climatic properties are substantially affected by long-term modes of climate variability (Section 9.1.3). Thus, a considerable fraction of any estimated model bias will simply be a random component. Long time series are therefore essential for a sensible model calibration.

Both natural and anthropogenic external forcing changes the climate system in time. Hence, also the climate distribution changes and climate model biases are in general time dependent. Some authors define a bias as the time-invariant component of a model error (Piani et al. 2010*b*). But this definition is itself time dependent, as it depends on the choice of the reference period. Such a bias definition is thus arbitrary. Please refer to Section 12.9 for a discussion of time-varying biases and their consequences.

12.3 Assumptions

As with the PP approach, the assumptions laid out in Section 4.3 can further be specified for the MOS approach. In principle, they are similar to those for PP. Some differences arise, however, due to the fact that MOS is calibrated between model and observations and corrects biases and that predictors of MOS are typically local and homogeneous:

- MOS predictors need to be credibly simulated. For present climate, simulations have to be realistic apart from correctable biases. In a climate change context, changes of the predictors have to be credibly simulated.
- MOS predictors need to be representative of the target variable, that is they have to represent the same spatial scale and location.
- The transfer function needs to have a suitable structure, which is applicable under changed climate conditions.

The first assumption corresponds to the PP assumption, but in the context of MOS, biases are explicitly allowed. The statement "apart from correctable biases" may at first sound like circular reasoning. It will be specified in more detail in the following sections. The second assumption replaces the informativeness assumption: MOS predictors are local and have the same physical dimension as the target variable. Thus, a selection of informative predictors is not required, it is by construction integrated into the predictor. Instead, the predictor variable should be representative of the target variable. Regarding the third assumption, we would like to point out that MOS transfer functions are not statistical models in a strict sense as, for example, a regression model (in the sense that observations are characterised by a probability distribution). Still, the transfer function has to be applicable in a future climate.

It might seem surprising that we do not list the stationarity or time-invariance ("biases have to be time invariant") assumption. The reason why we did not include this assumption is actually twofold: first, a bias correction might be sensible even if biases change in time; and second, the sensible aspect of the time-invariance assumption is implicit in the three other assumptions: a bias correction of future simulations makes sense if the climate model simulates a credible change, if the changes are representative and if the transfer function correctly post-processes the climate model output. The implications of all these assumptions will be discussed below.

12.4 Methods

In this section, we will give an overview of the main types of homogeneous MOS methods, that is bias correction methods. A widely used approach in hydrology – the delta change approach – is mathematically similar but conceptually very different from bias

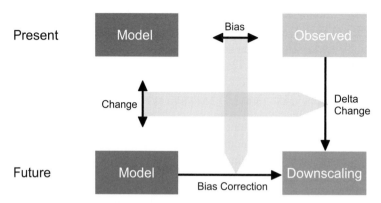

Figure 12.1 Direct and delta change approach. The delta change signal adjusts observations by a long-term change signal. Bias correction adjusts model output to remove the long-term bias.

correction. Instead of applying a bias correction to model data, this approach rather changes the observational record by a climate change signal. In the beginning of this section, we will briefly present the simplest implementation of this approach. Mathematically, one could apply any bias correction variant also as a delta change approach. To distinguish delta change from proper bias correction methods, the latter are often called direct methods; see also Figure 12.1.

12.4.1   Delta Change Approach

The simplest delta change method perturbs observed time series by the simulated climate change signal (Lenderink et al. 2007; see Figure 12.1). Two basic variants exist, depending on whether the climate change – and biases – is assumed to be additive or relative. For instance, projected temperature y^f_{i+T} at a time $i + T$ in the future is represented by observed temperature $x^p_{\text{obs},i}$ at time i in the observational record, perturbed by the additive climate change signal of mean temperature, that is the difference between mean future temperature $\bar{y}^f_{\text{mod}}$ and mean control run temperature $\bar{x}^p_{\text{mod}}$:

$$y^f_{i+T} = x^p_{\text{obs},i} + (\bar{y}^f_{\text{mod}} - \bar{x}^p_{\text{mod}})$$
(12.2)

Similarly, projected precipitation y^f_{i+T} at a time $i + T$ in the future is represented by observed precipitation $x^p_{\text{obs},i}$ at time i in the observational record, perturbed by the relative climate change signal of mean precipitation:

$$y^f_{i+T} = x^p_{\text{obs},i} \frac{\bar{y}^f_{\text{mod}}}{\bar{x}^p_{\text{mod}}}.$$
(12.3)

The delta method has been very popular, mainly because of its simplicity, and because in early downscaling studies the GCM simulations were too unrealistic even after a bias correction. The delta change method, however, utilises only changes simulated by a dynamical model. The statistical characteristics apart from the changes are still those of the unchanged observed time series. An underlying assumption is therefore, of course, also that the simulated climate change signal is representative of the target location. This and further issues will be discussed in Section 12.5.

In principle, the delta change approach is not limited to changing observed means, but other methods borrowed from bias correction could be used (Räisänen and Räty 2013, Räty et al. 2014). For instance, Willems and Vrac (2011) developed a quantile mapping variant, which applies changes individually for different quantiles of the observed distribution.

12.4.2 Additive Corrections and Scaling

The simplest bias correction methods are additive corrections and scaling. The former approach assumes an additive mean bias, while the latter assumes relative mean biases.

Mean temperature biases, in the meteorologically relevant range, can be considered additive. Assume that time series of observed present-day temperature, y_i^p, as well as simulated temperature in present climate, x_i^p, and in future climate, x_i^f, are given. Then a bias corrected temperature projection $y_{corr,i}^f$ at time i is given as the simulated future projection x_i^f at time i, corrected by the present-day mean bias, that is by the difference between present-day mean simulated temperature $\bar{x}^p$ and mean observed temperature $\bar{y}^p$:

$$y_{corr,i}^f = x_i^f - (\bar{x}^p - \bar{y}^p). \tag{12.4}$$

Higher-order moments such as variance and skewness are assumed to be correctly simulated.

To correct relative errors, Widmann and Bretherton (2000) suggested using a scaling factor. For instance, projected precipitation $y_{corr,i}^f$ at a time i in the future is similarly represented by the simulated future projection x_i^f at time i, corrected by the present-day mean bias, that is, by the ratio between present-day mean simulated precipitation $\bar{x}^p$ and mean observed precipitation $\bar{y}^p$:

$$y_{corr,i}^f = x_i^f / \frac{\bar{y}^p}{\bar{x}^p}. \tag{12.5}$$

The scaling method assumes a constant coefficient of variation, hence mean and variance are rescaled by the same factor. Higher-order moments are again assumed to be correctly simulated. As a variant, Leander and Buishand (2007) propose a power law transformation, where x_i^f is replaced by $(x_i^f)^b$, and b is an exponent to be estimated. Engen-Skaugen (2007) rescales mean precipitation and anomalies about the mean separately. Schmidli et al. (2006) proposed using separate corrections for precipitation occurrence and precipitation intensity.

12.4.3 Variance Correction and Quantile Mapping

As discussed in the previous sections, climate model biases often do not only affect the long-term mean but also higher-order statistical moments such as variance or the tail behaviour. For instance, Christensen et al. (2008) showed that temperature biases are higher for high temperatures, and precipitation biases might even change sign depending on precipitation intensity.

The simplest generalisation of the additive method would adjust discrepancies between modelled and observed variance, sd_x^p and sd_y^p respectively. The modelled anomalies with respect to present-day model mean, $x_i^f - \bar{x}^p$, would be rescaled by the present-day variance mismatch; finally, the observed present-day mean $\bar{y}^p$ would be added to the adjusted anomalies:

$$y_{corr,i}^f = \frac{x_i^f - \bar{x}^p}{sd_x^p / sd_y^p} + \bar{y}^p. \tag{12.6}$$

This implementation has probably never been used in practice, but it helps to illustrate potential problems of more sophisticated methods: a potential climate change signal is considered as anomaly in the same way as day-to-day, interannual or decadal variability; the rescaling affects all timescales in the same way.

Other variance correction methods approximately separate internal climate variability and forced trends (e.g. Räisänen and Räty 2013). Here, anomalies are considered with respect to the future mean signal $\bar{x}^f$, which is added again after rescaling; finally the present-day bias is subtracted:

$$y_{corr,i}^f = \frac{x_i^f - \bar{x}^f}{sd_x^p / sd_y^p} + \bar{x}^f - (\bar{x}^p - \bar{y}^p). \tag{12.7}$$

Variability and forced trend are of course only approximately separated, as the long-term means are still affected by variability. This implementation deliberately preserves the simulated climate change signal in the mean.

As a further generalisation, quantile mapping adjusts different quantiles individually (Panofsky and Brier 1968). From all modelled and observed values x_i^p and y_i^p over the calibration period, the corresponding cumulative probability functions $F_x(x)$ and $F_y(y)$ are estimated. Modelled values x_i^f, that is specific quantiles of the modelled distribution, are then mapped onto the corresponding observed quantiles as

$$y_{corr,i}^f = F_y^{-1}(F_x(x_i^f)). \tag{12.8}$$

Typically, climate models simulate too many wet days (the so-called drizzle effect; Gutowski et al. 2003). In this situation, quantile mapping automatically adjusts the number of wet days (as the wet-day threshold is a quantile of the distribution; Hay and Clark 2003). If a climate model simulates too few wet days, it has been suggested to randomly generate low precipitation amounts (Themeßl et al. 2012), although the inclusion of artificial wet days is somewhat arbitrary (Hewitson et al. 2014). A sketch of the method is given in Figure 12.2.

Different implementations differ in the model used to estimate the cumulative distribution functions, in particular for the tail. Some authors consider empirical, linearly interpolated quantiles (Themeßl et al. 2011), whereas others employ parametric models such as a normal distribution for temperature and a gamma distribution for precipitation (Hay and Clark 2003, Piani et al. 2010a). For extrapolated quantiles in the tail, Boé et al. (2007) use the correction derived for the highest available quantile. Wood

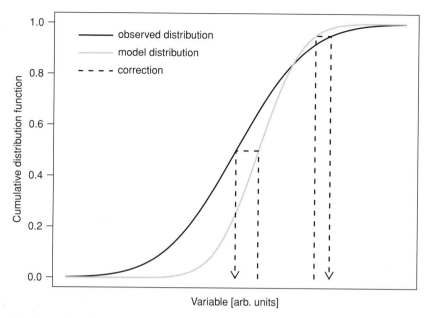

Figure 12.2 Quantile mapping. The simulated cumulative distribution function (grey) is mapped onto the observed cumulative distribution function (black). The correction function (dashed) depends on the actual quantile.

et al. (2002) take the empirical distributions, extended by different parametric distributions for high and low extremes; other authors employ specific extreme value models to model extreme precipitation intensities (Michelangeli et al. 2009, Kallache et al. 2011). Some quantile mapping approaches employ a further downscaling. These methods, such as BCSD (Wood et al. 2004), will be discussed explicitly in Chapter 14. Haas et al. (2014) have interpolated the transfer function between modelled and observed wind fields in space and thus created a full field of bias corrected wind distributions.

Quantile mapping is essentially a generalisation of the variance-adjusting bias correction method (Eq. 12.6). As such, it also modifies climate change trends (Hagemann et al. 2011, see Figure 12.3). These modifications are not restricted to the mean, since also trends in other aspects such as extremes may be affected (Dosio 2016). Whether and in which situations such a trend modification might be sensible will be discussed in Section 12.9. Nevertheless, several modifications of quantile mapping have been introduced which preserve raw model trends to different degrees. Haerter et al. (2011) suggest applying quantile mapping separately to different timescales, thereby very much reducing the effect on the climate change signal. Hempel et al. (2013) suggest removing the long-term trend prior to quantile mapping and adding it afterwards again. This variant is similar to the variance correction (Eq. 12.7) and preserves trends in the mean. In particular trends in low or high quantiles might still be modified. Li et al. (2010) developed a variant that preserves additive changes at all quantiles. Recently, a modification for relative changes has been proposed (Pierce et al. 2015). A new approach approximately preserves trends in all quantiles (Switanek et al. 2016).

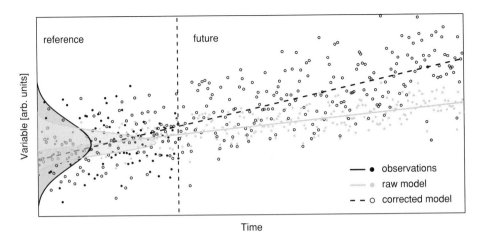

Figure 12.3 Trend modification by quantile mapping. The model distribution of day-to-day variability (light-grey shading) is stretched to match the observed distribution (dark-grey shading). As a result, also the long-term simulated trend (grey line) is inflated (dashed bold line).

12.4.4   Multivariate Bias Correction

Most bias correction approaches are univariate and only adjust the margins. Even though these approaches, applied separately to several variables, typically do not deteriorate the dependence between different input variables (Wilcke et al. 2013), they do not adjust it to better match observations. Over recent years, therefore, several multivariate bias correction methods have been proposed. Piani and Haerter (2012) introduced a conditional empirical bivariate correction. They first correct one variable based on a univariate empirical quantile mapping then group the resulting pairs of data points into bins of the first variable and finally apply univariate empirical quantile mapping in each bin to the second variable. A similar approach has been developed by Li et al. (2014).

Vrac and Friederichs (2015) propose a method that corrects the full multivariate, spatial and temporal dependence. This goal is achieved by first applying a univariate quantile mapping separately to all variables and locations and then by reordering all simulated values separately for each variable according to their ranks to match the temporal order of ranks in the observed reference record. In essence, this method takes the observed record and replaces each observed value with the corrected model value corresponding to the same quantile. It can thus be regarded as a multivariate quantile mapping delta change approach. It fully ignores the multivariable, spatial and temporal dependence simulated by the underlying climate model and just applies the simulated climate change to an observational field.

Cannon (2016) proposes an iterative method which conceptually lays between univariate bias correction methods, which approximately preserve the multivariate dependence of the driving climate model and the approach by Vrac and Friederichs (2015): in each iteration, univariate quantile mapping is first applied separately to each variable. Second, a linear multivariate bias correction is applied by rescaling the multivariate anomalies based on the Cholesky decomposition of the covariance matrix. The algorithm terminates when both the corrected margins and the dependence structure are sufficiently close to their observed counterparts. A variant is based on ranks rather than on the actual values. In a follow-up study, Cannon (2017) proposes an iterative procedure which provides a multivariate quantile mapping. It consists, in each iteration, of a random rotation of multivariate input data, a univariate quantile mapping on the rotated fields and the inverse rotation.

Multivariate bias correction by construction breaks consistency with the spatial-temporal structure of the driving climate model – in the most extreme case one ends up with a delta change approach (Vrac and Friederichs 2015). We will discuss the sensibility of such alterations in Sections 12.5.3 to 12.5.5.

12.5   Structural Skill and Limitations

In comparison to PP downscaling models (and non-homogeneous MOS models in numerical weather prediction), a bias correction model is easily built and calibrated to climate model data. But as we will demonstrate in Section 15.7 of the book, the evaluation of bias correction applications is far from trivial. Thus, it is essential to understand the structural skill and limitations of bias correction. As in the discussion about PP (Section 11.3), we will refer to the statistical climate aspects defined in Section 4.2. In MOS, the role of the model structure is typically less relevant than in PP downscaling. Two important issues related to model structure will be discussed separately: downscaling to finer scales (Section 12.7) and modifications of the climate change signal (Section 12.9).

12.5.1   Direct versus Delta Change Approaches

In the following, we will only consider direct approaches. A similar discussion, however, would also apply for the delta change approach. In principle, the direct approach should be used if trust in model simulations at the daily scale is high – then the full advantage of the dynamical model, including changes in dynamical properties such as the temporal dependence structure, can be taken. If trust in the model is low apart from the mean response to external forcing, the delta change approach might be preferable – then one assumes unchanged dynamics and relies only on the simulated changes in intensities and magnitudes, that is essentially thermodynamic changes. Often it is argued that the delta change method is ideal for downscaling, as it intrinsically represents local variability at the target scale. While this is true for present climate, it may not necessarily be true for the climate change signal, which is derived from a coarser-resolution climate model. Here the discussion in Section 12.7 applies equally to the delta change approach.

12.5.2   Marginal Aspects

In principle, bias correction can trivially correct any desired marginal aspect, as long as sufficiently many data points are available to calibrate the required transfer

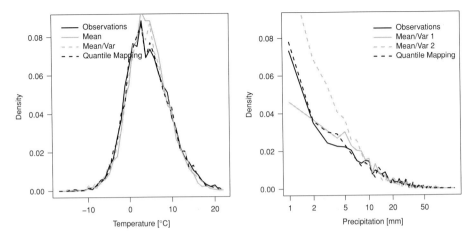

Figure 12.4 As Figure 11.2, but for MOS. Mean: additive correction of temperature (RaiRat-M6); Mean/Var: additional variance correction (RaiRat-M7), Mean/Var 1: separate scaling of mean precipitation and anomalies (Ratyetal M6); power law scaling of precipitation intensities (Ratyetal-M7); empirical quantile mapping (RaiRat-M9 for temperature; Ratyetal-M8 for precipitation). Local observations from ECA-D data (Klein Tank et al. 2002), for the period 1979–2008. For details of the methods, see Appendix A.

function. Figure 12.4 illustrates the representation of marginal aspects by bias correction methods for winter temperature (left) and summer precipitation (right) in Graz, Austria. All methods essentially capture the temperature distribution. The fact that this holds even for the simple additive correction just demonstrates that the predictor, ERA-Interim, captures the temperature variability well. In particular, the non-parametric quantile mapping closely follows the observed distribution. Differences arise simply due to the cross-validation. The case is different for precipitation. ERA-Interim grid box precipitation has different marginal characteristics as the observed data, which cannot be overcome by a simple rescaling of mean precipitation and anomalies (Mean/Var 1): low precipitation intensities are underrepresented, medium intensities over-represented. An attempt to adjust the variance by power law scaling (Mean/Var 2) just heavily over-represents low to medium intensities. Again, quantile mapping almost perfectly adjusts the marginal distributions. Given that the chosen quantile mapping method is non-parametric, this result is trivial and does not say anything about the skill of the bias correction.

Figure 12.5 illustrates the performance of different bias correction methods to capture extreme events at a rain gauge in the Harz Mountains, Germany. Predictors are taken from an RCM driven with perfect boundary conditions. Three bias correction methods have been considered: simple scaling of precipitation intensities at wet days (threshold 0.1 mm); parametric quantile mapping of intensities based on the gamma distribution including a wet-day correction; and non-parametric quantile mapping based on linearly interpolated empirical quantiles. The bias correction has been calibrated on the years 1961–1981 (left) and validated on the years 1982–2002 (right). For calibration as

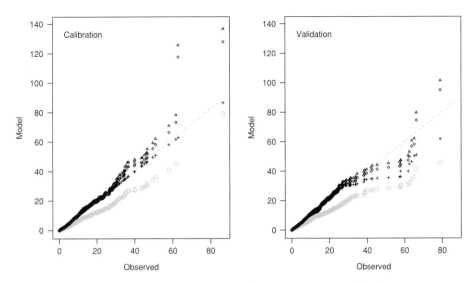

Figure 12.5 Comparison of bias correction methods for daily summer precipitation at Clausthal-Zellerfeld-Erbprinzentanne (data from German Weather Service). Left: calibration period; right: validation period. Grey circles: uncorrected RCM (KNMI RACMO); triangles: scaling; diamonds: parametric quantile mapping; crosses: non-parametric quantile mapping. For details on the methods see Appendices A, method CFE.

well as validation period, the dynamical model systematically underestimates precipitation intensities (grey circles). The non-parametric quantile mapping is trivially perfect during the calibration period (crosses). The transfer function derived from the scaling (triangles) and parametric quantile mapping (diamonds) methods are rather rigid, such that they cannot adjust the distribution flexibly enough. This effect is clearly visible for the highest quantiles. Overall, precipitation is underestimated – therefore the parametric models increase precipitation at all quantiles, causing a very strong amplification of the two most extreme simulated values. This problem would be mitigated by combining two distributions, for example, a gamma distribution for moderate intensities and a generalised Pareto distribution for the tail (e.g. Vrac and Naveau 2007).

The actual performance, however, has to be derived from the validation period (right panels). Here it becomes clear that an overly flexible method as the non-parametric quantile mapping might overfit: for all high quantiles, it performs worse than the parametric methods. In particular, the extreme tail should be modified with care. Here, the danger is high that non-parametric methods simply fit noise and thereby even decrease model performance. Conversely, inappropriate parametric models for the tail, even if they fit the observed range well, might be biased under extrapolation and create very unrealistic values (Volosciuk et al. 2017). Thus, care is needed for implementing corrections for the tails of a distribution.

In general, one should be careful correcting a simulated distribution that is very different from the observed one. It is likely that the difference is an expression of a fundamental model error or a representativeness problem (Maraun 2016).

12.5.3   Temporal Aspects: Persistence and Systematic Variations

In principle it is possible to adjust systematic temporal variations, for example, the seasonal or diurnal cycle. However, one should carefully consider the reasons for the mismatch. A pronounced seasonal dependence of biases might be an expression of model errors. For instance, season-specific weather types might occur systematically misplaced in time. Examples could be the occurrence of high-pressure systems over eastern Europe in Winter, which cause cold air outbreaks into central Europe or, even more regularly, the onset of the Monsoon. Errors in the timing of such events cannot be overcome by bias correction. Any attempt to do so will likely introduce major artefacts. Similarly problems at the sub-daily scale like the missing afternoon peak in precipitation cannot simply be bias corrected.

Most bias correction methods do only adjust marginal distributions and thus do not explicitly change the persistence, that is the residual temporal dependence structure of the predictor time series. Quantile mapping, which adjusts the number of wet days and thereby affects the dry–wet sequence, implicitly changes persistence. Similarly, multivariable correction methods implicitly (or even explicitly) modify the short-term persistence (see discussion below).

In fact, any major change in the local residual temporal dependence structure would induce strong inconsistencies between the bias corrected local weather and the large-scale dynamics of the driving model. For instance, if the length of long dry spells were extended to better match observations, the driving climate model would simulate synoptic weather conditions characteristic of rain, whereas the bias correction would force local weather to be dry. Major persistence errors reflect fundamental climate model errors, which cannot be overcome by bias correction (Haerter et al. 2011, Stocker et al. 2015, Maraun et al. 2017*b*).

Figure 12.6 illustrates artefacts one may face in an attempt to correct wet-day probabilities, if the driving model has major persistence errors. In the typical case (a), a wet-day correction might simply improve spells up to a length of about 20 days, longer spells are not produced by the correction (Rajczak et al. 2016). At best (b), the correction might even produce an almost perfect adjustment of spell lengths. But it is possible that a lack of long spells will be compensated by an over-representation of short spells (c), or even worse, that the correction produces artificially long spells which do not exist in the observational record (d). Thus, care is needed when correcting wet days; it might be worth discriminating between the pure drizzle effect and real persistence errors of the driving model, that is by conditioning on weather types.

Thus, most bias correction methods essentially inherit the persistence simulated by the driving climate model. Figure 12.7 illustrates the effects of bias correction on the ERA-Interim time series of daily maximum temperature (top) and daily precipitation (bottom) for summer 2003 in Graz, Austria. The temporal sequence of all three methods is essentially the same, namely that of ERA-Interim. Only at very close inspection might one find differences in the number of wet days.

Some methods attempt to bias correct multivariable aspects (Section 12.4.4). These approaches by construction modify the temporal dependence structure. For instance, if

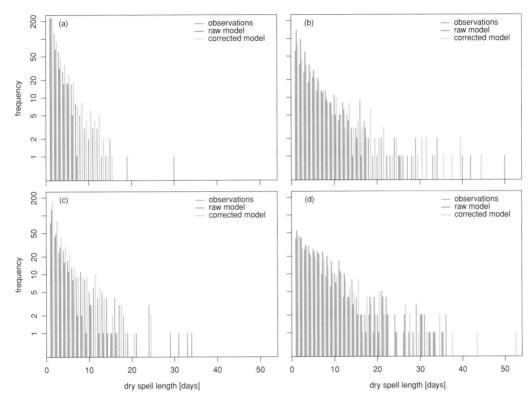

Figure 12.6 Effect of wet-day correction on spell length distributions for (a) Jokioinen-Jokioisten, (b) Lugano, (c) Sion, (d) Tortosa-Observatorio del Ebro, based on ECA-D station data, period 1979–2008 (Klein Tank et al. 2002).

the dependence of precipitation occurrence at two locations is bias corrected, at least at one of the locations the sequence of wet and dry days has to be modified. Even if the spell length distribution should be conserved under such modifications, the relationship between the large-scale weather sequence simulated by the driving climate model and the local weather sequence is altered. As an extreme example, the correction method by Vrac and Friederichs (2015) completely replaces the temporal sequence of the climate model with the observed sequence over the reference period. Again, major biases in the underlying temporal variability are likely an expression of severe climate model errors. Any attempt to correct such errors is questionable and cannot increase the credibility of the raw simulations.

12.5.4 Spatial Aspects

Bias correction by construction adjusts systematic spacial variations such as the influence of orography, distance to coast and latitude and longitude on the climatological fields. In doing so, even effects at a resolution higher than the model output can be imprinted onto the climate model simulation. This effect, however, should not be

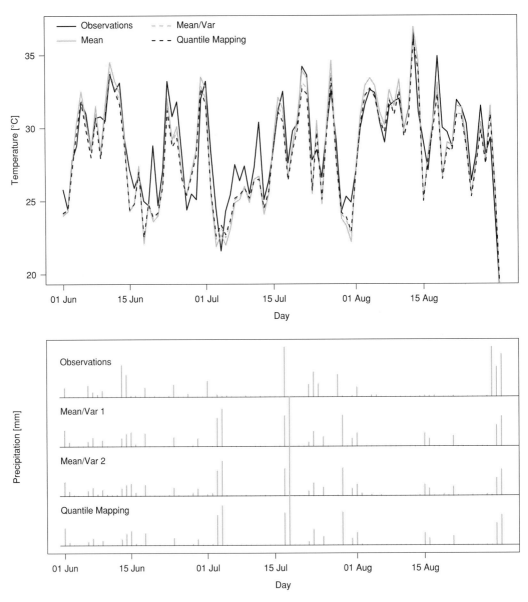

Figure 12.7 Observed and downscaled temperature (top) and precipitation (bottom) during the heat wave of Summer 2003 at Graz, Austria. Mean/Var: additional variance correction (RaiRat-M7), Mean/Var 1: separate scaling of mean precipitation and anomalies (Ratyetal M6); power law scaling of precipitation intensities (Ratyetal-M7); empirical quantile mapping (RaiRat-M9 for temperature; Ratyetal-M8 for precipitation). The different bias correction methods have taken their predictors from ERA-Interim and are cross-validated. Local observations from ECA-D data (Klein Tank et al. 2002), for the period 1979–2008. For details of the methods, see Appendix A.

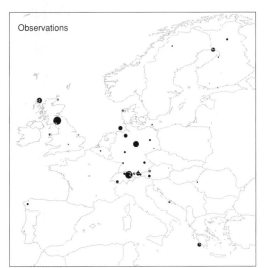 

Figure 12.8 As Figure 11.4, but for MOS. For details about the method, see Appendix A, method Ratyetal-M8.

mistaken as downscaling. Day-to-day variations at the higher resolution will still be missing, and also the climate change signal will be that of the coarser resolution. For a detailed discussion, see Section 12.7.

Since most bias correction methods only adjust marginal distributions, the effect on the residual spatial dependence is similar to that on the temporal dependence – in principle it is inherited from the driving model and only implicitly modified. Figure 12.8 illustrates the effect, as for the PP methods for 8 December 1993. ERA-Interim simulates the precipitation clusters well, and this structure is inherited by the bias correction. The discussion for multivariate bias correction methods (Section 12.4.4), if they are applied jointly to several locations, is essentially the same as that given for temporal aspects in the previous section.

12.5.5 Multivariable Aspects

Bias correction only moderately affects the overall inter-variable dependence (Wilcke et al. 2013): the ranking within the individual variables is not affected, such that measures such as correlations are only mildly changed. The effect is illustrated in Figure 12.9. Nonlinear effects, such as conditional threshold exceedances (e.g. precipitation occurrence for positive/negative temperatures), might however be crucially affected. Multivariate methods (Section 12.4.4) will improve the dependence but by construction alter the temporal dependence (see Section 12.5.3). A major bias in the multivariable dependence structure of the underlying model data is, again, likely an indication of several model errors. Similarly, such biases can arise from representativeness issues. For instance, when jointly correcting temperature and wind, the simulated grid-box temperature may well represent local conditions, the grid-box simulated wind may not. Also

---

Box 12.1 Bias Correction of RCM Input

It has been suggested to bias correct three-dimensional GCM fields prior to using them as boundary conditions for dynamical downscaling (e.g. Colette et al. 2012, White and Toumi 2013). The reasoning behind this proposal is that one would be able to correct large-scale biases and reduce the overall biases of the final RCM output.

But whereas the correction will remove spatial biases in the long-term mean field of the GCM, it will not remove spatial biases in GCM variability (Maraun et al. 2017*b*). For instance, assume that the region of strongest meridional temperature gradient and thus baroclinicity in the North Atlantic region is southward displaced. As a result also the storm track may be southward displaced. A bias correction would move the long-term temperature gradient into the correct position. The RCM would see this corrected mean field. Individual storms, however, would still enter the RCM at a wrong southward-displaced latitude. In fact, the storms would enter a region of reduced baroclinicity and weaken or even die out. Conversely, in the corrected region of strong baroclinicity even new storms may be generated. Thus, one could decouple the GCM and RCM and create unrealistic cyclogenesis over the Atlantic. These potential pitfalls need to be carefully studied.

---

here, a multivariate bias correction will not overcome these fundamental issues and remains questionable.

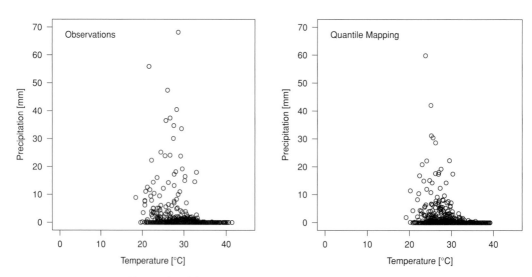

Figure 12.9 Observed and downscaled precipitation during summer at Madrid-Barajas, Spain (3°33'20"W, 40°28'00"N). Local observations from ECA-D data (Klein Tank et al. 2002), for the period 1979–2008. For details about the method, see Appendix A, method RaiRat-M9 and Ratyetal-M8.

12.5.6   Predictands

As standard MOS in a climate change context is homogeneous, predictand variables can only be those variables that are simulated by the climate model.

12.6   Predictors

The key advantage that makes homogeneous MOS easy to use is that one does not have to care about predictor variables, their location and domain size: it will be the same variable as the predictand, and it will be local. Unfortunately, this is a rather narrow and incomplete view of the issue. First of all, this statement ignores the credibility assumption: climate model output, which is not credible, will not become credible by a standard bias correction. Even credible output, if not representative, will not result in a sensible bias correction. Thus care is also needed to carry out a successful bias correction.

12.6.1   The Credibility Assumption

Bias correction cannot overcome fundamental errors of climate models. As discussed in Section 8.5, model simulations might be unrealistic for present climate or noncredible for future climate. Causes are misrepresented (or missing) processes on all scales, which affect regional climate change projections, from global climate sensitivity to large-scale phenomena such as planetary waves, ENSO or monsoons to local feedbacks involving snow-cover effects on the albedo or soil-moisture effects on temperature and precipitation. Standard bias correction methods cannot correct any such misrepresentations. In Section 12.9, we will discuss for which type of misrepresentations bias correction may still be sensible.

12.6.2   The Representativeness Assumption

A standard bias correction is not sensible if the simulated variable does not represent the observed variable. Consider a climate change projection with an energy balance model. Such models do not resolve the atmospheric and ocean circulation. Even though such conceptual models have been shown to simulate a credible response of global mean surface temperature to external forcing, one would not bias correct the resulting output against local temperature observations. The reason is not only credibility: the climate model simply does not represent temperatures in, say, Graz. Or, with less extreme illustrations: a climate model which does not resolve the main topographic features of a location does not represent local climate. A GCM might "miss" a full mountain range, an RCM an inner alpine valley. The problem may at best be limited to the representation of present climate, but it may involve also the response to climate change.

The representativeness assumption comprises two aspects: spatial (and temporal) scale and location. Both aspects cannot in every case be addressed by bias correction.

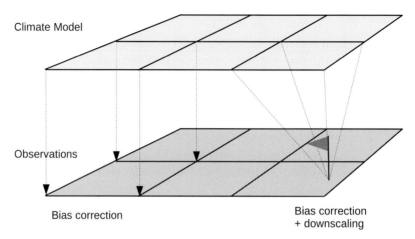

Climate Model

Observations

Bias correction

Bias correction
+ downscaling

Figure 12.10 Bias correction settings. A pure bias correction leaves the output resolution unchanged (left), whereas a bias correction to a higher resolution or even point scales includes downscaling.

Given their relevance, both aspects will be discussed separately in Sections 12.7 and 12.8.

12.7   Bias Correction and Downscaling

Bias correction may be calibrated between climate model and observational data of a similar resolution. Often, however, the observational data is of considerably higher resolution, for example, high-resolution gridded data or even station data – in this case, the bias correction also attempts to bridge a scale gap (see Figure 12.10). Whether such downscaling is sensible depends on the size of the gap, on the variable, on the location, on the relevant statistical climate aspects and on the chosen bias correction method.

- Is the scale gap so big that relevant processes are missing or severely mis-represented? Squall lines missing in a GCM or feedbacks missing in an RCM will not be represented by a bias correction.
- Has the variable strong residual sub-grid variability? Most bias correction methods cannot create non-resolved random fluctuations.
- Is the terrain complex, with relevant sub-grid variations? In such a situation, local processes (mountain breezes, snow-albedo feedbacks, orographic precipitation) may not be represented.
- Are only mean values relevant, or also the distribution's tail? A bias correction may provide a reasonable guess of local long-term means but may fail to represent local variability and extreme events.
- Is a variance-correcting method such as quantile mapping used? Such methods try to represent local variability by variance inflation and may introduce severe artefacts when applied for downscaling.

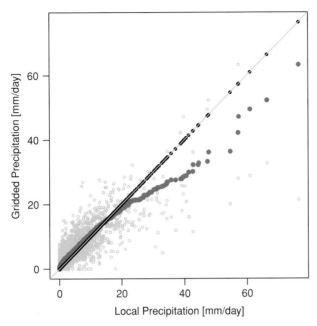

Figure 12.11 Quantile mapping and downscaling. Scatter and QQ plot of grid-box precipitation (E-OBS) against station observations for summer (JJA) in Rheinstetten, Germany (data from Germany weather service, DWD). Light grey: scatter plot; dark grey: QQ plot of uncorrected data; black: QQ plot of corrected data.

In the following, we will discuss these issues with some examples. As laid out in Chapter 4, the discussion has two aspects: the representation of present climate but also that of the climate change signal. The second aspect is part of a broader problem which will be discussed in detail in Section 12.9. In Chapter 14, we present methods that combine bias correction with explicit downscaling to at least represent local variability in present climate.

12.7.1 Example 1: Representation of Sub-Grid Precipitation

Climate models simulate area-average precipitation (Osborn and Hulme 1997). At daily and sub-daily scales, however, precipitation exhibits random fluctuations much finer than a typical climate model grid box. Figure 12.11 illustrates problems that may occur in such a setting. Instead of climate model output, gridded observational data sets (E-OBS; Haylock et al. 2008) are used. These gridded data sets should, by construction, be approximately bias free (see Chapter 7 for a critical discussion); differences will thus mainly be the result of the scale gap between grid box and station. The QQ plot between station and grid-box data very much resembles the QQ plot of an RCM against station data: high values are lower in the gridded data set, and a drizzle effect is visible at the lower end of the distribution (zero precipitation at the station, low intensities at the grid box). These effects are typical of a scale gap and not biases: extreme events tend to be localised, in particular during summer, and averaging over the grid-box area will automatically reduce intensities. Similarly, low-intensity events may only occur in parts of the grid box: it might be dry at a particular station, but the area average may have a low though non-zero value. Quantile mapping removes these effects, but still, the correction does not make sense. The scatter plot shows the daily pairs of grid box versus station values. A correlation is clearly visible but of course no deterministic one-to-one relationship: for a certain grid-box value of, say, 20mm, the observed station values range between about 7mm and almost 80mm. Quantile mapping, however, would always create the same local value of slightly more than 20mm. Thus, even though the marginal distribution of the bias corrected grid-box precipitation is identical to that of the station data, and the local-scale variability is not at all represented. Grid-box values are merely inflated, but no random variability is added. In other words: precipitation, quantile-mapped against station data, will have no sub-grid random variability, and the temporal variability will be that of a grid-box average. The effects might be severe in particular for extreme events: whereas a thunderstorm, simulated (i.e. parameterised) by an RCM will in reality only extend over part of the grid box, quantile mapping will essentially "copy" it across the whole grid box. The total grid box precipitation will then be severely over-represented. In hydrological models, this artefact might in turn produce unrealistically high flood peaks.

A simple scaling method would not produce such artefacts: only mean values would be adjusted, and variances would not be inflated.[2] Also such a correction would only adjust systematic spatial variations (caused by orography) but would not create any day-to-day sub-grid variations. Such a correction would thus produce reasonable estimates of mean values but not of variability around the mean and extremes. The situation is similar to deterministic regression models in PP (see Section 10.2 and Chapter 11). If only mean values (e.g. at the seasonal scale) are relevant, such a correction may nevertheless be very sensible (and definitely more than quantile mapping).

12.7.2   Example 2: Representation of Temperature in Complex Terrain

In contrast to precipitation, temperature varies rather smoothly in space; much of the sub-sub-grid variability will be systematic variations with topography. Such variations will fully manifest themselves in the observed climatological averages and may thus be bias corrected. In complex terrain, however, also substantial random sub-grid fluctuations might occur at the daily scale. For instance, narrow alpine valleys (The Rosental Valley in southern Austria, shown in Figure 12.12, has a typical width of less than 12km) are often not resolved by standard RCMs. A typical phenomenon occurring in such valleys, in particular in winter, are low-level inversions accompanied by low clouds or fog. During an inversion, temperatures up the slopes of the valleys will be higher than those down in the valley itself, that is, on some days the climatological temperature gradient will be reversed. In an RCM without sufficiently high resolution, the surface temperatures would be represented by one single value – a bias correction may be able to adjust

---

[2] Of course, also the variances are rescaled, but only to match the change in the mean, not beyond that.

Figure 12.12 Low-level inversion. Rosental Valley, Carinthia, Austria (photo: D. Maraun, 25 Dec 2016).

the climatological temperature distributions at each location, but it will not produce any day-to-day sub-grid variations: in the bias corrected RCM, temperatures in the valley will always be higher than those up the valley slopes. The lower the climate model simulation, the more severe such effects will be.

Note that one might be able to successfully create such temperature inversions even with standard bias correction methods if the inversion occurs at the grid box scale within the RCM. Then, instead of using only surface temperature as input for the bias correction, one could in fact also use higher-altitude temperature values (e.g. at 850hPa) as input for the bias correction of the mountain slopes.

12.7.3 Example 3: Representation of Temperature Changes in Complex Terrain

If the climate model resolution is too low to capture important feedbacks, also the resulting climate change signal might be implausible. Consider the Central Valley and Sierra Nevada in California, USA (Figure 12.13). A GCM does not resolve the complex terrain and simulates a very smooth spring (MAM) climatology (a). High-resolution gridded observations (b) reveal the influence of the coastal mountains, the Central Valley and the high mountain range of the Sierra Nevada. Bias correction with quantile mapping would perfectly match the GCM to the observations over the calibration period. A very–high-resolution RCM resolves the observed influence of the orography on the temperature field (c).

Forced according to the RCP8.5 scenario, the GCM simulates a smooth response to climate change (d). The quantile mapping would inflate local variances and thereby slightly modify the GCM climate change trends (e; see also Figure 12.3). Also the high-resolution RCM modifies the GCM trends but with a clear topographic finger-print (f). This spatially varying climate change signal is consistent with our physical

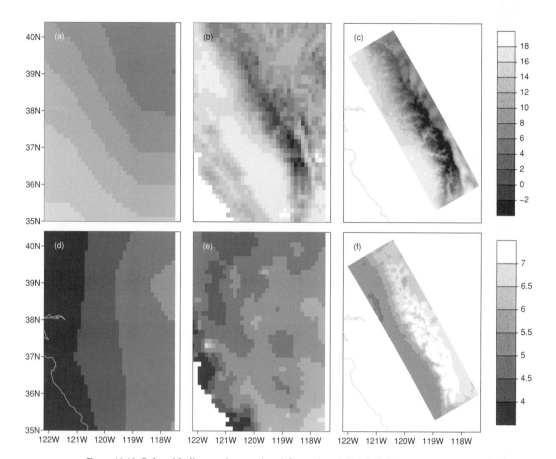

Figure 12.13 Sub-grid climate change signal for spring (MAM) daily mean temperature [°C] in California, USA. (a–c): present climate (1981–2000 average); (d–f): simulated change (2081–2100 average minus 1981–2000 average, RCP8.5 scenario van Vuuren et al. (2011). (a,d): GFDL-CM3 GCM, bilinearly interpolated to 8km grid; (b,e): corrected GCM (for present by construction identical with observations at 8km horizontal resolution; Maurer et al. 2002); (c,f): WRF RCM at 3km horizontal resolution, driven with GFDL-CM3 climate change signal (Walton et al. 2015). Adapted from Maraun et al. (2017b).

understanding: at high elevations, the snow-albedo feedback amplifies the warming signal. Of course, we don't know whether the RCM signal is correct, but it is (at least) plausible. But both the raw GCM signal and its modification by quantile mapping appear to be unconnected to any local physical processes, so they are highly implausible. Such implausible sub-grid climate change trends cannot be overcome by standard bias correction.

12.8   Bias Correction and Mislocations

As presented in Section 8.5.1, climate models exhibit substantial large-scale circulation biases. For instance, the North Atlantic storm track is too zonal in many GCMs and

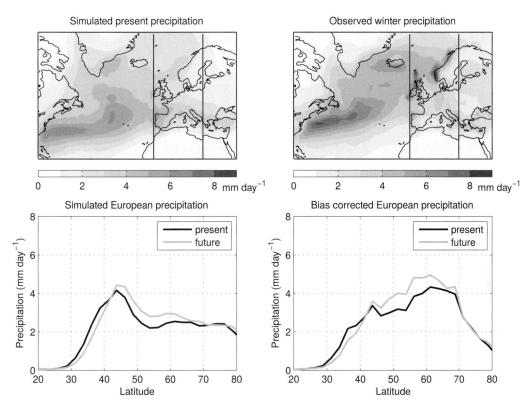

Figure 12.14 Effect of large-scale circulation biases. Winter (DJF) mean precipitation. Left: raw model (FGOALS-g2 GCM), right: observations (GPCP)/corrected. Top: climatological map; bottom: zonal average over 10W to 20E. Present: 1976–2005; future: 2070–2099, RCP8.5 scenario. Adapted from Maraun et al. (2017*b*).

often crosses the European continent too far south. Even if the response of the storm track to climate change is credibly simulated, a bias correction in the presence of such location biases may cause considerable artefacts. Figure 12.14 illustrates the problem: whereas the observed storm track, here measured by its imprint on seasonal winter (DJF) precipitation, passes across Ireland, Scotland and Scandinavia, the FGOALS-g2 GCM simulates a more zonal storm track entering the continent in Northern Spain and Southern France. The GCM simulates a slight northward shift of the stormtrack under climate change, and the precipitation peak at about 45N shifts northward by about 2–3°. A simple mean bias correction, calibrated on observed precipitation and applied to the simulated future precipitation field, would cause a southward shift of the precipitation peak by several degrees. Thus, even though the bias correction does not modify the local climate change signal, it modifies the large-scale climate change pattern: A northward migration of the storm track is modified into an implausible southward migration of the precipitation field. Depending on the context, such modifications may or may not be relevant: if one is interested in local precipitation only, one would not "see" the implausible large-scale pattern. If one is simulating the runoff over a large river catchment, such modifications may be important.

Figure 12.15 Meteorological divide. Kleinsölktal Valley, Styria, Austria (photo: D. Maraun, 25 Nov. 2016).

But even if the large-scale flow is unbiased, climate model simulations might suffer from systematic regional-scale location biases. The orography even of high-resolution RCMs is in general a coarse model of the true orography. In complex terrain, the simulated meso-scale flow might therefore substantially deviate from reality (Maraun and Widmann 2015). Mountain ranges often act as meteorological or climatological divides (see Figure 12.15). They may shield the climate south of the mountains from severe winter storms and cause inner alpine valleys to be rather dry. Climate models may misrepresent the strength of such divides: the topography may be too flat, such that no divide occurs; or the topography might be too compact, such that valleys connecting both sides of the range are missing.

To illustrate potential consequences, consider the example shown in Figure 12.16. The location of interest is the Maggia Valley in Ticino at the Swiss-Italian border (central grid box). The grid boxes show the correlation between winter (DJF) average simulated precipitation in the actual grid box and corresponding observed precipitation in the central grid box. Obviously, the interannual variability of simulated local precipitation is completely independent of observed local precipitation. But at distant grid boxes along the northern slopes of the Alps, correlations are almost perfect. That is, in reality the weather north of the Alps crosses the mountains into the valleys on the southern slopes. Yet in the climate model, the coarse orography unrealistically shields the southern valleys from the Northern influence.

Such a location bias is not correctable by a standard bias correction method. These methods can adjust local mismatches in the marginal distribution, but they cannot transform a Mediterranean climate into an Atlantic climate. A possible solution might be to identify a distant but representative grid box in the climate model and carry out a nonlocal bias correction. In perfect boundary conditions, such a grid box could be identified by the correlation at interannual timescales (Maraun and Widmann 2015).

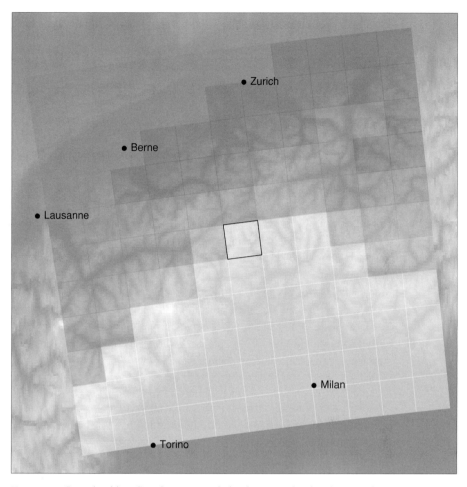

Figure 12.16 Location bias. Grey boxes: correlation between simulated seasonal mean summer precipitation in the particular grid box and observed precipitation in the central grid box (marked by a box). Grey background shading: orography. Observed data is taken from E-OBS, RCM data from RACMO2 driven with ERA-40. Adapted from Maraun and Widmann (2015).

12.9 Modification of the Climate Change Signal

A credibly simulated trend should not be modified by any bias correction. In such a case, the time-invariance assumption is fulfilled, and a trend-preserving bias correction is the method of choice. Several authors, however, have demonstrated that biases may be state-dependent and thus be different in a future climate, that is, biases may not be time-invariant (Christensen et al. 2008, Buser et al. 2009, Vannitsem 2011, Boberg and Christensen 2012, Maraun 2012). In other words: the simulated climate change is implausible, and a bias correction assuming time-invariant biases would fail. Reasons for state dependent biases are connected to processes at all spatial scales and may be related to credibility issues (Section 12.6.1; see Section 8.5 for a variety of

---

Box 12.2 Should Bias Correction Correct Large-Scale Trends?

As discussed throughout the chapter, bias correction is not designed to correct fundamental model errors. A major source of uncertainty in climate projections, also at the
regional scale, is the global climate sensitivity, which mainly results from the uncertain effect of clouds on the radiative budget of the atmosphere. Thus, biases in global
mean temperature – and linked to that in essentially all regional scale meteorological variables – will in general be time dependent. Bias correction cannot correct such
fundamental model errors in global mean temperature trends. As discussed earlier,
also trend errors due to misrepresented large-scale processes such as ENSO cannot
be corrected by bias correction. In other words: the assumption of time-invariant
regional biases is essentially worthless if not even global mean temperature biases
are time invariant.

  Thus, it is reasonable to interpret the time-invariance assumption conditionally on
the simulated large-scale climate change: do regional biases change in time, given
the (uncertain) simulated large-scale climate change? Any evaluation of bias correction methods, for example in a pseudo-reality, should therefore be conditional on
simulated large-scale climate, for example by expressing regional trends relative to
the large-scale trends. In a broader context, (bias corrected) regional climate change
projections should thus be interpreted as (at best) credible regional futures given a
simulated large-scale climate change signal.

---

examples) and representativeness issues (Section 12.6.2; see also the third example
in Section 12.7). In such a situation, the question arises whether bias correction may
improve simulated trends.

12.9.1   Trend Modifications by Quantile Mapping

It has been suggested that quantile mapping could be used to correct such implausible
changes (Boberg and Christensen 2012, Gobiet et al. 2015). This line of argument is in
contrast to the development of trend-preserving quantile mapping variants (see Section
12.4). As discussed in Section 12.4, quantile mapping is designed to correct quantile-
dependent biases: a value of, say 20°C may be corrected differently than a value of,
say, 10°C. The argument of Boberg and Christensen (2012) or Gobiet et al. (2015) is
that the value-dependent correction function would be invariant under climate change
conditions. This argument, however, is questionable: within a given month of the year,
day-to-day temperature variations are mostly spanned by the passage of different air
masses and the related cloud cover. In mid-latitude spring, 20°C may correspond to a
sunny day with a southerly airflow, 5°C to an overcast day with a westerly flow. Biases
are likely not only linked to temperatures but also to the general properties of the air
masses.

  Under climate change, however, the radiative balance of the atmosphere will change
and, in turn, the properties of air masses themselves will change: in a warmer future

climate, 20°C might be just a fair day, while 5°C may occur in an northeasterly cold air outbreak. It is thus likely that also the future biases associated with 20°C and 5°C will be different from those in present climate. Thus, biases will not only depend on the actual values but more generally on the state of the climate system.

A similar argument can be made from a timescale point of view: the distribution of daily weather values is mostly spanned by day-to-day variability – the passage of air masses. Quantile mapping is thus calibrated on short-term variability. The same correction is then applied to all timescales: to biases on day-to-day variability, to biases on interannual and decadal variability and to biases in long-term trends. Quantile mapping thus assumes that biases are independent of timescale. Haerter et al. (2011) showed that this assumption is not in general valid.

To summarise: trend modifications by quantile mapping would be justified if the value dependence of biases would directly correspond to a state dependence or – which is equivalent – if biases would be timescale-independent. Whether such a correspondence exists is not a statistical question; it has to be established on a case-by-case basis by physical arguments.

12.10   A Cookbook for Bias Correction

The following list summarises the most important steps in carrying out a successful bias correction (see Figure 12.17). Many steps are similar to those for developing a PP downscaling model, but the freedom to select predictors and model structure is much more limited.

12.10.1   Initial Considerations and Checks

As in PP downscaling, one should first get an overview of the relevant climate phenomena. As discussed in the Chapter 11, on the one hand, these phenomena depend on the given user problem (see also Chapter 5), and on the other hand, they determine characteristic space and timescales, and the relevant statistical climate aspects.

Second, one has to identify the relevant large-scale climatic phenomena; how realistically are they simulated for present climate, and how credible are their changes simulated by the driving climate model? Depending on the relevant statistical climate aspects, the evaluation for present climate may involve also non-marginal aspects, for example, regarding the temporal structure. Major errors in the residual temporal dependence are not correctable by bias correction (see Section 12.5). If the model simulates implausible changes of major involved large-scale processes, climate change projections are likely implausible (see the discussions in Section 12.6).

Third, it is important to check for representativeness issues:

- Does the model have major location biases? Large-scale phenomena such as the storm track but also local phenomena might be systematically displaced (see Section 12.8). In the former case, one should be alert and assess potential artefacts in the resulting climate change signal; in the latter case, a non-local bias correction might be feasible.

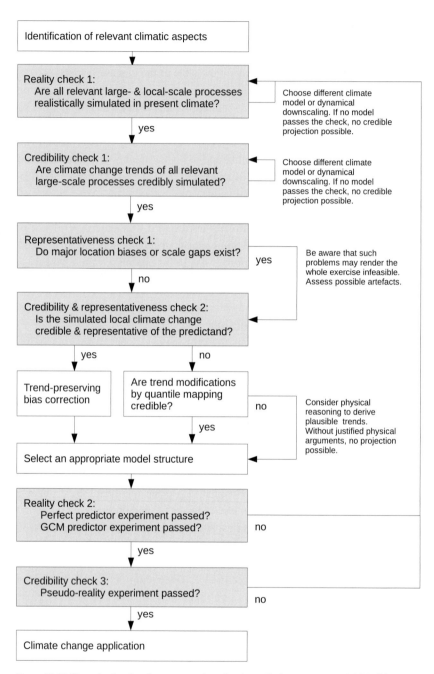

Figure 12.17 Steps in the development and evaluation of a homogeneous MOS (bias correction) model for climate change studies. The order of the individual steps is to some extent arbitrary.

- Is a substantial downscaling step involved such that relevant sub-grid variability, which is not represented by the climate model, exists? In this case, no variance-correcting bias correction (such as quantile mapping) should be used, and only long-term averages should be interpreted (see Section 12.7).

And finally, is the simulated regional climate change (conditional on the simulated large-scale climate change) credible (see Section 12.9)? If yes, one should choose a trend preserving bias correction method, that is, no standard quantile mapping. Different variants of quantile mapping preserve trends in the mean or even for different quantiles (see Section 12.4).

If the regional trend is not credible, one could check whether the trend modifications by quantile mapping are credible. This would involve demonstrating that a calibration on short timescales is valid on long timescales. In case of a negative result, one either has to develop a physically based bias correction which would intentionally modify the climate change signal (see Chapter 14 for some ideas) or one would have to combine the simulated trends with expert knowledge of the involved phenomena and regional conditions to come up with a modified best guess (see Section 18.5.2).

12.10.2 Model Development

Once a type of bias correction model has been selected, the actual model development is rather limited. If a quantile mapping-type approach is chosen, one has to consider the following issues:

- How should the bulk of the distribution be modelled? Non-parametrically (e.g. by linear interpolation) or parametrically (see Sections 12.5 and 16.5)?.
- In case of precipitation: how should wet-day occurrence be modelled? Should the overall number of wet days or the number of wet days conditional on specific weather types be adjusted (see Section 12.5)?
- How should the extreme tail be modelled? With a constant correction? Based on an extrapolation from the bulk of the distribution (e.g. gamma distribution for precipitation)? Or based on specific extreme value models?

The choice of the model structure, in particular for the extremes, may substantially alter the results.

12.10.3 Model Evaluation: Reality and Credibility Checks

The initial reality and credibility checks of the driving model should be complemented by an assessment of the bias corrected output. Such an assessment should involve an evaluation in a perfect predictor experiment, an evaluation with GCM predictors for present climate and an evaluation in a pseudo-reality (Maraun et al. 2015). A particular emphasis should be given to the assessment of artefacts, in particular implausible projections, resulting from the bias correction. For a detailed discussion, refer to Part III of the book.

12.11 Further Reading

The classical papers on MOS in weather forecasting are Glahn and Lowry (1972) and Klein and Glahn (1974). Recent reviews have been published by Teutschbein and Seibert (2012) and Maraun (2016).

13 Weather Generators

For many applications long weather time series are required (Richardson and Wright 1984): for instance, to drive impact models for deriving design values of hydraulic structures or to assess meteorological impacts on hydrology or agriculture. In practice, however, observational records are often short, suffer from gaps and inhomogeneities or are simply missing for some meteorological variables. Weather generators have been developed to synthetically generate weather time series of, at least in theory, infinite length.

DEFINITION 13.1 As weather generators we define stochastic models of meteorological variables that explicitly model their marginal distribution and temporal dependence.

Since the development of the first precipitation (Buishand 1977, Katz 1977) and weather generators (Richardson 1981), these models have become widely used to produce long surrogate time series, to impute missing data (e.g. Yang et al. 2005) and increasingly to downscale climate projections for impact assessment (e.g. Hulme et al. 2002).

In the following, we will first introduce the assumptions underlying the use of weather generators in climate change studies (Section 13.1), followed by an overview of the most widely used weather generator approaches (Section 13.2). Structural skill of weather generators will be discussed in Section 13.3. Finally, we will present approaches to incorporate climate change in weather generators in Section 13.4. The development of weather generators is – depending on how they incorporate climate change – similar to the PP and MOS approaches. For conditional weather generators, refer to the PP cookbook in Section 11.7; for change-factor weather generators, refer to the MOS cookbook in Section 12.10.

13.1   Assumptions

As will be discussed in detail in Section 13.4, weather generators can be used in two different settings to downscale climate projections: with so-called change factors – climate model–simulated changes in long-term climate statistics, which are used to modify the parameters of the weather generator – and with time series of meteorological predictors, taken from climate model simulations, that condition the weather generator parameters on a day-by-day basis and impose long-term changes.

The change factor approach closely resembles the delta change method; the underlying assumptions are therefore similar to those of the MOS approach (Chapter 12). The second approach is a specific variant of PP statistical downscaling (Chapter 11), and the underlying assumptions are therefore essentially the same. For both approaches, similarities and differences with standard MOS and PP will be discussed in Section 13.4.

13.2   Methods

Given the complexity of the problem, various weather generator approaches have been developed. Most of these approaches have in common that they first simulate a precipitation series and then simulate other variables conditional on this reference.

Precipitation itself has a highly skewed distribution on short timescales, with dry intervals in between. Different approaches have been suggested to model these characteristics. The most prominent ones, Richardson-type, truncated stochastic processes and Poisson cluster processes, will be described in detail in the following sections. The modelling of other meteorological variables is very similar across the different approaches and will be discussed jointly in Section 13.2.4. Weather generators have been conditioned on meteorological predictors – such stochastic PP statistical downscaling models will be discussed in Section 13.2.5. Finally, we will very briefly review multisite (Section 13.2.6), full field (Section 13.2.7) and subdaily weather generators (Section 13.2.8).

13.2.1   Richardson-Type Precipitation Generators

Richardson-type precipitation generators separate precipitation into an occurrence and an amount process, usually at the daily scale (Figure 13.1; for reviews see Wilks and Wilby 1999, Wilks 2010). The simplest model for the occurrence process was suggested in 1962 by Gabriel and Neumann; a sequence of wet and dry days is simulated by a Markov chain (see Equation 6.83 in Section 6.4.2). The transition between dry and wet states, K_d and K_w, at time t is given by a $2 \times$ matrix of probabilities

$$p_{kl}(t) = Pr(K_i = l | K_{i-1} = k), \quad k, l = (d, w). \tag{13.1}$$

This first-order Markov model assumes that the occurrence of precipitation depends only on the occurrence of the previous day. But such models tend to under-represent the length of dry spells. Therefore, higher-order models have been developed (Stern and Coe 1984, Mason 2004). As an alternative approach, instead of considering individual wet and dry days, Racsko et al. (1991) model the lengths of total spells following a geometric distribution.

Precipitation amounts are typically modelled independently for each day according to a chosen probability distribution $G(.)$ with parameter vector θ,

$$Y_i \sim G(\theta). \tag{13.2}$$

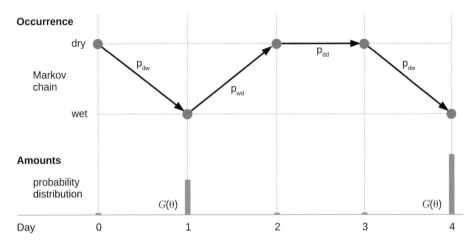

Figure 13.1 Richardson-type precipitation generator. At each day, a transition (arrows) between wet and dry states (grey dots) is simulated according to the transition probabilities p_{ij}. On wet days, amounts (grey boxes) are simulated according to a distribution $G(\theta)$.

A widely used model for $G(.)$ is the two-parameter gamma distribution (Section 6.1.6, e.g. Katz 1977); other authors use a mixed exponential distribution (Semenov and Barrow 1997, Wilks 1998). To account for seasonality, the parameters for both the occurrence and amounts process typically depend on the month of the year or the season.

13.2.2 Truncated Stochastic Processes

Instead of separately modelling precipitation occurrence and amounts, it has been proposed to jointly model both processes. The most common approach is based on a power-transformed and truncated normal distribution (Bárdossy and Plate 1992). Precipitation amounts Y_i at a day i are then modelled as

$$Y_i = \begin{cases} Z_i^{\beta} & Z_i > 0, \\ 0 & \text{otherwise,} \end{cases} \tag{13.3}$$

where Z_i is a normally distributed random variable and β is a transformation parameter. To account for temporal dependence, the Z_i are typically modelled as an autoregressive process (Bárdossy and Plate 1992, see also Section 6.4.2). Some authors suggest using more complex transformations (Bartholy et al. 1995, Glasbey and Nevison 1997, Allcroft and Glasbey 2003), whereas others suggested using different types of distributions (Dunn 2004).

13.2.3 Poisson Cluster Precipitation Processes

A different approach to rainfall modelling is Poisson cluster rectangular pulse models (Rodriguez-Iturbe et al. 1987); for reviews see Onof et al. (2000) and Wheater et al. (2005). These models are continuous in time and do not operate, for example, on the

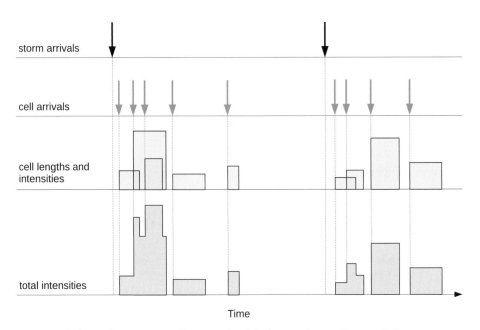

Figure 13.2 Poisson cluster process. Storms arrive (black arrows) according to a Poisson process. The arrival of rain cells (grey arrows) within a storm can follow either a Bartlett-Lewis or a Neyman-Scott process. Cell (grey rectangles) intensities and lengths are simulated according to chosen distributions. The total precipitation intensity is the sum over all cells at a given time.

daily scale. The basic assumption of these models is that precipitation arrives in rain cells, which cluster in rain storms (see Figure 13.2). A rain storm may be interpreted as the passage of a front or a squall line, a rain cell as an individual shower or thunderstorm or any structure at the meso-gamma scale (2–20km in space, typically up to 1 hour in time).

Typical timescales arise from the characteristic waiting times between storms and between rain cells and from the characteristic lengths of rain cells. The storms are assumed to arrive according to a Poisson process (Section 6.4.2). Basically two different point process models exist, which differ in their treatment of the rain cell arrival process, the Bartlett-Lewis and the Neyman-Scott process. The latter process is defined as follows:

- rain storms arrive according to a Poisson process, that is, with between-storm times assumed independent and following an exponential distribution with rate parameter λ;
- each storm contains a random number of C rain cells, which is typically Poisson distributed with mean ν. The arrival times of the rain cells relative to the storm arrival are exponentially distributed with parameter β;
- each rain cell has a duration following an exponential distribution with parameter η and an intensity X following an exponential distribution with parameter ξ;

- the total precipitation intensity $Y(t)$ at a time t is given by the sum of intensities of all rain cells "active" at time t;
- all random variables are mutually independent.

As stated, the Neyman-Scott and Bartlett-Lewis models differ only in the treatment of rain cell arrivals. In the Neyman-Scott model, the arrival times are modelled relative to the storm arrival (e.g. by exponentially distributed arrival times as given). In the Bartlett-Lewis model, the arrival times are themselves following a Poisson process, that is, the waiting times between two rain cells follow an exponential distribution (the first rain cell is generally assumed to occur when the storm arrives). Thus, in the Neyman-Scott model, there is a greater probability that rain cells cluster in the beginning of a storm.

No precipitation variability is modelled within a rain cell. Poisson cluster models should therefore not be interpreted at timescales smaller than a typical rain cell duration. To provide discrete time series at an aggregation scale s (e.g. hourly or daily), the instantaneous intensities $Y(t)$ can be aggregated to

$$Y_i^{(s)} = \int_{(i-1)s}^{is} Y(t)dt. \tag{13.4}$$

To characterise the stochastic process at a given time aggregation, Rodriguez-Iturbe et al. (1987) derived analytical expressions for relevant statistical moments such as expected value, variance or temporal auto-covariance.

Model calibration must be carried out at the available resolution of the observational reference data, which is often daily, sometimes hourly (Onof et al. 2000). The parameter estimation is usually carried out by the method of moments (Cowpertwait 1994, Wheater et al. 2005). Model parameters often vary throughout the calender year. Fitting different models for each month would result in an extremely high number of parameters, which are – given typical lengths of observational records – difficult to identify. Therefore, Cowpertwait (1994) suggested reducing the number of parameters by assuming a sinusoidal model for their seasonality.[1]

Several improvements to these standard variants of Poisson cluster processes have been proposed. To improve the representation of the proportion of dry periods, randomising the rain cell duration parameter η from storm to storm has been suggested (Rodriguez-Iturbe et al. 1988, Entekhabi et al. 1989). The representation of extreme events in the Bartlett-Lewis model has been improved by modelling the intensities of individual rain cells by the two-parameter gamma distribution rather than the exponential distribution and by adding jitter, that is high frequency random fluctuations within a rain cell (Onof and Wheater 1994). Finally, models have been developed to account for different precipitation processes, for example, for convective or frontal precipitation: Cowpertwait (1994) defined two different rain cell types with different duration and intensity parameters η and ξ, and Onof et al. (2000) suggested introducing dependence between intensity and duration.

---

[1] The complexity required for such a model depends, of course, on the seasonal cycle of precipitation and thus on the considered region.

13.2.4   Multivariable Weather Generators

The modelling of non-precipitation variables is similar for most weather generator approaches and typically based on a first-order autoregressive process (see Section 6.4.2 for a review see Wilks 2012):

$$\tilde{\mathbf{y}}_i = \mathbf{A}\tilde{\mathbf{y}}_{i-1} + \mathbf{B}\boldsymbol{\epsilon}_i. \tag{13.5}$$

The M dimensional vector $\tilde{\mathbf{y}}_i$ is standard normally distributed and – after a transformation that will be described next – represents the M different variables at time i. The $K \cdot K$ dimensional square-matrices $\mathbf{A}$ and $\mathbf{B}$ represent the lag-one auto- and inter-variable cross-correlations. For a single variable, $\tilde{\mathbf{y}}$, $\mathbf{A}$ and $\mathbf{B}$ are scalar.

The properties of precipitation and other variables are typically dependent. For instance, summer temperatures will tend to be higher on dry days, and lower on wet days. Therefore, non-precipitation variables are in general simulated conditional on the precipitation state. In the simplest case, the simulated $\tilde{y}_i$ are modified separately, depending on whether the day i is a dry $(k = d)$ or a wet $(k = w)$ day:

$$y_i = \tilde{y}_i \sigma_k + \mu_k, \quad k = d, w. \tag{13.6}$$

That is, the variables y_i have different means μ_k and standard deviations σ_k depending on the precipitation state k. Kilsby et al. (2007) condition these weather generator parameters not only on the state of the actual day but additionally on that of the previous day, that is, on the combined states dry-dry, dry-wet, wet-dry and wet-wet. Racsko et al. (1991) model temperature not only dependent on the state of day i but additionally on the length of the dry or wet spell and the position of the day i within that spell – this way, they may to some extent capture soil moisture feedbacks that amplify temperatures in long dry spells. All parameters may, again, depend on the time of the year.

13.2.5   Conditional Weather Generators

The parameters of Richardson-type weather generators have been conditioned on meteorological predictors, for actually two reasons: first, to improve the representation of long-term variability (Katz and Parlange 1993, Wilks and Wilby 1999), and second to create physically motivated precipitation generators that impose clustering, long-term variability and non-stationarities via the large-scale state of the atmosphere (Bárdossy and Plate 1991).

To improve the representation of longer-term variability, it has been suggested to condition weather generator parameters on atmospheric predictors (e.g. Katz and Parlange 1993). An alternative suggestion is to condition the parameters on total monthly precipitation taken from a monthly weather generator or the driving climate model (e.g. Wilks and Wilby 1999, Dubrovský et al. 2004).

One way to condition the weather generator parameters on the atmospheric circulation is essentially the weather type approach (see Section 11.2.2). Here, the weather generator model is calibrated to the precipitation data separately for each weather type.

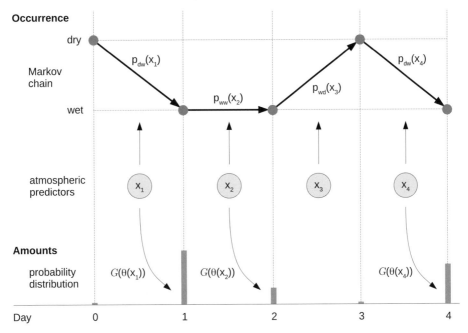

Figure 13.3 Richardson-type precipitation generator, conditioned on atmospheric predictors $\mathbf{X}_i$. Transitions (bold arrows) between wet and dry states (grey dots) are simulated according to transition probabilities $p_{ij}(\mathbf{x}_i)$, conditional on the state $\mathbf{x}_i$ of the atmosphere. On wet days, amounts (grey bars) are simulated according to a distribution $G(\boldsymbol{\theta}(\mathbf{x}_i))$, whose parameters $\boldsymbol{\theta}(\mathbf{x}_i)$ depend on the state $\mathbf{x}_i$ of the atmosphere.

For instance, Bárdossy and Plate (1992) conditioned a weather generator on atmospheric circulation states. Similarly, Fowler et al. (2000) conditioned the parameters of a Neyman-Scott rectangular pulse model on a set of weather types. Such conditional weather generators can be seen as PP statistical downscaling models (see Chapter 11). For a sketch of a conditional Richardson-type weather generator see Figure 13.3.

The occurrence and amounts process of the Richardson-type weather generator can be elegantly written as generalised linear models (GLMs; see Section 6.3.2). For instance, Coe and Stern (1982) and Stern and Coe (1984) modelled the Markov chain of the occurrence process by a logistic regression (Section 6.3.2). The probability of precipitation occurrence p_i at a day i is then given as

$$\ln\left(\frac{p_i}{1 - p_i}\right) = x_i\beta, \tag{13.7}$$

where $x_i = 1$ if precipitation occurred at day $i - 1$, and $x_i = 0$ when the previous day was dry, and β is a regression parameter. In this example, the predictor x_i is a scalar, but in general it might be a vector representing different covariates. For instance, a seasonal cycle is easily integrated into this model by simply adding a time-varying predictor such as a harmonic function. Meteorological covariates can be included in the same way as additional predictors (Yang et al. 2005). Similarly, geographical covariates such as longitude and latitude can be included (Rust et al. 2013); interactions between the seasonal cycle and the memory can be added by adding products of predictors as new predictors (Chandler and Wheater 2002). The amounts process can be written similarly with a gamma distribution with mean μ_i, where

$$\ln \mu_i = x_i \gamma. \tag{13.8}$$

Again, x_i is a predictor (or vector of predictors) and γ a (potentially vectorial) regression parameter. Even more generally, precipitation can be modelled using general additive models (Hyndman and Grunwald 2000, Beckmann and Buishand 2002, Underwood 2009).

A further approach uses hidden Markov models that are conditioned on the large-scale circulation (Hughes and Guttorp 1994, Bates et al. 1998, Hughes et al. 1999). These models have mainly been designed to model precipitation at multiple locations and will therefore be discussed in more detail in the following section.

13.2.6   Multisite Weather Generators

As there are many different types of weather generators, also a multitude of different extensions of these methods to a multisite context exists. A common problem is that multisite precipitation modelling is difficult, because few tractable models are available for multivariate distributions to capture the occurrence and amounts process in space.

Models Based on the Multivariate Normal Distribution
Many multisite precipitation models are based on transformations of the multivariate normal distribution. For the single-site case, this approach has been discussed in Section 13.2.2. For the multisite case, the univariate normal distribution Z_i in Equation 13.3 is replaced by a multivariate normal distribution, with a specific correlation structure to represent intersite dependencies (Bárdossy and Plate 1992, Stehlik and Bárdossy 2002). Multisite precipitation sequences are then generated by simulating correlated vectors z_i at each day i and then transforming the sequence according to Equation 13.3.

Similarly, the Richardson-type precipitation generator can be extended to a multisite model (Wilks and Wilby 1999). The transition probabilities between the occurrence states X_{i-1} and X_i (with $X_i = 1$ for a wet day and $X_i = 0$ for a dry day) determine the wet-day probabilities $p_{w,i}$ at day i as

$$p_{w,i} = \begin{cases} p_{dw} & X_{i-1} = 0, \\ p_{ww} & X_{i-1} = 1. \end{cases} \tag{13.9}$$

The occurrence state X_i at time i can then be determined by simulating from a univariate standard normally distributed random variable $w_i \sim \mathcal{N}(0, 1)$,

$$X_i = \begin{cases} 1 & \Phi(w_i) \le p_{w,i}, \\ p_{ww} & \text{otherwise.} \end{cases} \tag{13.10}$$

Here, $\Phi(.)$ denotes the standard normal cumulative distribution function, such that $\Phi(w_i)$ is uniformly distributed between $(0, 1)$.[2] The idea of this rather complicated representation of the Markov chain (Equation 13.1) is that the univariate w_i can straightforwardly be replaced by a correlated multivariate normally distributed random variable to describe the correlated occurrence at all sites. Wilks (1998) generates spatial dependence in the amounts process by conditioning the choice between two exponential distributions (representing medium- and high-intensity precipitation) on the latent Gaussian process w_i – since the w_i are correlated in space, also high-intensity and medium-intensity events will cluster in space.

Likewise, the GLMs presented in Section 13.2.5 can be extended to represent multiple sites (Chandler 2005, Chandler and Bate 2007). For instance, Yang et al. (2005) modelled the spatial dependence of precipitation by transforming daily intensities to approximate normal (by considering the third root of precipitation modelled with a gamma distribution) and then using a multivariate normal distribution. Alternatively, copulas can be used to model intersite dependencies (Bárdossy and Pegram 2009; see Section 6.1.7).

Hidden Markov Models

An alternative approach to model inter-site dependence resorts to (non-homogeneous) hidden Markov models (see Figure 13.4; Hughes and Guttorp 1994, Bates et al. 1998, Hughes et al. 1999). These models simulate a series of hidden weather states, which represent characteristic spatial weather patterns. If the transition probabilities between these weather states are conditioned on meteorological predictors X_i, the models are called non-homogeneous hidden Markov models. The weather state $\mathbf{S}_i$ at a day i itself controls the wet-day probabilities p_i:

$$p_{w,i} = p_w(\mathbf{S}_i). \qquad (13.11)$$

Hence the Markov model does not describe the transition between dry and wet days but a "hidden" weather state. The occurrence process is conditionally independent between consecutive days. Similarly, precipitation amounts can be simulated by a gamma distribution

$$Y_i = \begin{cases} \Gamma(\gamma(\mathbf{S}_i), \lambda(i)) & \text{if day } i \text{ is wet,} \\ 0 & \text{otherwise,} \end{cases} \qquad (13.12)$$

where the rate and shape parameters γ and λ depend on the weather state S_i (Bellone et al. 2000). The key advantage of introducing hidden weather states is the possibility to model spatially dependent fields with a simple univariate statistical model: if the same sequence of weather states is used to model precipitation at multiple sites, these are automatically dependent. Note, however, that all dependence between occurrence and

---

[2] A cumulative distribution function transforms the quantiles of a distribution to [0, 1], Section 6.1.3.

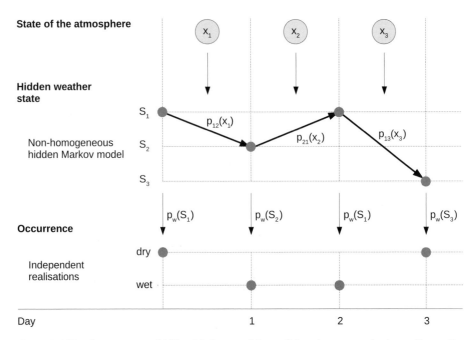

Figure 13.4 Non-homogeneous hidden Markov model, conditioned on atmospheric predictors X_i. A daily series of weather states S_i is simulated, with transition probabilities $p_{ji}(x_i)$ conditioned on the state x_i of the atmosphere. On each day, the occurrence of precipitation is simulated independently, conditional on the state S_i. Similarly, also amounts can be simulated. Even though the simulation of occurrence and amounts is conditionally independent in time and space, temporal and spatial dependence is imprinted by the weather states, which exhibit memory in time and determine occurrence and amounts distributions in all considered locations.

amounts on consecutive days is imposed only by the weather states; no residual temporal or spatial dependence is modelled. Vrac et al. (2007b) condition also the amounts distribution on atmospheric predictors, making this model in principle suitable for PP downscaling in a wetter future climate.

Multisite Poisson Cluster Models

Cox and Isham (1994) proposed a multisite version of Poisson cluster models. The basic underlying idea is that storms arrive to the considered area following a master Poisson process. Each storm is categorised according to the specific sites which experience precipitation during the storm. For a total set of k sites, $2^k - 1$ such storm types exist (if the case that no site is affected is not considered) – estimating the occurrence probabilities of all these different combinations is essentially infeasible. It has therefore been suggested to constrain the joint occurrence of precipitation at two sites (e.g. Onof et al. 2000, Wheater et al. 2000). One possibility would be to assume that the joint occurrence probability decays with distance. Another option is to explicitly model the geometry of storms (Wheater et al. 2005). As the latter approach simulates full spatial rainfall fields, it will be discussed in the following section.

13.2.7   Full-Field Weather Generators

The aim of full-field generators is to simulate gridded meteorological fields on a fine grid. These models are typically designed for precipitation, because it is very relevant for hydrological applications and at the same time exhibits very strong small-scale variability. Several types of approaches exist: (1) an extension of Gaussian-process-based multisite models to the full field by interpolating site-specific weather generator parameters. Variants of this approach are directly developed for full fields. (2) Spatial Poisson cluster models. And (3) random cascade models. A comparison of different types can be found in Ferraris et al. (2003). These models have rarely ever been used to downscale climate change projections (for an exception, see Perica and Foufoula-Georgiou 1996), but clearly have potential.

Transformed Gaussian Processes
In Section 13.2.6, we discussed multisite weather generators based upon transformed multivariate normal distributions. These models can be extended to a gridded field by extrapolating the model parameters to the grid (Wilks 2009). The extrapolation can in principle be implemented as a latent spatial process: the parameters are themselves modelled as a spatial stochastic process, conditional on geographical covariates such as longitude, latitude, elevation or climatic information available at the fine grid (Cooley et al. 2007).

Similar models are calibrated directly to full-field data: Gaussian random fields are generated with a prescribed covariance structure and nonlinearly transformed (Ferraris et al. 2003). Recently hierarchical variants have been developed that model the precipitation process in three stages, describing the storm arrival, the temporal evolution of areal mean properties and the temporal evolution of the storm structure (Paschalis et al. 2013). This model requires high-resolution radar data for calibration. Alternatively, Rebora et al. (2006) have developed a model that is calibrated entirely to the skillful scales of a climate model. The model is then extrapolated to smaller scales by assuming scaling laws of the precipitation power spectrum.

Full-Field Poisson-Cluster Models
As stated in Section 13.2.6, spatial extensions of Poisson-cluster processes exist. These models build upon an approach proposed by Cox and Isham (1988) and explicitly model the geometry of storms (Northrop 1998, Cowpertwait et al. 2002, Wheater et al. 2005). Here it is assumed that storms have a simple (e.g. circular or elliptic) shape and propagate along a storm axis at a specific velocity. Within these storms, rain cells occur, which propagate at the same velocity and in the same direction. The number, position and size of these rain cells is random. Their temporal clustering is modelled as a Neyman-Scott process.

Random Cascade Models
Finally, models have been constructed that assume fractal or multifractal relationships of precipitation characteristics across a range of time and space scales (Schertzer and

Lovejoy 1987, Over and Gupta 1996, Thober et al. 2014). A low-resolution precipitation volume is iteratively disaggregated into smaller spatial areas (and shorter time periods); the disaggregation is based on some underlying random process. Such models are typically constrained to observations only at the coarsest scale. Perica and Foufoula-Georgiou (1996) developed a cascade model based on wavelet decomposition that uses atmospheric predictors.

13.2.8   Subdaily Weather Generators

Many subdaily precipitation models are based on random cascade models (see the discussion in the previous section) that disaggregate precipitation in time (Olsson 1998, Güntner et al. 2001, Deidda et al. 2006). In general, the scaling characteristics of precipitation are more complex in time than in space (Lovejoy and Schertzer 2010), such that the temporal disaggregation of precipitation is more challenging than its spatial disaggregation (Thober et al. 2014). Poisson cluster models can be used down to hourly timescales (Onof et al. 2000, Wheater et al. 2005), although some recent research suggests it is possible to use Poisson-cluster models below the hourly timescales (Kaczmarska et al. 2014). Alternatively, a further layer of clustering can be introduced so that each cell is now a sequence of instantaneous pulses (Cowpertwait et al. 2007). Mezghani and Hingray (2009) developed a subdaily weather generator that combines a generalised linear model with a resampling approach. The generalised linear model is employed to simulate daily regional weather series conditional on the large-scale circulation. At a given day, the simulated regional weather characteristics and the large-scale circulation are used to select a suitable analog. The subdaily temperature and precipitation variations are then adjusted to match the simulated daily mean value.

13.3   Structural Skill and Limitations of Weather Generators

Standard PP methods derive much of their temporal and spatial variability from atmospheric predictors. Weather generators, by contrast, may include predictors but often only to represent long-term variability; for change-factor weather generators, no predictors are required at all (see Section 13.4). In other words, weather generators are complex statistical models that explicitly model essentially all desired properties. This sophistication has advantages and disadvantages: on the one hand, weather generators can in principle simulate any desired statistical climate aspect (Section 4.2), as long as sufficiently long data series and a suitable statistical model are available. But on the other hand, unconditional weather generators do not incorporate any physical knowledge. As a consequence, the fact that an unconditional weather generator works well in present climate is trivial and sheds no light on its performance in a future or simply different climate. In fact, their method specific assumptions should be seriously tested prior to any application in a different climate (Onof et al. 2000). The use of weather generators in future climate will be discussed in Section 13.4; here we discuss the skill of (mostly only) unconditional weather generators to represent present climate.

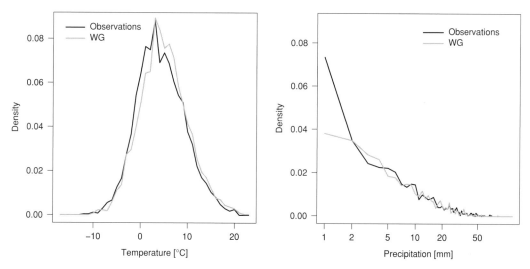

Figure 13.5 As Figure 11.2, but for a Richardson-type weather generator. For details on the method, see Appendix A, method SS-WG.

13.3.1   Marginal Aspects

As marginal distributions are a calibrated aspect , weather generators are – trivially – able to accurately represent them, as long as suitable models are available (see Figure 13.5 for an example of a Richardson-type weather generator). The choice of distribution depends, of course, strongly on the climate of interest. For instance, Wilks and Wilby (1999) demonstrate for a Richardson weather generator that a mixed exponential distribution better captures precipitation intensities in the state of New York, USA, compared to the exponential and gamma distributions. But Kilsby et al. (2007) find a convincing performance of a Neyman-Scott rectangular pulse model with exponentially distributed rain cell intensities to capture daily precipitation intensities in the UK. They argue, however, that heavy-tailed distributions may be required in different climates.

The representation of extreme events may be improved by specific extreme value models for the distribution tails (e.g. Vrac et al. 2007a). The inclusion of predictors may in principle improve the representation of marginal aspects. For instance, Wilks (1998) suggests the use of a mixed exponential to account for sampling from different exponential distributions at different days. Such a behaviour might be physically represented by conditioning the parameters of an exponential distribution on meteorological predictors.

13.3.2   Temporal Aspects

Richardson weather generators explicitly model the 1-day dependence of the occurrence process – it should therefore be perfectly represented. The amounts process, however, is typically modelled independently in time. Potential clustering of precipitation intensities will therefore not be captured. Poisson process models simulate precipitation intensities as overlapping intensities of finite-length rain cells. On timescales up to typical rain cell lengths, these models should therefore be able to also represent clustering of precipitation amounts. In fact, Kilsby et al. (2007) find a superior representation of autocorrelations of daily precipitation by a Neyman-Scott model compared to a Richardson-type model. Figure 13.6 illustrates the effect. The chosen Richardson-type weather generator is not conditioned on predictors, so no synchrony with observations is to be expected. But still, the behaviour of observed and modelled precipitation is evidently different. Whereas high-intensity events tend to cluster in reality, this behaviour is not represented by the model. Hidden Markov models that simulate occurrences and amounts independently (conditional on the hidden state) will imprint some temporal dependence. Clustering within a hidden state, however, is by construction not captured.

Residual temporal dependence on longer timescales is typically not explicitly modelled by any weather generator, such that interannual variability is in general underrepresented (Katz and Parlange 1998, Wilks and Wilby 1999). A pragmatic solution to overcome this problem was proposed by Racsko et al. (1991), who explicitly model the lengths of spells, or similarly by Bárdossy and Plate (1991), who simulate the length of hidden weather states. Some authors suggested implementing higher-order dependence (e.g. Katz and Parlange 1998, Wilks and Wilby 1999), although with limited success. A key question in this context is the origin of longer-term variability: in case of substantial moisture recycling, one would indeed expect a dependence between precipitation events, for instance in continental summer climates. Yet in situations of prevailing winds and moisture advection, the temporal structure of precipitation is not caused by memory in the precipitation process but rather by large-scale processes such as planetary waves and blocking. In this latter case, higher-order Markov chains are not suitable models. Long-term variability is somewhat better captured by models conditioning the weather generator parameters on atmospheric predictors (e.g. Katz and Parlange 1993) or on total monthly precipitation (e.g. Wilks and Wilby 1999, Dubrovský et al. 2004).

13.3.3   Spatial Aspects

Systematic variations in climatological properties (e.g. lower precipitation in the wind shadow) are of course automatically calibrated. But the representation of residual spatial dependence strongly depends on the choice of the multisite or full-field weather generator (for an illustration of a multisite weather generator, see Figure 11.4). Single-site weather generators trivially do not capture residual spatial dependence: weather sequences at different locations will be completely independent. If – as in the presented multisite implementation of the Richardson weather generator – only dependence in the occurrence process is explicitly simulated, the spatial clustering of amounts will be underestimated. The behaviour of hidden Markov models in space is similar to their temporal behaviour: the regional-scale spatial clustering on a given day is likely underrepresented. Models based on Gaussian random fields, random cascades and Poisson

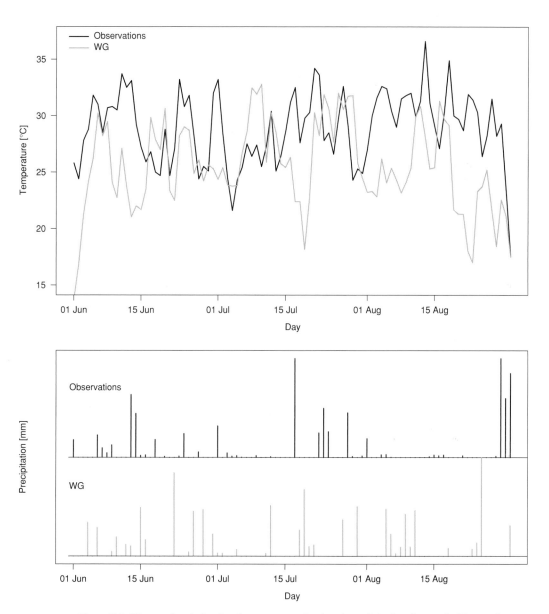

Figure 13.6 Observed and simulated temperature (top) and precipitation (bottom). Observations are from the heat wave of summer 2003 at Graz, Austria. The simulated data is a realisation of an unconditional Richardson-type weather generator, that is, no synchrony is to be expected. Observations from ECA-D data (Klein Tank et al. 2002). For details on the method, see Appendix A, method SS-WG.

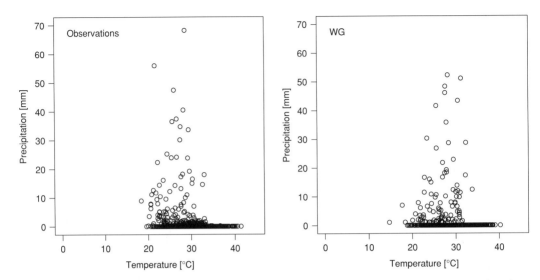

Figure 13.7 Observed and simulated precipitation during summer at Madrid-Barajas, Spain. The simulation is based on a Richardson-type weather generator. For details on the method, see Appendix A, method SS-WG.

cluster processes all satisfactorily represent the spatial autocorrelation of precipitation (Ferraris et al. 2003). Nevertheless, their performance differs depending on their sophistication, for example, on whether they account for possible anisotropies or topographic effects. For the representation of extreme events, it is important whether the underlying stochastic model allows for the representation of tail dependence, that is, the joint occurrence of extreme events. For example, models based on the multivariate normal distribution by construction cannot represent tail dependence and may therefore fail to represent the spatial clustering of extreme events.

13.3.4   Multivariable Aspects

A key rationale for the use of weather generators is the representation of multivariable dependence. It is typically explicitly modelled and thus well represented for mean conditions. Again, a potential dependence of extreme events will not be captured by models based on the multivariate normal distribution. For an example of the joint occurrence of precipitation and temperature in a Richardson-type weather generator, see Figure 13.7.

13.4   Using Weather Generators for Climate Change Projections

As discussed in the introduction to this chapter, weather generators can be used for creating local-scale climate change projections in two different ways. The more popular variant is arguably based on so-called change factors, the other is to condition the weather generator on atmospheric predictors that implicitly impose future changes.

13.4.1   Change-Factor Weather Generators

Change-"factor" weather generators are similar to the delta-change approach discussed in Section 12.4: no climate model information at the temporal resolution of the weather generator is employed. Only simulated changes between present and future long-term climatological properties are used to modify the weather generator parameters. That is, change-factor weather generators are unconditional weather generators; they are not synchronised with a climate model.

The approach is as follows (e.g. Wilks 1988, Bartholy et al. 1995): in addition to calibrating the weather generator to observations, it is also calibrated to climate model simulations of present and future climate. The "observed" weather generator parameters are then modified by the change in the simulated parameters. Parameters $\theta^p_{a,obs}$, that may respond additively to forced changes, such as mean temperature, are modified by an additive term:

$$\theta^f_a = \theta^p_{a,obs} + (\theta^f_{a,mod} - \theta^p_{a,mod}),$$

(13.13)

where $\theta^p_{a,mod}$ is the parameter obtained from the simulation of present climate, $\theta^f_{a,mod}$ that from the simulation of future climate. Equivalently, parameters $\theta^p_{a,obs}$, that may respond with relative changes to forcings, such as precipitation intensities, are modified by a multiplicative term:

$$\theta^f_a = \theta^p_{a,obs} \frac{\theta^f_{a,mod}}{\theta^p_{a,mod}}.$$

(13.14)

Alternatively, instead of applying the change factors to the weather generator parameters, they can be applied to the calibrating statistics (Kilsby et al. 2007).

As discussed, the underlying assumptions of the change-factor approach are similar to the MOS assumptions:

- the change factors have to be credibly simulated by the chosen climate model;
- the simulated meteorological variables, from which the change factors are derived, have to be representative of the corresponding observed variables;
- the structure of the weather generator has to be suitable to capture the relevant statistical climate aspects, and change factors for all relevant aspects have to be included.

The second assumption is a crucial limiting factor: often weather generator change factors are derived from coarse resolution GCMs and then used to modify station-calibrated weather generators. Even if the GCM would credibly simulate, say, daily precipitation on the resolved GCM scales, a big scale-gap would exist. Because of orographic influences, it would not in general be justified to assume that the changes in area-average precipitation are identical to the local changes. Similarly, local feedbacks might modify large-scale temperature changes. For an in-depth discussion of this issue, see Section 12.7.3. Finally, one, of course, assumes that all relevant possible changes in the weather generator parameters are captured.

Change factors can also be informed rather than determined by climate models (Wilby et al. 2014). For instance, one can distil a number of representative future evolutions from a multimodel ensemble projection or may even define change factors beyond the range of simulated changes (see also the discussion in Section 17.2).

13.4.2   Conditional Weather Generators

Conditional weather generators can in principle be used for climate change studies: a relationship between large-scale predictors and weather generator parameters is derived from observed data and then applied to climate model simulations. As mentioned, such models can be considered as stochastic perfect prognosis (PP) downscaling methods (Chapter 11). Thus, all considerations that apply for PP essentially also apply here, although with some crucial differences. The main underlying assumptions for the applicability of conditional weather generators under climate change are thus very similar:

- the predictors have to be realistically and bias free simulated in present climate, and their response to external forcing needs to be credibly represented (the PP assumption);
- informative predictors need to be selected that determine future changes in the relevant statistical climate aspects of the simulated local meteorological variables;
- the structure of the weather generator has to be suitable to capture the relevant aspects, and the predictors have to be included in a suitable way.

For the discussion of the PP assumption, please refer to Section 11.5.1. The informativeness assumption is slightly different from that of standard PP methods (see Section 11.5.2): weather generators explicitly model the short-term temporal variability. Thus, predictors are only required to represent long-term variability and climate change. Crucially, these predictors have to capture not only dynamical but also thermodynamic changes. Weather generators conditioned only on circulation patterns may be very useful to represent internal climate variability, but they will therefore likely fail to correctly represent forced future changes.

Finally, again, it is assumed that all relevant statistical climate aspects that may change in response to climate change are conditioned on predictors. For instance, a spatial dependence model that assumes a constant decorrelation length will not capture changes in the clustering of organised convection. For all aspects relevant for the predictor choice, refer to Section 11.5. For the selection of predictors for climate change projections, refer to Section 11.6.1.

In principle, also MOS variants (Chapter 12) of conditional weather generators are conceivable (Eden et al. 2014, Wong et al. 2014): a relationship between climate model simulated predictors and "observed" weather generator parameters is established, and used to drive the weather generator with simulations of that climate model. Such models, however, are difficult to calibrate, as climate models and observations are typically not in synchrony (for an exception see Eden et al. 2014).

13.5   Further Reading

Comprehensive recent reviews of different weather generator approaches can be found in Onof et al. (2000), Wheater et al. (2000), Ferraris et al. (2003), Wilks (2010) and Wilks (2012).

14   Other Approaches

In addition to dynamical downscaling, and the PP and MOS statistical approaches presented in Chapters 11 and 12 there are a number of alternative downscaling methods that combine different approaches. These will now be outlined, and some examples will be given.

14.1   Statistical-Dynamical Downscaling and Emulators

There are combinations of dynamical and statistical downscaling that are different from the MOS post-processing. They aim at using the physically based small-scale information from RCMs but without performing computationally expensive RCM runs for every timestep and ensemble member of the GCM simulations to be downscaled. In contrast to standard PP downscaling, which is also an alternative to dynamical downscaling, these approaches are not restricted by the availability of observations for model fitting and provide values for the same region and grid as the RCM. The output is thus spatially complete and can be produced for regions without any observations.

14.1.1   Statistical-Dynamical Downscaling

The first type of this approach is known as statistical-dynamical downscaling. It exploits the fact that many of the large-scale atmospheric states that usually would be used to drive an RCM are similar. The idea is that running the RCM for each situation in a set of similar states creates redundant information and can thus be avoided. Instead, the RCM is only run for typical weather situations, and then the final output is generated by a weighted average of the RCM outputs for the typical situations, with weights determined by the frequency of these weather types in the target period. The approach goes back to wind field studies by Wippermann and Gross (1981) and Heimann (1986) and has been introduced for downscaling of climate models by Frey-Buness et al. (1995). Fuentes and Heimann (2000) have improved the definition of the weather types and demonstrated skill comparable to full RCM simulations for winter precipitation in the Alps for the period 1981–1992. The approach has been used for instance for downscaling windstorm impacts over Germany from climate change simulations (Pinto et al. 2010) and near-surface wind fields from decadal hindcasts and climate change simulations, which in turn were used for wind energy estimates (Reyers et al. 2015).

Statistical-dynamical downscaling is conceptually similar to PP–weather-type down-scaling, but with the observations for a given weather type replaced by an RCM simulation, which however may be biased. The successful application in a climate change context is therefore subject to the same assumptions, namely that the future change in the predictand is mainly determined by changes in the frequency of weather types rather than by changes in the character of weather types. This in turn means that the method can be expected to work better for predictands for which the climate change signal is dominated by dynamical rather than by thermodynamic changes. The wind-related studies mentioned earlier fall in the former category; studies related for instance to the climate change signal in precipitation would not. As the predictands are simulated values, the performance of dynamical-statistical downscaling with respect to the climate change signal can be determined straightforwardly in the model world, that is by comparing the change obtained from the dynamical-statistical downscaling with the climate change in a full dynamical downscaling. Good performance in this pseudo-reality setup is a necessary but not a sufficient condition for a credible outcome, as the climate change signal in the full RCM might be biased (cf. Chapters 9 and 15).

14.1.2   RCM Emulators

The second type is RCM emulators, which are statistical models that aim at producing output similar to RCMs. They are fitted by using RCM–simulated data as predictands and variables from the global model that drives the RCM as predictors. The global model can be a reanalysis or any other GCM simulation; usually only condensed GCM information is used. As a consequence, also the RCM output is typically condensed. In contrast to the RCM, emulators usually also provide only one meteorological variable as output. If the RCM is driven by a reanalysis the RCM output can be viewed as replacing observations, again with potential differences due to RCM biases as for statistical-dynamical downscaling. In this case the approach is similar to PP or MOS downscaling, depending on whether biased or unbiased predictors from the reanalysis are used.

One example for an emulator is Haas and Pinto (2012), who use wind data over Europe based on simulations with a 7km resolution RCM driven by the ERA-Interim reanalysis for 100 storms as predictands and wind at the 16 closest gridcells from the ERA-Interim reanalysis as predictors for a downscaling based on linear regression. Prior to fitting the downscaling model, the RCM-simulated winds are transformed into approximately unbiased estimates for wind gusts using friction velocity as a predictor for turbulence. A comparison with actual observations showed good skill.

Instead of emulating an RCM for every timestep, Walton et al. (2015) presented an emulator for only the climate change signal, which often is the only quantity of interest. For fitting the emulator the authors dynamically downscaled 5 GCMs for the mid-21st century with a 2km resolution RCM for the Los Angeles (US) area. The warming patterns were calculated for each month of the year, leading to 60 (5×12) warming patterns. The warming was then separated into the mean warming over the area and the main spatial anomaly pattern determined through PCA. Finally the amplitude of the mean warming and the spatial anomaly pattern were linked through a linear statistical model to GCM–simulated warming. The locations from which the GCM input was taken were determined from the correlation maps of grid cell GCM temperature changes and the RCM regional mean change or the RCM spatial change pattern expansion coefficient. The study has shown that this emulator outperforms purely statistical downscaling methods, in particular for the ensemble mean of the downscaled climate change signals. In order to save computational effort, a reanalysis-driven RCM simulation was used to define the reference climate. To be consistent, the GCM climate change signals for each GCM were then added to the reanalyis to provide the boundary conditions for the RCM, using different patterns for each month of the year. The emulator was applied to downscale temperature in the target area for the mid-21st century from 32 GCMs (Walton et al. 2015) and the for the mid- and end 21st century from a comprehensive set of CMIP5 simulations (Sun et al. 2015).

Emulators have been pioneered outside the downscaling context, for instance to represent intermediate complexity atmosphere–ocean models and estimate climate change from a large range of forcings for which it would have been not feasible to run the dynamical model (Holden and Edwards 2010) or to approximate the output from forced GCM simulations based on computationally cheaper simulations with an energy and moisture balance model (Tran et al. 2016). Emulators can be more general than the regression models applied for downscaling so far (e.g. Tokmakian et al. 2012).

14.2   Separating Bias Correction and Downscaling

It has been discussed in Section 12.7 that applying bias correction methods is problematic when the predictand is on a smaller spatial scale than the predictor. The reason is twofold: first, the temporal variability on the small scale is usually not fully determined by the variability on the larger scale, in particular for variables with high small-scale variability such as precipitation. Here the bias correction can lead to variance inflation, unrealistic sub-grid variability and in turn unrealistic spatial extremes and long-term trends (Maraun 2013a). Second, local processes and feedbacks modifying the climate change signal may not be correctly represented. To address the first problem, it has been proposed to apply bias correction on the same spatial scale and perform the downscaling step separately, and three specific approaches are discussed in what follows. Note that potential local modifications of the grid-scale climate change signal will not be captured by any of these methods (see the in-depth discussion in Section 12.7).

14.2.1   Bias Correction Spatial Disaggregation

Wood et al. (2002) and Wood et al. (2004) presented a two-step approach (see also Maurer and Hidalgo 2008) in which monthly mean GCM or RCM output was first bias corrected and then further downscaled in space and time. The bias correction was calibrated against observations on the same grid as the climate model output, based on empirical quantile mapping and a parametric model for the tails. For future temperature simulations, the mean climate change signal was removed prior to the bias correction and subsequently added again. The monthly means were then disaggregated in time and space by resampling and amplitude-adjusting high-resolution gridded observations. To this end, daily observations from a randomly selected month (for, say, a simulated January from all observed Januaries) were adjusted (additive for temperature, rescaled for precipitation) such that the monthly total anomaly (compared to a long-term reference), averaged across the climate model grid-box area, matched the simulated anomaly. These adjusted daily observations served as a spatial-temporal downscaling of the coarse-resolution bias corrected model output. To obtain spatially coherent fields, the random selection of observational data was carried out jointly for all considered coarse-scale grid boxes. This method, coined bias correction spatial disaggregation (BCSD), produces high-resolution daily fields. The random selection of observational fields, however, is not linked to the simulated large-scale fields. It might well be that, say, a dry month is selected and rescaled to represent a wet month or vice versa. The effects of this structural limitation on, in particular, the temporal structure and daily intensities of precipitation should be analysed in detail. A variant of BCSD that is fully based on daily values has been published by Thrasher et al. (2012).

14.2.2 Bias Correction Constructed Analog

As an alternative to BCSD, Maurer et al. (2010) proposed the bias correction constructed analog (BCCA) method. This method combines the bias correction of the BCSD method with the constructed analog method (Hidalgo et al. 2008) presented in Chapter 11. A more recent, related method is the localised constructed analogues (LOCA) method (Pierce et al. 2014). A key difference compared to BCCA is that the local downscaled field is not a weighted average of analogs but a single analog. This difference avoids smoothing of extreme events. Additionally, future analogs are calculated not as anomalies with respect to present climate but with respect to a future 30-year climatology. This feature allows one to better simulate, for example, future heat waves.

14.2.3 Quantile Mapping Plus Stochastic Regression

Another approach that addresses the potential problems of bias-correcting to smaller spatial scales has been suggested by Voloschuk et al. (2017). It produces downscaled values that have a link with the RCM–simulated predictors for daily values and has been applied to precipitation over Europe. It also consists of two steps and separates a pure bias correction from the downscaling. The first step is performed by using a parametric quantile mapping to correct precipitation simulated by an RCM, with the target given by gridded observations on the same 0.44 deg grid as the observations. As no unexplained small-scale variability is influencing the bias correction, no variance inflation occurs. The second step is a pure PP downscaling step, implemented by using a stochastic linear model to downscale from the gridded observation onto precipitation station records. The stochastic component accounts for the unexplained local variability. The first step only requires simulated and observed distributions and can thus be performed independent of the setup of the RCM simulation, while the pairwise setup required to fit the linear downscaling model is naturally given by using observed grid cell predictors and local predictand records.

14.3   Further Reading

Recently, Chandler (2013) presented an interesting approach to stochastically simulate predictors for perfect prognosis statistical downscaling. By combining information from different GCMs with observations, this approach may arguably be capable of correcting location biases.

Pattern scaling is a technique with potential for downscaling applications (e.g. Mitchell et al. 1999, Huntingford and Cox 2000, Osborn et al. 2016). Such methods are based on the assumption that different rates of global warming only change the amplitude of a change pattern but not the shape of the pattern. A simulation of one global warming scenario can thus be rescaled to yield any desired scenario. More recently pattern scaling has been also applied to RCM–simulated patterns (e.g. Cabré et al. 2010).

Part III

Downscaling in Practice and Outlook

15 Evaluation

As discussed in Chapter 5, two key information requirements for users of climate information are credibility and salience. A proper evaluation is key to establish credibility of regional (in fact: any) climate projections: by analysing the realism of the chosen, potentially statistically post-processe, climate model simulations and by assessing the credibility of regional future projections. A proper evaluation has to be designed in a way to provide salient information: statistical aspects need to be evaluated that are relevant to users.

Barsugli et al. (2013) and Hewitson et al. (2014) highlight the practitioner's dilemma: users of downscaled information are faced with a plethora of different regional climate projections, based on different GCMs, downscaling methods and approaches, realisations and forcings, with widely varying and often contradictory results. Key to a sensible evaluation is thus a common framework to be able to trace differences between the individual simulations.

For global climate models, the CMIP framework (Meehl et al. 2007a, Taylor et al. 2012) provides the basis for broad intercomparison studies. Regional climate models have been intercompared within the PIRCS (Takle et al. 1999), PRUDENCE (Christensen and Christensen 2007), ENSEMBLES (van der Linden and Mitchell 2009), NARCCAP (Mearns et al. 2009) and most recently the CORDEX initiatives (Giorgi et al. 2009). The first broad intercomparison of statistical downscaling methods was carried out within the European STARDEX project (Goodess et al. 2010). Similar projects have been carried out for Australia (Frost et al. 2011), China (Hu et al. 2013) and the US (Gutmann et al. 2014). Recently, the VALUE network carried out the most comprehensive intercomparison of statistical downscaling and bias correction methods for European climate (Maraun et al. 2015). Ongoing activities are the BCIP intercomparison of bias correction methods, and the CORDEX-ESD evaluation for South Africa and the La Plata basin in South America. Guidelines for the evaluation of regional climate projections have been published by Hewitson et al. (2014) and Maraun et al. (2015).

As discussed in Chapters 2, 8 and 9, climate is a complex and high-dimensional system, and climate models are always substantial simplifications of the real-world system; thus no climate model or downscaling method can be expected to reproduce all aspects of the system perfectly, and an evaluation of all aspects would be practically impossible. But in a given context only a small part of the system will be relevant: specific weather and climate phenomena at specific space and timescales in a specific region. An evaluation of regional climate projections should therefore be user focused. Such perspective not only reduces the amount of work but also helps provide salient information that is relevant for the user.

A user-focused approach to evaluation must first identify the phenomena and scales of interest; with respect to these, it must seek to identify the key strengths and weaknesses of a method: does a method perform well for a given application, how does it perform relative to other methods, and where is it likely to fail? The details of a user-focused evaluation will be application dependent, but some general requirements can be formulated (Maraun et al. 2015): the evaluation framework should (1) be transparent and provide relevant and defensible guidance for users; (2) assess the performance of the method under climate change as far as possible; and (3) allow, in principle, for a comparison of all different types of dynamical and statistical downscaling approaches. The latter requirement is key to provide salient information: in principle, a user is not interested in a particular type of downscaling but in the most suitable approach for a given application.

In this chapter, we discuss how the evaluation of regional climate projections could be designed; the actual performance wil be discussed in Chapter 16. An evaluation exercise consists of a proper choice of evaluation diagnostics (Section 15.2) as well as a suitable choice of experiments. Apart from routinely carried out standard experiments (Section 15.3), one should conduct experiments to assess the representation of long-term trends (Section 15.4). How to implement a suitable cross-validation for these experiments will be discussed in Section 15.5 and the question of added value will be discussed individually in Section 15.6. The evaluation of bias correction methods will be presented separately in Section 15.7.

Every evaluation is of course limited by the quality and availability of observational reference data – either as local references or to serve as predictors. Please refer to Chapter 7 for an in-depth discussion.

15.1   The Regional Climate Change Context

The evaluation of numerical weather forecasts is rather straightforward: past forecasts are compared with actual observations. By construction, the forecast time series is in synchrony with real-world weather, and a pairwise evaluation is possible. Forecast verification scores measure different aspects of forecast skill such as overall accuracy, bias or reliability to forecast individual events (Jolliffe and Stephenson 2003, Wilks 2006). Such an evaluation is in general not possible for climate simulations, because the simulated weather sequence is not synchronised with the observed sequence. But moreover, such an evaluation may not be useful in a climate change context, and it is definitely not sufficient.

As discussed in Chapter 4, the setting of climate change projections is very different from numerical weather prediction. The latter intends to predict individual states in a cloud (the climate attractor) of possible states under present conditions. The former, however, is inherently about how the cloud of possible states changes and moves away from present conditions. Thus, evaluation in a climate change context cannot fully rely on skill established by comparing simulations with observations. Instead, it has to rest upon credibility arguments and thus physical understanding (see the discussions about credibility in Chapters 11 and 12).

For a given emission scenario, the uncertainty of regional climate change projections arises mainly due to model errors and internal climate variability (see Chapter 9). The latter source of uncertainty is a property of the climate system and essentially irreducible for multidecadal projections. It should thus be thoroughly assessed and communicated. It does, however, not limit the usefulness of a climate projection (see Sections 9.2 and 18.5.3).

To evaluate model errors in a downscaling context, one has to address the following questions (Maraun et al. 2015):

1. How well does the downscaling method perform for present and future climate?
2. How well does the driving climate model simulate the input for the statistical downscaling for present and future climate?

These questions can both be separated into reality checks, that is, comparisons with observational reference data and credibility checks of future changes. Many evaluation studies only address the comparison with observations, although the credibility checks are at least as important in a climate change context.

15.2   Evaluation Diagnostics

For a quantitative model evaluation, one derives specific indices from the model output and compares these indices with indices derived from the reference data. The mismatch between simulated and reference indices is quantified by suitable performance measures. The combination of index and performance measure is often called a model diagnostic.

As discussed in the introduction to this chapter, evaluation indices need to be user specific and well selected to derive relevant information for a given context. A useful starting point for selecting indices might be the set defined by the Joint Expert Team on Climate Change Detection and Indices (ETCCDI; Klein Tank et al. 2009, Sillmann et al. 2013) or the evaluation indices defined by VALUE (Maraun et al. 2015). Nevertheless, more specific indices might be relevant: for instance, an agricultural crop may tolerate both a slow decrease in temperature and also short drops below a certain threshold. But sharp persistent temperature drops, where temperature stays below the same threshold for a longer period, may severely damage the crop. Thus, an evaluation of cold temperatures or cold spells alone may not be sufficient – indices need to be defined that capture both the strong drop and the persistent low temperatures.

In some cases, region-specific indices have to be designed. For instance, seasonality has very different characteristics in different climates, in particular for precipitation. Whereas in many areas of the mid-latitudes, precipitation has one seasonal peak only, it may have two peaks in other regions. In some regions, the onset and length of the rainy season may be relevant.

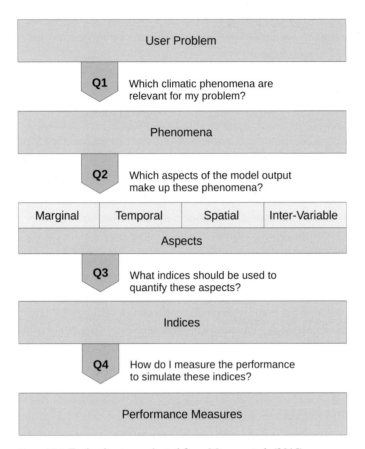

Figure 15.1 Evaluation tree, adapted from Maraun et al. (2015).

To guide the selection of indices for different user problems, a decision tree has been developed as part of the VALUE framework (Maraun et al. 2015); see Figure 15.1. A user should first identify the relevant weather and climate phenomena in a given context. Phenomena could be mean temperature, extreme rainfall events or the average length of the growing season. In principle, phenomena can be compound, that is, comprise different variables or aspects such as in the example given. In the second step, one would identify which statistical climate aspects (Section 4.2) are involved in these phenomena: marginal aspects such as intensities, temporal aspects such as the autocorrelation, a mean spell length or the seasonal cycle, spatial aspects such as the typical event size or multivariable aspects such as the interplay between temperature and precipitation. In the third step, one would select (or define) indices to quantify the different aspects of the relevant phenomena. Finally one would select appropriate performance measures, such as biases or skill scores, to compare indices derived from model data with those from observations.

A list of selected diagnostics for different statistical climate aspects is given in Table 15.1.

Table 15.1 List of example diagnostics. A more comprehensive list can be found on www.value-cost .eu/indices.

Index	Performance Measure
Marginal Aspects	
mean	bias/relative error
variance	relative error
20-season/-year return level	bias/relative error
number of threshold exceedances	bias
Temporal Aspects	
time series	mean squared error/correlation
ACF lag 1, 2, 3	bias
mean of spell length distribution	bias
90th percentile of spell length distrib.	bias
minimum/maximum of annual cycle	bias/relative error
Spatial Aspects	
decorrelation length	bias
variogram range	bias
decay length of tail dependence	bias
Multivariable Aspects	
Pearson/rank correlation	bias
probability of joint exceedances	bias
indices conditional on (no) exceedance	as above

Additionally, model evaluation can address the representation of specific relevant processes. In statistical downscaling, such an evaluation is of course limited, because physical processes are not explicitly implemented. One idea is to assess whether the response of the predictand to relevant meso- to synoptic-scale meteorological phenomena is realistically simulated. For instance, one can calculate indices and performance measures conditional on large-scale phenomena such as blocking or the passage of a Genoa low or mesoscale phenomena such as the Mistral or Bora winds (Soares et al. 2017). In this context, one has to ensure that the local predictand and the considered phenomenon do not experience different realisations of internal variability. For instance, the Mistral wind may be calculated from a reanalysis, whereas the local predictand may stem from a bias corrected RCM, driven by the reanalysis. Well inside the RCM domain, the realisations of internal variability may differ substantially. Thus, in this example, the Mistral wind should be calculated based on the RCM.

15.3 Standard Evaluation Experiments

The questions raised in Section 15.1 can be addressed for present climate by the following specific experiment. These experiments may be further tailored to investigate even

> **Box 15.1** Are Significance Tests Required for Evaluation?
>
> Sometimes, significance tests are applied to assess whether a model simulation is different from observations. Whereas such an assessment might be useful in specific contexts, we do not generally recommend the use of significance tests in model evaluation.
>
> The main reason is that significance is not necessarily the same as relevance. The power of a significance test, that is, its ability to detect a violation of a null hypothesis, depends crucially on the number of data points and the chosen test. For a short observational record, a high deviation is required to be significant; for a long record, a tiny deviation might be sufficient. Thus, one would have to choose a suitable number of data points to get significant results only in case of relevant mismatches (in medical science studies this is regularly done). But of course one would be interested in using as many data points as possible to get as robust an estimate of the performance as possible. Therefore, we recommend to employ a sensible performance measure in the first place. Such a measure would typically also be better interpretable as an abstract p-value. Of course it is useful to calculate confidence intervals for the estimated indices and performance measures. These would give a measure of the sampling uncertainty and thus the robustness of the evaluation result.
>
> In addition, one should always visually inspect the data instead of just considering aggregated statistics (see Section 6.3.5): one should, for example, investigate whether the residuals have a temporal structure, whether a scatter plot indicates a relationship or whether a quantile–quantile plot supports the chosen model.

more specific questions. For instance, to disentangle the effects of biases and scale gaps, one would have to compare results for gridded and station reference data. To separate the influences of model structure and predictors on downscaling skill, one would have to systematically define a matrix of possible model structures and predictors.

15.3.1   Perfect Predictor Experiment

In a perfect boundary setting, that is with (quasi) observations as input, one can assess how well a simulation of a downscaling model compares with local observations. In principle, such an evaluation can also address forecast skill, that is the pairwise correspondence between downscaled simulation and observations. Even if climate models are not used for forecasts of individual events, such an evaluation might in principle be very useful: if the forecast skill is high, the downscaling method simulates a realistic local response to a given large-scale weather situation. Here, internal climate variability would be a limiting factor: large-scale weather does in general not fully determine local weather: real climate and RCMs, but also stochastic statistical downscaling methods, typically develop different realisations of local weather. Thus, observed and simulated local weather randomly differ at the daily scale: a particular precipitation event may occur a day later or earlier or at slightly different locations. In weather forecasting, this effect would be correctly interpreted as a reduction of predictive skill. But for a climate projection this effect is irrelevant and has to be eliminated when assessing skill.

Artificial skill should be eliminated by a suitable cross-validation (see Section 6.3.6). Note that it is difficult to establish skill of a bias correction by cross-validation. A detailed discussion is given in Sections 15.5 and 15.7. Of course, perfect predictor experiments can be used to assess the representation of trends. This issue will be discussed in Section 15.4.

15.3.2   GCM Predictor Experiment

In a GCM predictor experiment, the downscaling model is driven with GCM simulated predictors to assess the overall skill of the chosen model – GCM plus downscaling/bias correction – to represent present-day regional climate. PP methods are first calibrated with real-world predictors and predictands and the predictors are then replaced by GCM predictors. MOS methods are directly calibrated between modelled predictors and observed predictands. For PP methods, this experiment provides an overall assessment of the PP and informativeness assumptions for present climate, that is, whether the predictors are realistically simulated and carry information to represent regional climate. To isolate to what extent the PP assumption is fulfilled, it is useful to separately assess skill of the driving GCMs (Brands et al. 2012, Jury et al. 2015). For bias correction, one has to be aware that one cannot gain any insight about calibrated aspects but only about non-calibrated aspects (see discussion in Section 15.7).

The value of any GCM predictor experiment is fundamentally limited by internal climate variability: GCMs, driven with observed radiative forcings, will simulate their own realisations of internal climate variability. Observed climate, however, is itself a combination of forced changes and a realisation of internal climate variability. Modelled and observed realisations are not synchronised, and simulated climate variations will thus not follow observed climate on decadal scales. Even on multidecadal scales, regional climate might be substantially modulated by random fluctuations (Latif and Park 2012; see also Chapter 9). Thus, a considerable fraction of the mismatch between the downscaling and the observation will be purely random and not a bias. Averaging multiple initial condition simulations of the same GCM will reduce the influence of internal climate variability on the bias estimation, but the randomness in the observed statistics cannot be removed.

15.4   Experiments to Evaluate Trends

The experiments discussed in the previous section are routinely carried out to evaluate regional climate projections. But in a climate change context, it is of course of equal importance to assess whether regional climate projections are able to represent trends in both present and future climate. Such an assessment can partly be formalised – how well does the downscaling method reproduce local trends? How well does the driving model reproduce simulated trends? – but additionally has to rest upon credibility assessments of future GCM trends.

15.4.1　Perfect Predictor Experiments for Trend Analysis

Perfect predictor experiments can of course also be carried out to assess the representation of long-term historical trends. Such an evaluation assesses whether the structure and predictors of the downscaling model are suitable to translate changes in the large-scale predictors or boundary conditions into local changes. As laid out in Chapter 7, the evaluation is limited by the quality of both the predictor and the predictand data: if the perfect predictor data does not correctly capture long-term trends in the large-scale climate, a downscaling model cannot be expected to correctly simulate local trends. Similarly, if the local observations contain inhomogeneities, they cannot serve as a reference for long-term trends (see Chapter 7).

15.4.2　Pseudo-Reality Experiments

Pseudo-reality experiments (Charles et al. 1999, Frías et al. 2006, Vrac et al. 2007c, Gutiérrez et al. 2013, Maraun et al. 2015), also called perfect model experiments, are essential to assess the performance of PP statistical downscaling models under future climate conditions: the downscaling model is calibrated to large-scale predictors and local-scale predictands in present climate simulations of a given high-resolution climate model (typically a dynamically downscaled GCM). The downscaling model is then applied to future simulations of the predictor with the same climate model and evaluated against the local-scale future simulation of the climate model. This setting allows one to address the validity of key PP assumptions: are all relevant predictors informative of climate change included? Is the model structure suitable to extrapolate?

For MOS, one of course cannot use the same climate model for predictors and predictands: a climate model is bias free compared to itself, and all MOS assumptions would be trivially fulfilled. One could, however, restrict the perfect model condition to the driving GCM and take different RCMs for predictors and predictands. Such a setting would represent a situation in which a climate model perfectly simulates large-scale climate, including the large-scale response to external forcing, but regional biases exist and may vary in a future climate. In such a setting one could assess how regional biases may change in a future climate. Moreover, one could address the issues discussed in Section 12.9: is, for example, quantile mapping suitable to capture time-varying regional biases?

For a pseudo-reality experiment to make sense, the reference model should simulate a credible (in particular regional) response to climate change. This requirement is crucial for assessing, for example, extreme events or the effect of local feedbacks. For instance, there is evidence that climate models with parameterised convection do not capture changes in sub-daily extreme precipitation (see Chapter 8.5.2). That is, using such models as a reference one would merely test whether a certain downscaling method captures the behaviour of a questionable semi-empirical model (the parameterisation).

Similarly, if one aims to assess the influence of soil moisture feedbacks on local temperature changes, reference models with sufficiently complex land-surface models are required that credibly simulate these feedbacks.

Also in pseudo-reality experiments, it is crucial to account for the effects of internal variability. Recall that a pseudo-reality would usually be constructed by a GCM, dynamically downscaled with an RCM. Typically, the predictors for a PP method would be defined at synoptic scales (Section 11.5), and it would be quite natural to take them from the GCM. But at the daily scale, internal variability could create a substantial discrepancy between GCM and RCM fields even at the synoptic scale. Local climate simulated by the RCM, however, is driven by the RCM synoptic fields, not the GCM fields. Part of the correlation between large and local scales would thus be destroyed by choosing GCM predictors. Therefore, predictors for a statistical downscaling model should not be taken from the GCM but rather from the RCM – even for assessing PP methods.

15.4.3　GCM Predictor Experiments for Trend Evaluation

GCM predictor experiments can in principle be used to assess the representation of historical long-term trends at the regional scale by the downscaled GCM (of course, one could carry out a similar experiment at larger scales with the raw GCM). In this context, internal variability is a key limiting factor. As discussed in Chapter 9, both simulated and observed trends will be substantially affected by random fluctuations, even over several decades. At multidecadal timescales, internal climate variability is even fundamentally unpredictable, such that also improved initial conditions will not synchronise internal variability on these scales. As a consequence, a GCM, even if forced with observed radiative forcing, cannot be expected to closely follow observed trends, in particular for noisy variables such as precipitation at the regional scale.

Thus an evaluation of simulated historical trends, even over several decades, is far from trivial.[1] The only reasonable suggestion to evaluate the representation of historical trends is therefore to evaluate the consistency of observed and modelled trends (see Section 8.5.3 for an overview of such studies). One would thus ask the question: is an observed trend distinguishable from an ensemble of simulated trends (Bhend and Whetton 2013)? In this context, the use of a climate model ensemble is essential to represent the uncertainty range caused by internal variability. If an observed trend is indeed indistinguishable from the ensemble the reason could either be that the ensemble as a whole does not realistically simulate the long-term trend or that the ensemble as a whole simulates too weak internal variability. The assessment of individual model trends, however, is essentially impossible.[2]

---

[1]　At centennial timescales, trends should dominate internal climate variability, but dense observational networks with homogeneous data are typically only available since the 1960s.

[2]　Of course one could try to identify outliers, which simulate trends that are compatible with neither forcings nor internal variability.

15.4.4   Credibility of Climate Model Projections

Even if we could fully assess whether the driving climate model reproduces forced historical trends, we could not conclude that this model would simulate credible future projections: the trends might be correct because of compensating errors, and important feedbacks may kick in only at a higher level of global warming. Thus, additional credibility checks are required that refer to process understanding.

For PP methods, which rely only on large-scale predictors, one should address the following question related to model performance (Section 8.5):

- What are the relevant large-scale climatic phenomena, such as the ITCZ, El Niño/Southern Oscillation, mid-latitude cyclones?
- How credible are these processes simulated by the driving climate model in a future climate?

For bias correction methods, one additionally has to address the following questions:

- What are the relevant regional-scale phenomena, such as regional circulation patterns, feedbacks and forcings?
- Are sub-grid scale processes relevant for the variable of interest?
- How credible are these processes simulated by the driving climate model in a future climate?

A careful climate model choice is thus essential for a successful downscaling. This choice requires a broad interdisciplinary collaboration with experts in climate modelling and in the relevant large-scale climate phenomena and climatologists familiar with the local climate in the target region. A useful basis can be obtained already by a literature review on GCM and, if necessary, RCM performance. The idea of these credibility checks is not so much to choose a small subset of credible models. Rather, it is to identify and discard models that are obviously not credible, because they misrepresent – or do not at all represent – crucial processes. Such analyses may also help researchers better understand – and potentially resolve – contradictions between different regional climate projections.

15.5   Cross-Validation

Statistical model predictions have to be evaluated on independent data that has not been used for calibration; otherwise overfitting by overly complex models cannot be identified. A standard approach to eliminate artificial skill is k-fold cross-validation (see Section 6.3.6).

For a k-fold cross-validation, the statistical model needs to be calibrated k times. In principle, also the model selection (which predictors should be included, how should they act upon the predictand) should be carried out individually for each fold of the cross-validation. To reduce the computational burden, in particular for complex statistical models, and to be able to interpret the sensitivity to specific predictors, we suggest deriving the model structure once from the whole data set and then applying the cross-validation keeping the model structure identical for each block. But note that such a simplified approach may lead to apparent cross-validated predictive skill (Wilks 2006).

One should always consider whether cross-validation is sensible in a given context. Cross-validation is of course necessary for perfect predictor experiments. Pseudo-realities are experiments in which the data is automatically separated into a calibration (present climate) and validation period (future climate). The setting for GCM predictor experiments is quite peculiar: in case of PP methods, the simulated predictors are independent from the observed predictors used for calibration, so artificial skill is thus eliminated. Some authors suggest using cross-validation when assessing the performance of bias correction in a GCM predictor experiment. We argue that this is a potentially misleading use of cross-validation, for an in-depth discussion, see Section 15.7.

15.6   Assessing Added Value

As discussed in Chapter 4, added value may arise from reducing model biases at the resolved scale and from better representing small-scale variability. Obviously, the former contribution has to be quantified at the resolution of the driving model, the latter contribution at the target scale.

Whether a downscaling model adds value depends on the considered aspects (see Section 4.2). Ultimately, the answer to this question is thus user specific. Assume one

---

Box 15.2 Are Split-Sample Tests Useful to Evaluate the Representation of Climate Change?

It has been suggested to split the observational record into two climatically distinct subsets, for instance into the 50% warmest and the 50% coldest years. The aim of this test is to evaluate whether the statistical model captures non-stationarities in climate and thus long-term climate change. This approach, however, assumes that the predictors capturing and the processes governing interannual variability are the same as those capturing and governing long-term climate change. This is not in general true: interannual variability is mainly caused by dynamical processes (the occurrence of certain weather types in a year), whereas long-term trends are also caused by thermodynamical changes. (This discussion is equivalent to the question whether downscaling and bias correction methods calibrated to short-term variability can capture long-term changes; see Chapters 11 and 12.) Such a test therefore mainly assesses whether downscaling methods capture interannual variability. A sensible split-sample cross-validation could, however, split a long observational record into two consecutive blocks. The first would represent a period with weaker, the second one with stronger radiative forcing. Note, however, that such a test is not sensible if substantial long-term internal variability is present and not synchronised between model and observations (Section 15.7.2).

quantifies the performance of a model to simulate a specific index A by a performance measure $d(.,.)$, then added value can be defined as (Di Luca et al. 2015)

$$AV = d(A_{cm}, A_{obs}) - d(A_{ds}, A_{obs}), \tag{15.1}$$

where cm denotes the coarse(r) resolution driving model, ds the downscaling model and obs the observational reference.

In statistical downscaling, added value arises trivially by calibrating the model to observational data (in case of bias correction additionally by removing model biases). Thus, the assessment of added value should always consider non-calibrated aspects.

Even if added value (also trivially added value) in present climate might be essential for users, it does not imply any added value for climate change simulations (Castro et al. 2005). In fact, the latter can only be assessed by evaluating whether the climate change signal simulated by the downscaling model is more credible than that simulated by the coarse resolution model. An obvious example is the representation of elevation-dependent warming, as discussed in Section 12.7.3.

15.7   Evaluating Bias Correction Methods

Here we illustrate two key points that limit the evaluation of bias correction methods: first, cross-validation is not a feasible approach for a GCM predictor experiment. And second, the assessment of calibrated aspects cannot establish bias correction skill.

15.7.1   Illustrating the Evaluation Problem

To illustrate the problem, we begin with an artificial and exaggerated example (Figure 15.2). Consider an attempt to bias correct simulated temperature against observed precipitation for boreal winter (DJF). Of course, both variables are locally correlated: winter precipitation, for example, would tend to come along with higher temperatures, while dry winter days would typically be colder. Still, one would not seriously attempt to represent local precipitation simply by transformed temperature values. To take the example to the extreme, we consider a target area over central Europe and a predictor area just opposite the globe over the Southern Ocean. This choice ensures that any correlation between temperature and precipitation vanishes on daily to multidecadal scales. The setting is thus very similar to a GCM predictor experiment, in which simulation and observations are not synchronised. The bias correction – a parametric quantile mapping – is calibrated over the period 1961–1980 and validated over the period 1981–2000. If cross-validation would be suitable to assess the skill of a bias correction, it should identify the nonsense character of this example.

Figure 15.2 shows the results for the validation period, the top row (a–c) for winter daily mean precipitation, the middle row (d–f) for the 95th percentile of daily winter precipitation. The left panels (a, d) show the uncorrected model data, that is Southern Ocean temperature fields; the centre panels (b, e) show the observed fields. As expected, the temperature fields do not at all match the central European precipitation fields. The right panels (c, f), however, show the bias corrected temperature fields (still

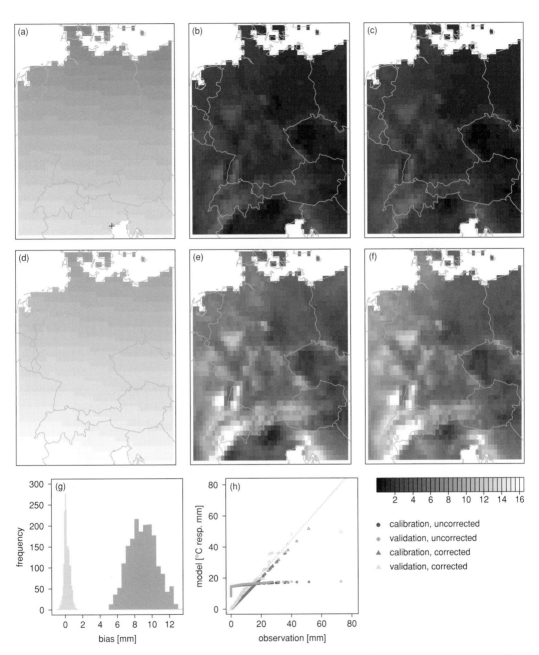

Figure 15.2 Cross-validation problem. Quantile mapping from ERA40 daily temperature over the Southern Ocean [°C, 45S-55S, 175W-163W] to E-OBS daily precipitation over Central Europe [mm/day, 45N-55N, 5E-17E], calibrated over 1961–1980. Mean (a–c) and 95th percentile (d–f) over validation period (1981–2000). Uncorrected ERA40 (a, d), observations (b, e), corrected ERA40 (c, f). Histogram of biases across all grid boxes (g). QQ plot for grid box close to Venice (h; see cross in panel a). A QQ plot plots the quantiles of two distributions against each other, that is for two time series, and the values are sorted separately and then plotted against each other. Adapted from Maraun et al. (2017*b*).

for the validation period): they look strikingly similar to the observed fields, with only minor differences which are typical of practical bias correction applications; the nonsense character of the example is not detected. The bottom panels further illustrate the problem: the spatial histogram of biases (g) is strongly narrowed and shifted by the bias correction, and the QQ plot of an example location (Venice) is much improved also for the validation period. Thus by all measures typically employed to assess the quality of a bias correction, this application looks sensible – even though it is meaningless. In other words: cross-validation is essentially useless to establish bias correction skill.

The reasons for the failure of cross-validation in this setting are threefold: the evaluation is limited to calibrated statistics (such as the mean, or the 95th percentile), quantile mapping can flexibly adjust distributions and the biases are large compared to the climate change between the validation and the calibration period. The bias correction perfectly maps the predictor distributions onto the predictand distribution during the calibration period. The trends in predictor and predictand (i.e. Southern Ocean temperatures and central European precipitation) are rather different, in particular their spatial patterns. But still, the resulting validation plots look reasonable, as the biases over the validation period remain small.

15.7.2   Sensibility of Cross-Validation

It has been argued that a positive cross-validation is a necessary but not sufficient criterion to establish bias correction skill (Stocker et al. 2015). But whereas cross-validation is sensible for perfect predictor experiments, it is misleading for GCM predictors experiments, that is in any context in which the long-term internal climate variability of the driving model is not in synchrony with observations.

The issue is particularly obvious for the hold-out method, but can be generalised for cross-validation. Assume that simulation and observation are split into two consecutive blocks; one is used for calibration, the other for validation. Over the calibration period, a calibrated statistic (say, the 95th percentile) is perfectly matched. In the validation period, modelled and observed statistics have slightly diverged, and a residual bias emerges. The only reason why a residual bias exists is thus the fact that modelled and observed changes are slightly different – this reason is not necessarily linked to the sensibility of the bias correction. In fact, four results are possible (see Maraun and Widmann 2017 for further details):

- Case 1: the bias correction is sensible, and the (bias corrected) climate model simulates a trend closely resembling the observed trends. A cross-validation would thus yield a true negative result.
- Case 2: the bias correction is sensible, but due to internal climate variability, the (bias corrected) climate model simulates a trend different from the observed trend. A cross-validation would yield a false positive result.
- Case 3: the bias correction is not sensible, but the (bias corrected) climate model for some reason simulates a trend similar to the observed trend. A cross-validation would yield a false negative result. This case corresponds to the example earlier.

- Case 4: the bias correction is not sensible, and the (bias corrected) climate model simulates a trend different from the observed trend. A cross-validation would yield a true positive result.

Thus, the outcome – positive or negative – does not depend on the sensibility of the bias correction but on the agreement of simulated and observed trends. Recall that these are, at local scales and in particular for precipitation, strongly influenced by internal variability. In a GCM predictor experiment, internal climate variability is not synchronised with the real world. Here the outcome of a cross-validation is thus essentially random and misleading.

In a perfect predictor setting, the long-term internal climate variability is synchronised. Here, a cross-validation is in principle a feasible approach. One could, for example, assess to what extent different bias correction methods capture different climatic conditions. Of course, also in such a setting, calibrated aspects will look reasonable in a validation period, such that it is essential to evaluate uncalibrated aspects.

15.7.3 Recommendations for Evaluating Bias Correction Skill

We have shown that a cross-validation fails to establish skill of a bias correction. Key to a sensible evaluation of bias correction is therefore the choice of appropriate evaluation diagnostics. If only marginal, that is calibrated, indices are considered, any evaluation will likely give a positive result (as long as climate change trends between model and observations are not too different, as discussed). Therefore, it is essential to evaluate also non-calibrated indices: does the temporal dependence match? Are spatial dependence characteristics similar? How is the performance conditional on relevant climatic phenomena? A simple but very efficient test is a visual comparison of simulated and observed time series: a major difference in the time series characteristics points to either a credibility or a representativeness issue. Similarly, a QQ plot between simulated and observed quantiles may reveal an unphysical bias correction: again, if the uncorrected simulated and the observed marginal distributions are very different, one likely faces credibility or representativeness issues. Both tests should be carried out at least for some selected locations.

15.8 Further Reading

The need for downscaling evaluation has been discussed in depth by Barsugli et al. (2013), Hewitson et al. (2014) and Maraun et al. (2015). A good summary of forecast verification approaches can be found in Wilks (2006) and Jolliffe and Stephenson (2003). A classical experiment to evaluate the performance of dynamical downscaling is the big brother experiment (Denis et al. 2002).

16 Performance of Statistical Downscaling

As laid out in the previous chapter, the quality of regional climate projections depends on the performance of the driving GCM (and potentially RCM) and the downscaling model in both present and future climate. GCM and RCM performance has been captured in Section 8.5 – this chapter focuses on the performance of different statistical downscaling methods. In Part II we presented the PP, MOS and weather generator approaches. Along with these, we also explained how the structure of these approaches determines their skill and limitations (Sections 11.3, 12.5 and 13.3). Here we will discuss their de facto skill to reproduce the statistical climate aspects (Section 4.2) of observational data. A broad synthesis, however, will be given in the end of this chapter.

The results that follow are to a large extent based on the comprehensive VALUE evaluation study (Maraun et al. 2015), that is mainly on the papers by Gutiérrez et al. (2017) on marginal, Maraun et al. (2017a) on temporal, Widmann et al. (2017) on spatial and Hertig et al. (2017) on extreme aspects. Other studies have investigated the marginal and often temporal performance of statistical downscaling (e.g. Semenov et al. 1998, Wilby et al. 1998, Schoof and Pryor 2001, Schmidli et al. 2007, Teutschbein and Seibert 2012, Gutmann et al. 2014, Huth et al. 2015, Rajczak et al. 2016). Some of these additionally considered spatial aspects (e.g. Ferraris et al. 2003, Frost et al. 2011, Hu et al. 2013), some focused on extremes (e.g. Haylock et al. 2006, Goodess et al. 2010, Bürger et al. 2012). To our knowledge only one study addressed multivariable aspects (Wilcke et al. 2013). We will refer to these studies whenever they complement our results.

The results we present are all from perfect predictor experiments, that is, they isolate the skill of the downscaling method – in present climate with reanalysis predictors, in pseudo-reality studies with perfect model predictors. Many of the investigated PP methods where driven with grid-box predictors. It is likely that the PP assumption is not fulfilled in several cases, such that these methods would fail when driven with GCM predictors. To fully assess the performance in a practical setting, that is, with GCM predictors (or from an RCM driven by a GCM), one has to assess the overall performance: of the downscaled GCM/RCM, and of the plausibility of the GCM/RCM predictors in a future climate (see Chapter 15). These results of course depend strongly on the chosen GCM and are thus not discussed here.

Note also that many of the results for bias correction in a perfect predictor setting are more or less trivial. The large-scale spatial-temporal variability (at least up to interannual variations) of a reanalysis is essentially perfect (at least over Europe); arte-facts caused by GCM errors, as discussed in Chapter 12, will therefore not occur.

Most evaluation studies are limited in that they do not systematically attempt to disentangle the effects of method choice and predictor choice. In fact, the options to evaluate – different experiments, indices, stations and seasons for different methods and predictors – are so diverse that a comprehensive study is almost impossible (Huth 1999, Wilby and Wigley 2000, Dayon et al. 2015, for some very focussed studies). Further-more, in practice the available methods span an ensemble of opportunity, determined by the participating groups, such that any systematic understanding is already difficult. In some studies, differences in performance have been attributed to the choice of methods, where in reality the methods where structurally very similar, but the choice of predic-tors was very different. Thus, all inter-comparison studies should be interpreted very carefully.

A key question is the added value of both the RCM and the statistical post-processing (see discussion in Chapter 15). In the plots of the following sections, a first guess of the value added by a downscaling method can be obtained as the difference in performance of the driving ERA-Interim data compared to the downscaled results. Results for ERA-Interim are in many cases given for both the 2° resolution, which resembles a typical GCM, and the native 0.75° resolution, which is used as input for the bias correction. Similarly, the performance relative to a good RCM (here: RACMO2, van Meijgaard et al. 2008) can be evaluated. Of course, this inspection does not distinguish between added value and added detail. Also, statistical methods are calibrated and thus trivially reduce biases in many indices. In some cases, however, non-trivial added value becomes evident. This holds for all non-calibrated aspects, where the improvement results from a careful predictor selection.

16.1   Representing Marginal Aspects

Marginal aspects can in principle be calibrated – to what extent depends of course on the chosen model. VALUE conducted a broad analysis of marginal aspects (Gutiérrez et al. 2017). The main results will be discussed here for a selection of methods, and complemented by other studies.

Long-term temperature means (Figure 16.1, top row, for T_{max}) are in general cali-brated and are consequently well represented. The analog method performs comparably badly, as it does not calibrate the mean.

As expected (Section 11.3), deterministic regression (MLR) strongly underestimates the day-to-day variance (Figure 16.1, middle row); randomisation creates the missing variance (MLR-ASW). Also, inflation (MLR-ASI) produces a result with the right vari-ance value, but see the discussion in Box 10.3 about the ill-design of this approach (also see below how inflation deteriorates other aspects). Analog methods well rep-resent day-to-day variance. For MOS, the representation of variance simply depends on whether this aspect is calibrated: simple additive methods (RaiRat-M6-E) inherit the – often wrong – model variance, variance correcting approaches (M7/M9) such as

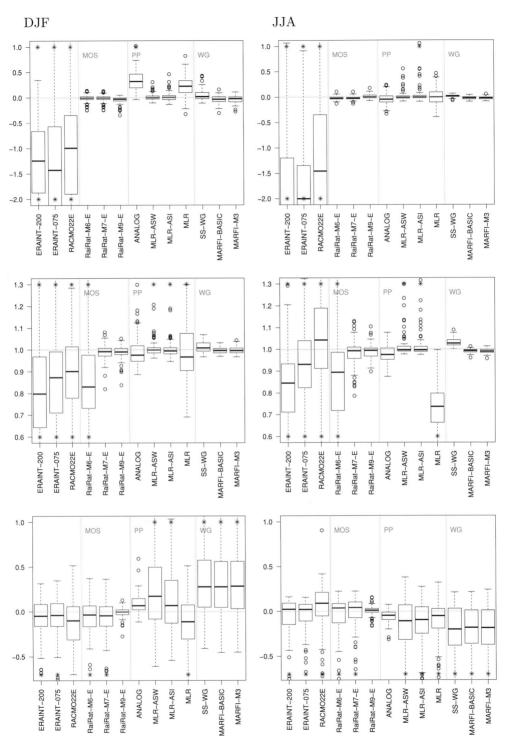

Figure 16.1 Selected marginal diagnostics for T_{max}. Top: mean (bias [K]); middle: variance (relative bias); bottom: skewness (bias). For details on the methods, see Appendix A.

quantile mapping reproduce the observed variance. But note again that this correction may induce inflation artefacts. In weather generators, variance is calibrated and therefore well represented.

Skewness, in case of temperature asymmetric deviations from a Gaussian distribution, typically arises from rare weather patterns that are locally amplified – blocking causing heat waves in summer, amplified by a loss of soil moisture and a subsequent increase in sensible heating of the lower atmosphere. In winter, blocking causes cold spells, which are amplified by radiative cooling (Figure 16.1, bottom row). In case of PP methods, a good predictor choice should capture the circulation effects but not necessarily the amplification. In fact, the analog method performs rather well, and also some regression models have comparably low biases – the Gaussian white noise randomisation is the worst-performing regression model. Standard MOS reproduces the skewness of the driving data and quantile mapping further adjusts it. Weather generators – mostly based on normal distributions – do not represent the observed skewness.

Simple linear regression models are not suitable to downscale daily precipitation: they do not separate the amounts from the occurrence process and do not randomise, that is, they inherit the distribution from the predictors (which, according to the central limit theorem, should be close to normal). Zero precipitation is generated by setting all negative predictions to zero. As a consequence, even though these methods are calibrated to the long-term mean, they fail to represent any marginal aspect of precipitation (Figure 16.2, MLR methods): wet-day probabilities are strongly overestimated and intensities underestimated; the calibrated mean precipitation is over-represented, because negative values are set to zero. Simple randomisation (MLR-ASW) and inflation (MLR-ASI) do not improve the results. Only more advanced generalised linear models (GLM) which separate the occurrence from the amounts process and model both stochastically reproduce the observed marginal aspects. Frost et al. (2011) obtained a similar result for GLIMCLIM (Chandler and Wheater 2002, Yang et al. 2005). The analog method performs surprisingly well (see the discussion in the following section), although Frost et al. (2011) find a rather bad representation of intensities for Australia. Both MOS and weather generators typically calibrate wet-day probabilities and intensities and thus represent marginal precipitation aspects well. For the US, Gutmann et al. (2014) found that BCCA and BCSDd (see Chapter 14; Wood et al. 2002, Maurer et al. 2010) both simulate far too many wet days.

16.2   Representing Temporal Aspects

The three approaches – PP, MOS and weather generators – differ fundamentally in their treatment of residual temporal dependence: PP typically inherits it from large-scale predictors and may or may not modify it. MOS methods inherit it from the simulated variable and, for most variants, do not explicitly alter it. Unconditional weather generators explicitly model temporal variability – aspects which are not explicitly modelled (or trivially result, such as short wet spells from transition probabilities) will not be represented by unconditional weather generators. The following discussion is based on the results from Maraun et al. (2017a).

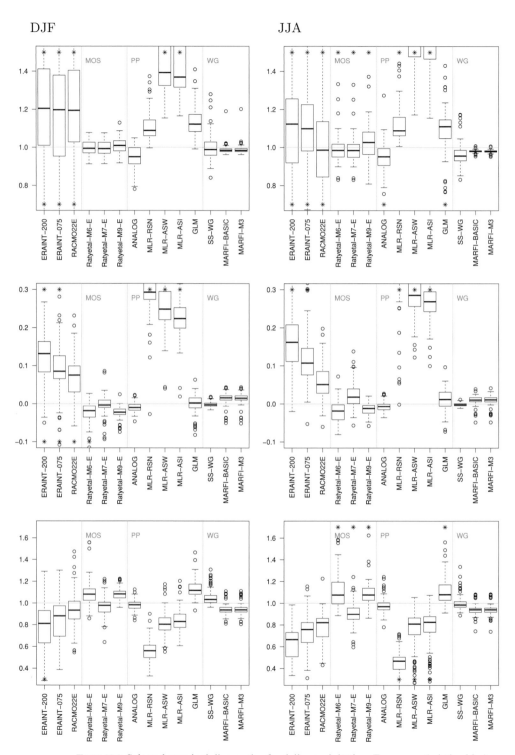

Figure 16.2 Selected marginal diagnostics for daily precipitation. Top: mean (relative bias); middle: wet-day probability ($\leq$ 1mm) (bias [%]); bottom: wet-day intensity (relative bias). For details on the methods, see Appendix A.

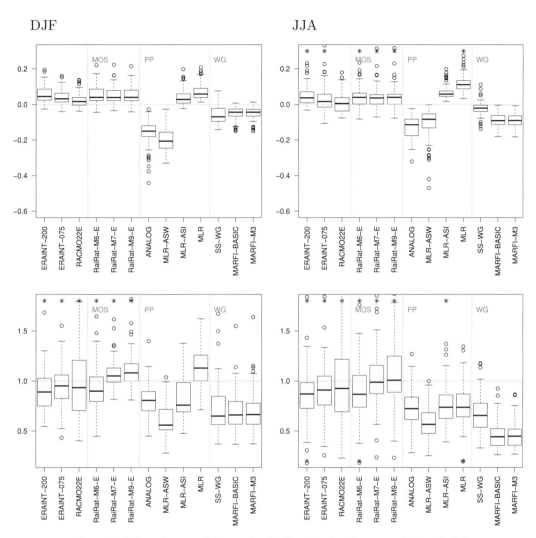

Figure 16.3 Selected temporal diagnostics for T_{max}. Top: lag-1 autocorrelation (bias); bottom: variance at interannual scale (relative bias). For details on the methods, see Appendix A.

Short-term temporal dependence of daily temperature, as expressed by the lag-1 auto-correlation (Figure 16.3, top), is too strong for deterministic regression models (MLR-AAN, MLR-AAI, MLR), in particular during summer, because small-scale noise unexplained by the predictors is not simulated. Randomisation can in principle account for this problem, but simple white noise randomisation (i.e. a noise model without temporal dependence, e.g., MLR-AAW) will destroy too much of the inherited dependence. Also, analog methods simulate too low a temporal dependence for temperature. Given that wet–dry transition probabilities (see below) are well simulated, this result seems at first sight surprising. But multivariable analog methods use the same analog for all variables, that is also the same predictors for the selection of analogs. If the predictor variables determining precipitation (circulation and humidity fields) have stronger spatial-temporal variability than the large-scale temperature fields, they would thus dominate the construction and selection of analogs. Such predictors would therefore not well constrain temperature variability. Another reason could be that large-scale analogs are rather poor at the station scale. A sequence of analogs may thus not be able to correctly represent the rather smooth temporal variability of temperature. As laid out already, standard MOS does not explicitly alter temporal variability. But some implementations of quantile mapping may slightly improve short-term temporal dependence by adjusting the marginal distribution. Weather generators by construction reproduce short-term variability.

Interannual variability of temperature, as expressed by the variance of seasonal means (Figure 16.3, bottom), is underrepresented by most PP methods. Thus, either an important fraction of long-term variability is not determined by the large-scale circulation (this may in particular be the case for summer, when local land–atmosphere feedbacks might play an important role), or important predictors are missing or not suitably included. MOS methods inherit the – here well-simulated – interannual variability from the driving data. All unconditional weather generators under-represent interannual variability, because it is not explicitly modelled. These results are in line with earlier findings (e.g. Wilks 1998). Including predictors in general improves but still underestimates interannual variability (Wilks 2010).

The treatment of wet-and-dry transition probabilities for precipitation (Figure 16.4 top and middle) illustrates again that deterministic linear models are not suitable for precipitation (MLR-RSN, MLR-SI). By construction they overestimate wet–wet and dry–wet transitions and the length of wet spells, while stochastic methods adding white noise underestimate these aspects (MLR-ASW). Only more advanced stochastic generalised linear models capture these aspects (GLM). Also, GLIMCLIM has been shown to produce realistic short-term variability (Hu et al. 2013). Analog methods perform quite well for short-term precipitation dependence. Simple MOS based on rescaling does not alter the wet–dry sequence. Yet wet-day probability corrections as implemented in quantile mapping indirectly affect transition probabilities and short spells (see also Rajczak et al. 2016): the drizzle effect is mitigated, and short spells merge into longer spells. Gutmann et al. (2014) shows that BCSDd strongly overestimates wet spell lengths in summer. Weather generators are calibrated to represent transition probabilities. If only one-day memory is explicitly modelled, dry spell lengths tend to be too short. Hu et al. (2013) also find that a weather generator based on a nonhomogeneous hidden Markov model, in which the transition between weather states is conditioned on atmospheric predictors, has difficulties representing dry spell characteristics.

As in the case of temperature, all regression-based PP methods under-represent interannual variability (Figure 16.4 top and bottom). Only inflated regression simulates too-strong interannual variability: the inflation affects not only short-term variability but also long-term variability. As the latter is better explained by predictors than the former, inflation overcompensates. The same effect occurs for quantile mapping: the correction of day-to-day variance affects variance at all timescales and thus inflates interannual variability. Unconditional weather generators generally under-represent interannual

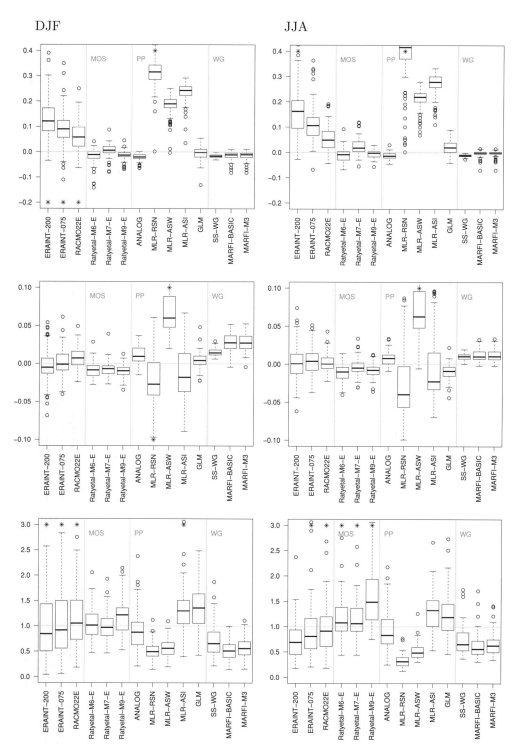

Figure 16.4 Selected temporal diagnostics for daily precipitation. Top: wet-wet transition probability (bias [%]); middle: dry-wet transition probability (bias [%]); bottom: variance at interannual scale (relative bias). For details on the methods, see Appendix A.

variability – it is not a modelled aspect (see also the earlier discussion on temperature). Frost et al. (2011) find that for their study region also, conditional weather generators based on non-homogeneous hidden Markov models and Richardson-type weather generators under-represent interannual variability with the chosen predictors.

Maraun et al. (2017*a*) find that the annual cycle of temperature is well simulated by all methods; for precipitation only the linear regression models underestimate the relative amplitude. For temperature, the positive result even holds for annually calibrated methods without a model of the annual cycle – the reason, of course, is that the predictor–predicand relationship for temperature is very close to actual physics (see the discussion in Section 11.5). But also for precipitation, a model with carefully chosen predictors descrbing humidity and circulation will capture seasonality. Almost all MOS methods improve the representation of seasonality. But note that most methods are explicitly trained for each calender month or season. Seasonally varying biases, however, cast doubt on the applicability in future conditions, since the annual cycle to some extent resembles different future climates.

16.3   Representing Spatial Aspects

Almost all statistical downscaling methods reproduce systematic spatial variations, that is the spatial pattern of the climatological mean, because the mean is calibrated for each site individually. The representation of residual spatial dependence, that is the day-to-day variations of weather patterns, differs strongly across different approaches and individual methods (see Sections 11.3.3, 12.5.4 and 13.3.3). Standard PP deterministic linear regression models imprint the spatial coherence of the large-scale predictor fields and hence overestimate the spatial dependence of temperature variability. White noise randomisation produces independent realisations for each station and thus destroys spatial coherence – the lower the variance explained by the predictors, the stronger the effect. The analog method by construction produces spatially coherent fields. MOS essentially inherits the residual temporal dependence and only slightly modifies it by adjusting the marginal distribution. Weather generators differ widely in their implementation of spatial models. In case of single-site weather generators, no residual spatial dependence is modelled, neither based on predictors or based on a model of the residual temporal dependence. Some approaches explicitly model residual spatial dependence based on multivariate distributions; hidden Markov models only simulate coherent fields of distribution parameters but then simulate from these distributions independently for each location. Accordingly, the performance to simulate residual spatial variability differs strongly from case to case, in some cases from region to region.

Figure 16.5 illustrates this discussion. For temperature (left), the decay length of daily maximum temperature is shown (left). The deterministic linear regression model (MLR) overestimates the spatial dependence of temperature variability, and white noise randomisation (WT-WG) destroys it. The analog method produces spatially coherent fields. Note that the actual performance is lower if the analogs have been defined individually rather than in common for several stations. The single-site weather generator (SS-WG)

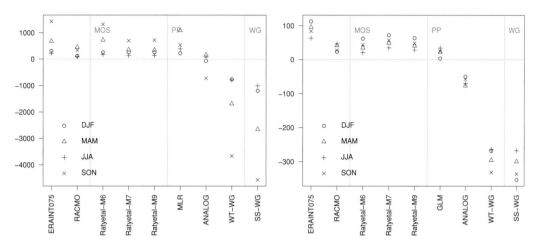

Figure 16.5 Decay length for (left) correlation of T_{max} and for (right) Brier skill score for joint wet-day occurrence (bias [km]). For details on the methods, see Appendix A.

illustrates a situation in which no spatial dependence is modelled. The different MOS methods either inherit residual temporal dependence (Ratyetal-M6) or slightly modify it indirectly (M7/M9). For precipitation, a characteristic length describing the joint occurrence of precipitation is shown (right). The actual values are much lower than for temperature, because precipitation is much more variable in space. Yet the qualitative behaviour of the different methods is very similar. Interesting, however, is the case of the GLM model. It simulates precipitation occurrence independently for each location, conditional on atmospheric predictors. The nevertheless good performance – and the comparison with a pure white noise model (SS-WG) – indicates that the joint occurrence is well captured by the predictors. Note that the ERA-INT and RACMO biases might to a large extent be due to the different statistics of precipitation on different spatial scales. Although these biases therefore do not necessarily indicate shortcomings of the models they do quantify to what extent the raw simulated values can be used to represent local values.

Based on the VALUE perfect predictor experiment, Widmann et al. (2017) have investigated a range of additional spatial diagnostics including dependence of inter-site correlations on distance, the dependence of joint threshold exceedances on distance and the spatial degrees of freedom. The results corroborate the findings for regression-based PP statistical downscaling methods. As expected, the use of large-scale predictors without adding regional noise leads to overestimations, while underestimations can be caused for instance by adding local noise that is uncorrelated between different locations. Analog methods based on local analogs, which often turn out to be different for different locations, underestimate correlations, while those with common analogs for multiple sites lead by construction to realistic correlations. The direct RCM output considered in the VALUE study has moderately positive correlation biases for precipitation, which get consistently reduced when it is post-processed by deterministic MOS methods. Many quantile mapping methods have very low correlation biases. The spatial degrees of freedom, which are a measure for the complexity of the spatial fields, have biases consistent with the correlation analysis, showing a too-low (high) number of degrees of freedom for methods that overestimate (underestimate) the correlations.

When evaluated on shorter spatial scales in a high-density network for Germany, the method is almost bias free for correlation and occurrence length scales (Widmann et al. 2017). Based on the same network, the analysis of joint exceedances of high precipitation thresholds (Widmann et al. 2017) shows an overestimation of the associated length scale for almost all methods. The only methods that perform well are the ANALOG method and some methods that combine an analog method with various transfer functions. All these findings essentially corroborate results from earlier studies (e.g. Easterling 1999, Kettle and Thompson 2004, Huth et al. 2008, Frost et al. 2011, Hu et al. 2013, Huth et al. 2015).

As mentioned, weather generators differ substantially in their characterisation of spatial dependence. Unfortunately, only single-site methods participated in the VALUE experiment. Important insight, however, is provided by the studies of Frost et al. (2011), Hu et al. (2013), Ferraris et al. (2003) and Paschalis et al. (2013). Frost et al. (2011) evaluated correlations of occurrence and amount of daily precipitation at different locations. They found that a nonhomogeneous hidden Markov model for occurrence combined with conditional multiple regression for amounts, and GLIMCLIM, a conditional weather generator based on a generalised linear model, substantially underestimated correlations between station timeseries. In contrast, Hu et al. (2013) had similar results for GLIMCLIM but found that a nonhomogeneous hidden Markov model performed well for inter-site correlations. The difference can be a result of the predictor choice or the specific regional climate. Disaggregation methods for precipitation investigated in Ferraris et al. (2003) show substantial over- and underestimations of correlation lengths with no method performing systematically better than others. However, advanced stochastic models for precipitation that include a disaggregation step based on two-dimensional, latent Gaussian fields showed realistic spatial characteristics (Paschalis et al. 2013).

All these studies confirm the structural skill and limitations of downscaling methods for representing spatial variability, which have been discussed in Sections 11.3.3, 12.5.4 and 13.3.3. When the fraction of variability unexplained by the predictors is low, statistical downscaling methods tend to yield realistic spatial variability. When the local, unexplained variability is high, only MOS, if it inherits a realistic spatial dependence from the driving model, or common analogs provide realistic results. Further systematic studies comparing the range of different weather generators are, however, required to come to a conclusive statement regarding multisite weather generators.

16.4   Representing Multivariable Aspects

Wilcke et al. (2013) analysed the effect of univariate quantile mapping on intervariable correlations, and some studies compared multivariate bias correction with univariate

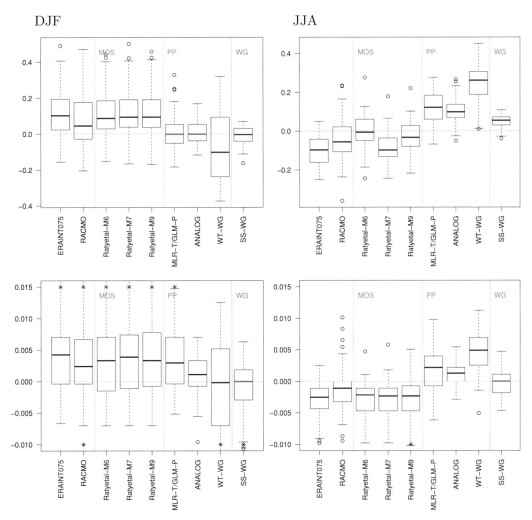

Figure 16.6 Selected diagnostics for multivariable aspects of T_{max} and daily precipitation. Top: Spearman correlation (bias); bottom: probability of jointly exceeding the 90% percentile (bias). For details on the methods, see Appendix A.

methods (Piani and Haerter 2012, Li et al. 2014, Vrac and Friederichs 2015, Cannon 2016, 2017). But no broader intercomparison studies about the representation of multivariable aspects exist. The following illustrations are therefore exclusively based on VALUE results.

Figure 16.6 shows selected multivariable diagnostics for temperature and precipitation. During winter, univariate MOS essentially inherits (and slightly overestimates) the Spearman rank correlation (top panel, Ratyetal-M6/M7/M9 for precipitation, RaiRat-M6/M7/M9 for temperature) from the driving ERA-Interim. During summer, ERA-Interim correlations are slightly too low; MOS may indirectly improve these values by adjusting the wet-day probability (M6 and M9). The regression model (MLR/GLM, independent for temperature and precipitation) well simulates dependence in winter (note, however, that the rank correlation does not "see" the unrealistic marginal distribution for precipitation). For summer, the dependence is too weak (note that the observed rank correlation is negative in summer), probably because precipitation occurrence is randomised. The analog method by construction captures the inter-variable dependence. In weather generators, the dependence is explicitly modelled and thus also well represented. The results for the joint occurrence of rare events are qualitatively similar (bottom panel).

Several studies have demonstrated that multivariate bias correction improves the representation of intervariable dependence (Piani and Haerter 2012, Li et al. 2014, Vrac and Friederichs 2015, Cannon 2016, 2017) compared to the driving model and univariate bias correction. But these studies have not investigated potential impacts on the temporal dependence.

16.5   Representing Extremes

Extreme events may have marginal, temporal, spatial and compound aspects. The former two are discussed in this section. For a brief discussion of spatial extremes, refer to Section 16.3. To our knowledge, the representation of compound extremes has not been systematically assessed. The following results are based on Hertig et al. (2017).

The representation of marginal aspects of extreme events depends both on the implementation of the tail in the marginal distribution itself but also on the choice of predictors: Extreme events that are not merely random but explained by the predictors will be well represented even by a simple model. Deterministic PP linear regression models have problems capturing cold winter extremes (Figure 16.7, top left); both randomisation and inflation improve the representation of these phenomena. The analog method, MOS and weather generators represent low winter temperatures well. Summer cold extremes are better represented by all methods (top right). Warm extremes of daily maximum temperatures are better – in general well – represented than cold extremes of daily minimum temperature (middle). In general, the performance of MOS and weather generators depends on the implementation of the tail. Note, however, that warm and cold extremes might occur in spells – this relationship is not implemented in weather generators, an explicit tail model for the marginal distribution might thus lead to unrealistic time series of extreme temperatures (see also the discussion regarding skewness in Section 16.1).

Again, PP linear regression models – also those using inflated regression – cannot be used to downscale daily precipitation. Stochastic generalised linear models (GLM) instead provide a reasonable representation, although with an overestimation in winter and an underestimation in summer. Frost et al. (2011) found in their study of Australia that the choice of distribution is important for GLMs. For instance, GLIMCLIM based on a gamma distribution overestimates monthly maxima of daily precipitation. The reason is likely that precipitation in Australia is heavy tailed; the gamma distribution might compensate for the missing heavy tail with a lower rate parameter, which would

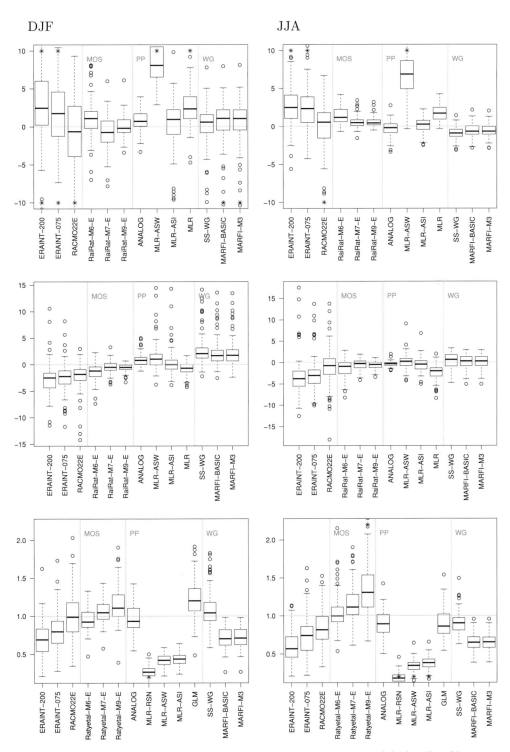

Figure 16.7 Selected extremal diagnostics for temperature and precipitation. Top: 20-season return level of T_{min} (lower tail, bias [K]); middle: 20-season return level of T_{max} (upper tail, bias [K]); bottom: 20-season return level of daily precipitation (upper tail, relative bias). For details on the methods, see Appendix A.

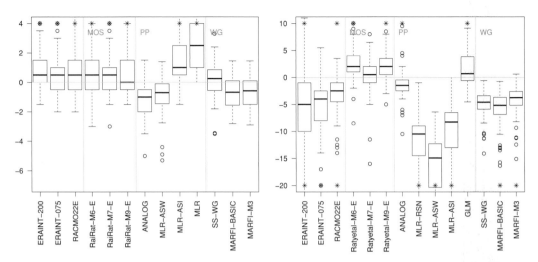

Figure 16.8 Average annual maximum heat wave (consecutive days of T_{max} larger than the 90th percentile of the daily distribution, left) and dry spell (right) length (bias [days]). For details on the methods, see Appendix A.

dominate over the light tail for moderate extreme events. The analog method produces rather good results. The representation of heavy precipitation by MOS methods depends on the implementation of the tail. Most overestimate high intensities, but for the delta change approach Frost et al. (2011) find an underestimation. Similarly, weather generators represent heavy precipitation according to their model of the tail.

The 20-year return level is typically still within the range of observed values. The representation of even more extreme values depends strongly on the extrapolation by the tail model. The analog method will likely under-represent such high extremes, as it cannot produce unobserved values. In case of quantile mapping as well as stochastic PP and weather generator methods, it is still an open question how the distribution tails should be modelled. Hertig et al. (2017) found evidence that methods based on simple parametric models (such as a Gaussian for temperature and a gamma distribution for precipitation) capture the overall behaviour well but may have difficulties representing extreme events. A properly selected parametric distribution, however, typically performs as well as empirical methods in the observed range of extreme events (see discussion in Section 12.5.2). For extrapolations beyond the observed range, some kind of model has to be assumed. (Volosciuk et al. 2017) found that parametric models (in particular heavy-tailed models) may produce extremely high unphysical values for extreme return values. Thus some constraints should be included.

Also temporal aspects can be extreme, such as heat waves or drought (Figure 16.8). The duration of long heat waves is essentially prescribed by the driving reanalysis and thus well represented by most methods. Only some linear regression models overestimate it, and methods using white noise randomisation underestimate it. The analog method, having problems in representing the temporal dependence of temperature, also struggles representing long heat waves. Weather generators by construction, though only slightly, underestimate the duration of long heat waves.

For long droughts, again PP linear regression models are not suitable. Again, a stochastic GLM performs well. Similarly, GLIMCLIM, based on general linear models, yields realistic representations in studies for Australia (Frost et al. 2011) and China (Hu et al. 2013). The analog method performed well in the VALUE experiment but simulated too-long droughts in the study by Frost et al. (2011). MOS methods perform all well, and weather generators by construction simulate too-short droughts. Similarly Hu et al. (2013) report that a nonhomogeneous hidden Markov model under-represents dry spell lengths because it assumes conditional independence of precipitation within a weather state.

16.6   Representing Climate Change

Evaluating the performance to represent regional climate change requires – in the case of statistical downscaling – two analyses (see Chapter 15): evaluating the performance of the driving dynamical model(s) and evaluating the performance of the statistical downscaling model. Of course, this evaluation is inherently limited, as it cannot fully rely on observations – it is an extrapolation problem. GCM performance has been discussed in Chapter 8, and possible experiments to evaluate how well GCMs reproduce observed trends have been briefly sketched in Section 15.4. Here we will focus on the downscaling performance.

As discussed in Section 15.4, two experiments should be carried out, which complement each other:

- evaluating the reproduction of historical observed trends in a perfect predictor experiment;
- evaluating the reproduction of simulated future trends in a pseudo-reality experiment.

16.6.1   Reproducing Observed Trends

Several studies have been conducted that assess the skill of RCMs to reproduce observed trends when driven with boundary conditions from reanalysis data (e.g. van Oldenborgh et al. 2009, Lorenz and Jacob 2010, Bukovsky 2012, Ceppi et al. 2012, van Haren et al. 2013a). Only a few similar studies exist for statistical downscaling.

Widmann and Schär (1997) showed that observed precipitation trends cannot be explained by changes in the atmospheric circulation alone. Benestad et al. (2007) analysed seasonal precipitation trends in Scandinavia in a perfect predictor study. Observed trends were weakly positive and often not reproduced by statistical downscaling, driven with ERA-40 predictors.[1] Yet predictions with circulation-based predictors were closest to the observed trends. These findings demonstrate that historical precipitation trends are still dominated by internal climate variability (which manifests in variability of the atmospheric circulation) rather than radiative forcing (which would manifest in a

---

[1] The mismatch between observed and simulated trends may at least partly result from mis-represented trends in the ERA-40 predictors (see Chapter 7).

relevant influence of humidity). In other words: analyses of historical precipitation trends do not bear information about whether a downscaling method credibly represents future climate change.

In a similar perfect predictor experiment, Huth et al. (2015) analysed the representation of seasonal temperature trends in central Europe. They found similar performance for all statistical downscaling methods, indicating that the choice of predictors plays a dominant role. Simulated trends were too strong for winter, too low for summer; simulations were better for T_{min} than for T_{max}.

Recently, Maraun et al. (2017a) analysed how well the methods participating in the VALUE perfect predictor experiment reproduced observed temperature trends. The MOS methods reproduced the trends simulated by the driving model, regardless of the quality of these simulations. Quantile mapping methods mostly inflated and in fact deteriorated these trends. The performance of PP methods, in line with the findings by Huth et al. (2015), depends very much on the predictor choice and the predictor domain. Temperature trends were only well captured by those models including surface temperature as predictor. It is, however, questionable whether this variable fulfils the PP assumption.

16.6.2   Pseudo-Reality Studies

Also only a few perfect model (or pseudo-reality) studies exist that assess the performance of statistical downscaling methods to represent simulated future climates. Perhaps the first such study was carried out by Charles et al. (1999), who demonstrated that changes in the occurrence of precipitation could not be simulated by predictors representing specific humidity but rather by predictors representing relative humidity (in their case: dew point temperature depression). In a study for the US, Vrac et al. (2007c) analysed the performance of a nonhomogeneous hidden Markov model with predictors influencing both the occurrence and the amounts process. They showed that a model with predictors representing circulation and specific and relative humidity was capable of reproducing RCM simulated future trends. Similarly, in a study for France, Dayon et al. (2015) showed that predictors without the combined effect of sea-level pressure, surface temperature, moisture flux, stability and specific humidity were not able to reproduce RCM precipitation trends. In general, downscaling methods based on weather types and the analog method are limited in representing warmer and moister future climate states. For instance, Gutiérrez et al. (2013) demonstrated that downscaling based on analogs and weather types strongly underestimates warming trends.

16.7   Synthesis

In the following, we attempt to summarise what is currently known about the performance of statistical downscaling methods. Table 16.1 provides an overview of the discussion throughout this chapter. Note that the table attempts to give an overview of

Table 16.1 Performance of different statistical downscaling methods for daily resolution. BC: bias correction, QM: quantile mapping, REG: (generalised) linear model, det: deterministic, infl.: inflated variance, stoch: white noise randomisation; WT: weather typing; ANA: analog; RI: Richardson-type; POI: Possion clustering; HM: hidden Markov; SS: single-site; MS: multisite; U: unconditional; C: conditional; "+": should work reasonably well based on empirical evidence and/or expert judgement; "○": problems may arise depending on the specific context; "−": weak performance either by construction or inferred from empirical evidence; "?": not studied. The categorisation assumes that predictors are provided by a well-performing dynamical model simulating informative and representative predictors, also for climate change. Statements about extremes refer to moderate events occurring at least once every 20 years. This table is necessarily to some extent subjective.

Aspect	MOS				PP				WG[a]			
	BC	QM emp.	QM para.	QM extreme	REG det.	REG infl.	REG stoch.	ANA SS/MS	RI U SS/MS	RI C SS/MS	POI SS/MS	HM U/C
Temperature, marginal												
mean	+	+	+	+	+	+	+	○	+	+	+	+
variance	○	+	+	+	−	○	+	○	+	+	+	+
extremes[b]	○	+	+	+	−	○	+	+	+	+	+	+
Temperature, temporal variability												
autocorrelation	+	+	+	+	+	+	−	−	+	+	+	+
mean spells	+	+	+	+	○	○	−	−	+	+	+	+
extreme spells	+	+	+	+	+	+	−	○	+	+	+	+
interannual variance	+	○	○	○	−	○	−	−	−	○	−	−/○
climate change	+	○	○	○	+	−	+	−	+	+	+	+
Temperature, spatial variability												
means	+	+	+	+	○	○	−	−/+	−/?	−/?	−/?	?
extremes	+	+	+	+	−	−	−	−/+	−/?	−/?	−/?	?
Precipitation, marginal												
wet-day probabilities	+	+	+	+	−	−	+	+	+	+	+	+
mean intensity	+	+	+	+	−	−	+	+	+	+	+	+
extremes[b]	○	+	○	+	−	−	+	+	○	○	○	○
Precipitation, temporal variability												
transition probabilities	○	+	+	+	−	−	+	+	+	+	+	+
mean spells	○	+	+	+	−	−	+	+	○	+	○	○/+
extreme spells	+	+	+	+	−	−	+	+	−	○	−	−/○
interannual variance	+	○	○	○	−	○	○	○	.−	○	−	−/○
climate change	+	○	○	○	+	−	+	○	+	+	+	+
Precipitation, spatial variability												
means	○	+	+	+	−	−	−	−/+	−/○	−/○	−/○	○
extremes	○	○	○	○	−	−	−	−/+	−/?	−/?	−/?	?
Multivariable												
bulk	+	+	+	+	−	−	−	+	+	+	+	+

[a] We consider standard WGs with Gaussian distribution for temperature and, for example, gamma distribution for precipitation.

[b] We consider extreme events within the range of observed values. No extrapolation is assessed.

isolated downscaling performance. We assume that the predictors are informative, representative and credibly simulated. Of course, no downscaling can overcome errors in the large-scale atmospheric circulation, and no downscaling can correct errors in large-scale climate trends. In many cases, the performance will depend on the specific implementation, the predictor choice, the region and the season. A generic ranking is thus impossible. Still, we believe it is useful to give at least rough guidance.

Based on the results from VALUE and other evaluation studies cited throughout the chapter and considering structural limitations of different models and our own judgement, we therefore try to classify each method for a range of statistical aspects into three categories: methods that should work reasonably well based on empirical evidence and/or expert judgement; methods for which problems may arise depending on the specific context; and methods that perform weakly either by construction or inferred from empirical evidence. Of course, this table is to some extent subjective. When selecting a method based on this table, please also refer to the corresponding sections throughout this chapter and to the discussions of structural skill in Sections 11.3, 12.5 and 13.3. In the following, we will briefly discuss the table.

16.7.1 Model Output Statistics

Bias correction performs well in general when driven with credible predictors. Moderate extremes within the previously observed range are well adjusted by quantile mapping, independent of the implementation. Care is needed when addressing very rare events. Empirical quantile mapping will suffer from sampling variability – the correction function will be very noisy. Extrapolations from the bulk (e.g. with a Gaussian or gamma distribution) might be substantially biased for severe events. In general, specific extreme value models might prove successful, but there is a danger that the transfer function, which for extremes is a combination of the tails of two distributions, might also produce unphysical adjustments to very extreme events. A careful adaptation to specific stations is thus required.

Temporal variability is inherited from the driving model and may be improved by adjustments of the marginal distribution. It should be investigated, however, whether wet-day corrections induce artefacts by distorting the spell-length distribution (and in the worst case create unrealistically long extreme spells). A general problem of variance correction and quantile mapping is the inflation of interannual to multidecadal variability and forced trends. In particular for precipitation, this artificial amplification might be substantial.

Bias correction also inherits spatial variability from the driving model and may improve it slightly by adjustments of the marginal distribution. For local extreme events, it is likely that the driving climate models have limitations simulating the spatial structure; bias correction cannot correct for that. Also, multivariate aspects are typically not deteriorated by bias correction.

Bias correction should be very carefully applied when used in a downscaling context, that is when mapping from a model grid to a much finer grid or even point scales. Bias correction is – in most cases – deterministic and cannot create sub-grid variability.

Quantile mapping will instead inflate grid-box variability and create and unrealistically smooth sub-grid temporal and spatial variability; extreme events will be inflated in space. The problem will be substantial for precipitation but may also be relevant for temperature in complex terrain. Feedbacks will not be represented below the grid scale; if these locally influence the climate change signal, this modulation will not be correctly represented by bias correction.

16.7.2   Perfect Prognosis

Deterministic linear regression models represent mean precipitation and its short-term variability fairly well. Long warm or cold spells, controlled by the large-scale circulation, are captured as well and interannual variability is substantially underestimated. But these models are unsuitable for modelling temperature extremes and fail to represent daily precipitation. Inflated regression improves some aspects but creates artefacts and potentially modifies the climate change signal. Many aspects of daily precipitation are successfully represented by stochastic generalised linear models, even without an explicit model of temporal dependence. For temperature, however, stochastic regression models with white noise randomisation fail: they do not capture the unexplained temporal dependence and thus misrepresent spells. Here, including a Markov component, for example by using previous values as predictor, might proof useful. Such models are conditional weather generators. Regression models have severe limitations describing spatial fields. Mean temperature is fairly well described, while extremes and precipitation fields are not: deterministic models are too smooth, single-site stochastic models do not simulate spatial dependence. Thus, spatial models are required.

The analog method performs well for precipitation, but is of limited skill for downscaling temperatures: temporal dependence is underestimated, and even mean values are misrepresented. Moderate extreme events are well sampled, but classical implementations fail to represent unobserved extremes. The analog method is also likely to substantially underestimate warming trends, because no present-day analogs exist for weather in a much warmer world. The method does, however, by construction capture spatial and multivariate dependence.

A key issue with PP methods is the selection of suitable predictors that credibly capture climate change. This issue comprises both representing all relevant influences by credibly simulated predictors and constraining the predictor–predictand relationship for long-term timescales based on comparably short time series.

16.7.3   Weather Generators

Weather generators explicitly model marginal, short-term temporal and inter-variable and sometimes also spatial aspects. As such, they trivially represent these aspects in present climate. The representation of marginal extreme events may be improved by specifically designed marginal distributions. Only interannual and longer-term variability as well as long precipitation spells are substantially underestimated. The latter issue might be mitigated by including meteorological predictors.

Single-site weather generators cannot represent residual spatial dependence. Multisite weather generators have, to our knowledge, not yet been studied systematically enough to allow for any robust conclusions. For instance, the findings for hidden Markov models, which attempt to link the variability at different locations through underlying common weather states, differ between studies, with some showing realistic spatial variability whereas others do not.

Weather generators can in principle capture climate change by including change factors from climate model simulations. Here two issues have to be ensured: first, the simulated change factors have to be representative of the real-world local weather; in particular, they have to represent changes at the target spatial scale. For instance, a climate model that is too coarse to represent the local climate change signal will provide noncredible change factors. Second, all relevant weather generator parameters that respond to climate change have to be modified by change factors.

16.8   Further Reading

For literature on dynamical model performance, see Chapter 8. Comprehensive evaluations of statistical downscaling are given by Frost et al. (2011), Hu et al. (2013), Gutmann et al. (2014) and Huth et al. (2015), as well as the special issue dedicated to the VALUE perfect predictor experiment (Gutiérrez et al. 2017, Maraun et al. 2017a, Hertig et al. 2017, Widmann et al. 2017, Soares et al. 2017).

17 A Regional Modelling Debate

Along with the increasing amount of research in climate downscaling, a critical debate about the limitations and suitability of downscaling has arisen. The debate revolves around essentially two questions: first, are climate models skillful to provide user-relevant information about climate change? Some researchers argue that GCMs cannot provide skillful input to downscaling, and downscaling itself cannot add value to GCM simulations. Second, is downscaling necessary, or could it be avoided by bottom-up approaches to decision making? We will review these two discussions and critically comment on the issues raised.

17.1 Are Climate Models Fit for Purpose?

Kundzewicz and Stakhiv (2010) discuss whether climate models are "ready for prime time" in climate impact research. They argue that GCMs have originally been developed to advise mitigation policies but are more and more applied also to inform adaptation decisions. Whereas for the former purpose, a broad representation of global climate change is sufficient, the latter requires accurate projections of regional changes, in particular of highly uncertain processes such as the hydrological cycle. The authors argue that climate models are not (yet) skillful for direct application in adaptation planning. This view has been shared by other authors (e.g. Pielke and Wilby 2012); it is based on a series of claims (Kundzewicz and Stakhiv 2010, Pielke and Wilby 2012): first, GCMs do not skillfully include all first order forcings and feedbacks; second, GCMs do not skillfully simulate relevant regional processes such as El Niño; third, GCMs do not reproduce observed trends; and fourth, downscaling cannot improve GCM simulations. The first three claims are related to GCM skill, the fourth to downscaling skill and added value. In the following we will discuss these issues.

17.1.1 Skill of GCMs

Pielke et al. (2009) argue that current GCM projections do not consider all first-order forcings that determine future climate, such as the effect of aerosols or changes in land use and land cover. Over recent years, these aspects have more and more been in the focus of climate research and featured prominently in the most recent IPCC assessment report (Boucher et al. 2013, Myhre et al. 2013). In fact, cloud feedbacks and the interactions of aerosols with radiation and clouds are a major source of uncertainty in climate sensitivity estimations (Bony et al. 2006, Flato et al. 2013, Myhre et al. 2013, Bony et al. 2015). The impact of land use change on climate is becoming more researched with dedicated scenarios. Still, it is useful to recall that climate change projections are themselves based on scenarios that should be kept separate from other anthropogenic forcings. An intercomparison of the relevance of different forcings, in particular at the regional scale, is of course a highly useful undertaking.

In Section 8.5, we discussed the limitations of GCMs to simulate important regional climate processes. Of course, wherever these are dominating regional climate, they need to be skillfully simulated. If a GCM does not credibly represent changes in ENSO variability, this model will not be useful to simulate regional climate changes in, for example, Peru (even after downscaling). Teleconnections additionally complicate the issue. For instance, El Niños have a weak effect on European winter climate – such effects of course limit the value of regional climate projections, but they do not necessarily make them useless. We will further discuss this issue in Section 17.1.2. The argument that climate models do not reproduce observed trends is mainly based on a study by Anagnostopoulos et al. (2010); these authors compare simulated grid-box as well as continental-scale trends with observations, essentially by correlating them at the 30-year timescale. A similar line of argument has later been used by Racherla et al. (2012) and Shindell et al. (2014). These authors compared simulated and observed climatic changes between the 1970s and the 2000s for selected regions of North America.[1] Such comparisons of course only make sense if modelled and observed trends can be expected to be largely caused by the same mechanism, namely some external forcing.

Over recent years, however, the role of internal climate variability has more and more been understood. Large-scale internal modes of variability (see Section 2.3) such as the Pacific Decadal Oscillation (PDO) or the Atlantic Multidecadal Oscillation have intrinsic timescales of several decades (Latif and Park 2012) and have a strong impact on surface climate (Mantua and Hare 2002, McCabe et al. 2004, Knight et al. 2006, Sutton and Dong 2012). It has been demonstrated that internal variability is responsible for about 30% of the uncertainty of 30-year projections of decadal mean temperature at the regional scale (Hawkins and Sutton 2009) and even for 50% for precipitation projections at the continental scale (Hawkins and Sutton 2011). Deser et al. (2012) showed that internal variability might reverse forced 50-year precipitation trends at the sub-continental scale, and Maraun (2013b) demonstrated that seasonal precipitation trends may not emerge from internal variability within the next decades in many mid-latitude regions. Freely evolving climate model simulations that are tied to observations only by radiative forcing will by construction simulate realisations of internal variability that are different from that observed. They will therefore not correlate with observations, even

---

[1] Racherla et al. (2012) have also compared the 30-year GCM trends trends with RCM trends to show that RCMs do only add marginal added value – because they do not improve these trends compared to observations. Laprise (2014) has demonstrated that, because of the reasons that will be laid out in the following, also this reasoning was not correct.

at climatic timescales. Trend analyses like those by Anagnostopoulos et al. (2010) and Racherla et al. (2012) are thus ill designed.

Also, initialising climate models with observations will not overcome our limited ability to evaluate GCM–simulated trends. Large-scale modes of internal variability may be predictable on decadal scales (Smith et al. 2007, Keenlyside et al. 2008, Meehl et al. 2009), but the complex nature of the climate systems fundamentally limits predictions beyond that time horizon (Collins et al. 2006). It is sometimes argued that, therefore, internal climate variability makes climate projections essentially useless. But as discussed in Section 5.2, an assessment of the signal-to-noise ratio itself provides useful information: even if a projection, because of internal variability, spans opposite changes, this projection is useful (as long as it is created with credible models), as it demonstrates that climate change is less relevant and internal variability is expected to dominate.

As discussed in Section 15.4, the only way to proceed is to assess whether observed trends are indistinguishable from simulated trends – in the presence of strong internal variability, no stronger tests are possible. Several such studies have been carried out, with broadly consistent results (see Section 8.5.3). For instance, Bhend and Whetton (2013) find, for both the CMIP3 and CMIP5 ensembles, that observed precipitation trends are still dominated by internal climate variability, such that observations and model simulations are consistent across the globe. In other words: observed precipitation trends can thus not be used to constrain the response of climate models to radiative forcings. Temperature trends are consistent in many regions of the world but, crucially, not everywhere. For instance, western Europe is warming much faster during summer than simulated; similarly winter temperatures over central Asia are increasing faster than modelled. It is not yet fully understood whether these discrepancies result from missing forcings, from a wrong response to forcing, from an under-representation of internal variability or whether they are simply a random difference. In these regions, indeed research is needed to understand the causes of these deficiencies and whether they will also manifest in future projections.

In general, it needs to be acknowledged that climate projections are not probabilistic but at best credible simulations of future climates (see also the discussion in Section 9.2.2).

17.1.2 Added Value of Downscaling

Pielke and Wilby (2012) argue that downscaled climate projections cannot have greater skill than interpolated GCM output, because (among the other reasons listed) they inherit biased large-scale weather fields (either as predictors or boundary conditions) from the driving GCM and cannot improve these. In the literature, this is referred to as the garbage-in, garbage-out problem (e.g. Hall 2014).

Here, it is useful to recall the discussion from Section 8.5: GCMs do not produce garbage everywhere, and not for all properties of the climate system. As discussed, Shepherd (2014) argues that most of our robust knowledge about climate change stems from thermodynamic responses (such as an increase in moisture in a warmer atmosphere) to external forcing, but changes in the atmospheric circulation are highly uncertain. In Section 8.5, we have used this line of argument to identify where downscaling may indeed add value (Hall 2014): regional phenomena, where dynamical changes are important, are highly difficult to project. In several cases, GCMs will share common biases, such that even a multimodel ensemble would not help span a credible range of uncertainty. Here we might have a garbage-in, garbage-out situation: regional climate model simulations could potentially simulate a physically credible regional response, but this response would at worst just refine an unrealistic large-scale climate. But for many phenomena, thermodynamic effects might dominate the large-scale response. Here, regional climate change simulations may indeed improve global climate change simulations by simulating regional-scale dynamical effects and feedbacks, which are not represented in a coarse-resolution GCM. Kendon et al. (2010) show that the relative importance of dynamic and thermodynamic changes are, for a given region and aspect, a function of season and scale. In fact, dynamical uncertainties will be most relevant in regions of strong climatic divides, for example, linked to the position of the polar front. As discussed in Section 8.5, this issue can be nicely illustrated by projections of European precipitation changes: whereas the RCM ensemble from the ENSEMBLES and EURO-CORDEX projects both simulate consistent precipitation changes in northern and southern Europe, these model ensembles simulate contradicting precipitation changes in many central European regions (Maraun 2013b, Jacob et al. 2014).

To summarise: if large-scale dynamical processes control the regional phenomenon of interest, uncertainties will be high. Here a careful model evaluation is required to select GCMs that credibly simulate the relevant large-scale processes. If no such GCMs exist, of course, no credible regional climate projections can be conducted. However, if regional climate is controlled mainly by large-scale thermodynamic processes, regional climate modelling is in principle a sensible approach that can add substantial value. Which case applies depends on the phenomenon of interest, the region of interest, and the season of interest. For any given downscaling application it is thus crucial to understand the relevant influences on the particular regional climate phenomena and how credibly they are simulated.

17.2   Can Bottom-Up Approaches Reduce Uncertainties?

Over recent years, impact modellers have developed approaches called response surfaces (Prudhomme et al. 2010) or decision scaling (Brown et al. 2012) to study the sensitivity of impact systems to climatic changes. Two major ideas behind the development of these approaches were (1) to decouple the impact modelling from the climate modelling and thereby gain flexibility in repeating impact assessments with new climate model generations or under different scenarios; and (2) to carry out stress tests of the impact system, that is to better understand how the system reacts upon forcings (Brown and Wilby 2012). But it has also been implied that these approaches reduce uncertainties compared to "the uncertainty-laden scenario-first perspective" (Brown and Wilby 2012). The argument is that in traditional top-down approaches, from GCMs to RCMs to impact models, "the range of outcomes offered to the decision maker is simply bewildering (or worse, spans changes of opposite sign)" (Prudhomme et al. 2010). Whereas Prudhomme et al. (2010) apply this approach in conjunction with downscaling, Brown et al. (2012) and Brown and Wilby (2012) apply it directly to GCMs, hence suggesting it as an alternative to downscaling.

The approach works as follows: an impact model is driven with a range of inputs resembling different climatic conditions (decreased mean rainfall, increased interannual variability, etc.), and for each input, the output of a desired target variable (e.g. river flood flows) is recorded. If many such different climatic conditions are sampled, the sensitivity of the impact model to climatic changes can be explored. Each aspect of the climate input that is changed would span one dimension of the input space, and a hyper-surface in the input-output space would describe the response of the target variable. A flat surface (with respect to a particular input variable) would indicate no sensitivity, a steep slope a strong sensitivity. This approach can be used to guide adaptation decisions in a "bottom up meets top down" approach (Brown and Wilby 2012): users might define a certain decision-relevant threshold (e.g. runoff higher than that threshold would require a higher dam), and information about future climate change can be included to evaluate whether the response to that change would be below or above the critical threshold.

The strengths of the approach are obvious: the response surface allows one to better understand the behaviour of the impact model. In regions where the surface is flat, the system will be robust to changes in the studied climate drivers; as discussed already, one can easily update climate impact studies by plugging in new climate model data, and one can carry out stress tests of the system under consideration. In fact, the approach allows one to study the sensitivity of the modelled system to climatic changes beyond the range simulated by the available climate model ensembles (Wilby et al. 2014). For instance, one can easily investigate whether even stronger changes might push the system across a tipping point.

Yet the approach cannot reduce uncertainties. First, uncertainties in regional climate projections exist – one can account for them or one can ignore them, but they will not go away without considerable improvements of our observational database and climate models. If a climate projection includes changes of opposite sign, and none of the underlying simulations can be ruled out as implausible, this is – potentially critically – relevant information for users. Withholding them might result in maladaptation. Moreover, also the impact model and thus the response surface itself is subject to uncertainties.

Furthermore, the approach is not an alternative to downscaling. As stated by Brown et al. (2012), it relies on credible climate information: if the slope of the response surface is steep, the adaptation decision will sensitively depend on the climate model output and thus crucially rely on the adequacy of the chosen climate models. Indeed, the source of information is not prescribed by the approach. It can be information from climate models, from observational and potentially paleoclimatic data, or from expert judgement. Note that this choice might crucially influence the outcome of the assessment (Kay and Jones 2012). But if the choice is to use model-based information, the specific application defines whether downscaled climate simulations are necessary or not (and whether they are sufficient or not). A GCM may skillfully simulate mean precipitation and its response to climate change in relatively flat terrain, but it will fail to simulate the response of precipitation intensities in a mountain area (see discussion in Section 12.7.3).

Finally, we would like to highlight an issue which – to our knowledge – has not yet been discussed in the literature: response surfaces have to include all climatic changes relevant for the system of interest and have to precisely include them; otherwise they might be dangerously misleading. Consider a system – for example a small catchment – that responds to changes in daily precipitation intensity or extremes. A response surface including only mean precipitation may produce a wrong response: higher intensities are linked to higher mean precipitation via the number of wet days. Thus, changes in mean precipitation might be different from changes in intensities – for instance, intensities might increase, the number of wet days might decrease, such that the mean remains constant. In such a case, the response surface would not be informative. Moreover, the system might respond to climatic phenomena that are assumed to remain constant – and are thus not included in the response surface. To avoid ill-designed response surfaces, detailed understanding of the range of possible climatic changes is thus equally important as an in-depth understanding of the behaviour of the impact model. We therefore believe it is necessary to construct response surfaces starting from a classical ensemble of climate model simulations, driving the impact model. This ensemble could then guide the choice of climatic aspects spanning up sensitivity space. Of course, the simulated range of changes could then be extended to assess sensitivities to non-simulated worst-case changes.

17.3   Further Reading

Key papers to understanding GCM performance at the regional scale are Hall (2014) and Shepherd (2014). The reproduction of observed regional trends across the globe has been analysed by, for example, Bhend and Whetton (2013) and van Oldenborgh et al. (2013).

18 Use of Downscaling in Practice

Users of climate information, as broadly discussed in Chapter 5, require credible, defensible and actionable information as a basis for decision making. This chapter intends to close the circle and comes back to these initial requirements.

In Part I of the book, we discussed the broader context of climate projection uncertainties in the light of observational limitations as well as the skill and remaining shortcomings of dynamical climate models. In Part II, we introduced the range of statistical downscaling approaches and methods and discussed their structural skill and limitations as well as the assumptions underlying their use in climate projections. In Part III, finally, we have reviewed the performance of many of these methods in practical applications.

In the following sections, we synthesise the relevance of these issues for generating and providing useful regional climate projections. We assume a stance in the spirit of Mastrandrea et al. (2010), who highlight the need of bottom-up/top-down vulnerability assessments that bring together bottom-up knowledge of existing vulnerabilities with top-down climate-impact projections.

Hewitson et al. (2014) call for considering statistical downscaling in a wider landscape of climate information. In fact, users of climate information do not care about which method has been used to generate this information as long as it meets their requirements. We therefore explicitly assume a method-agnostic stance: we do not restrict ourselves to discussing statistical downscaling options but also present situations in which dynamical downscaling might be required or in which no downscaling might be needed at all.

Regional climate modelling might be purely curiosity driven. In this chapter, however, we assume a user-relevant context and thus a collaboration between regional climate modellers and stakeholders such as impact modellers or decision makers. Section 18.1 highlights the relevance to first assess whether climate and climate change are important factors in a given context. Sections 18.2 to 18.4 then summarise the discussions throughout the book about how to select suitable climate models and downscaling methods for a given application. The actual interpretation of the results in the context of projection uncertainties is the subject of Section 18.5. Section 18.6 emphasises the need for interdisciplinary collaboration but also critically discusses the difficulties of such efforts. Suggestions for providing data for users via online portals are made in Section 18.7. Sections 18.8 and 18.9 briefly present both the importance of distilling useful information out of climate model ensembles and the ethical responsibility of regional climate modellers.

18.1   Assessing the Relevance of Climate Change

Already in the design of a collaborative project, the partners should consider whether climate change information is relevant in the given context.

- Is climate change a relevant factor?
  First of all, socio-economic or environmental factors might dominate over the influence of climate change in a given context. For instance, land-use change may have a stronger effect on landslide risk than climate change. Furthermore, the planning horizon might be short compared to climatic timescales: to adjust the premium of a harvest insurance, seasonal to decadal predictions may be required. For their long-term strategic planning, however, the company may need multidecadal climate projections. In some situations, of course, the aim of the project might be to first establish whether climate change is indeed a relevant factor.
- Is climate information relevant?
  In some situations, decision makers may be able to sufficiently reduce vulnerability to climate change by implementing low-regret measures (Dessai 2009, Wilby and Dessai 2010). In some cases, adaptation measures can be designed to be easily extendable, such that – depending on the actual future evolution of climate – measures can be corrected and updated. For instance, a dike may be built such that it can easily be increased further in height. In other cases, however, defensible climate change projections may be crucial to inform decision making.

18.2   Assessing the Climate Information Requirements

As a basis for selecting a suitable climate modelling approach, it is important to understand both the impact system and the relevant climatic phenomena.

18.2.1   Understanding the Spatial-Temporal Scales of the Impact

As discussed in Chapter 5, most impact processes have characteristic time- and spacescales, which might be different from the input resolution of the chosen impact model. Crucially, the relevant target scales for the downscaling are the process scales – not necessarily the impact model scales. For instance, a hydrological model may require hourly input. If monthly base flow in a small catchment is the impact of interest, daily precipitation input, disaggregated in a simple way, may well be sufficient. If flash floods in the same catchment are sought, one may indeed require credibly represented hourly precipitation in space.

18.2.2   Understanding the Relevant Climatic Phenomena

Different meteorological and climatic phenomena, spanning a wide range of scales, might control the considered impact. Examples of large- and regional-scale phenomena have been discussed in Chapter 2.

Typical large-scale phenomena are the Hadley cell, monsoonal circulation, ENSO, or the mid-latitude storm track. When addressing large-scale influences, one should also consider whether the regional climate phenomenon of interest is controlled by thermodynamic or dynamic processes. For instance, the precipitation climate of a region depends both on the moisture-bearing circulation (dynamic) and the moisture content of the atmosphere (thermodynamic). In northern and southern Europe, precipitation changes are relatively robust, and dynamical uncertainties are severe but manageable. In some regions of central Europe, however, the position of the jet stream and storm track might be of such importance that precipitation changes are highly uncertain. Regional to local phenomena are, for example, convection; influences of topography such as orographic precipitation, Föhn winds, mountain breezes, land-sea breezes, coastal jets and lake effects; and surface feedbacks with soil moisture and albedo.

18.3 Choosing the Driving GCMs

One must choose GCMs that credibly simulate the relevant large-scale phenomena. This selection requires process understanding rather than an evaluation of surface variables in present climate. For instance, many state-of-the-art GCMs simulate ENSO-like interannual variability (Flato et al. 2013). Yet many of these models have difficulties in realistically simulating ENSO feedbacks, undermining the credibility of ENSO projections (Bellenger et al. 2014, Chen et al. 2015). GCMs that evidently misrepresent first-order processes or feedbacks should be discarded. If no GCM is found that credibly simulates the key large-scale processes for the considered impact, model-based projections should be interpreted very carefully. In such a case, a model-informed expert assessment might be the only possible option (see Section 18.5.3).

Furthermore, if the location of large-scale atmospheric circulation features plays an important role for local climate, GCMs should be avoided that simulate these features in a substantially wrong position. For instance, many GCMs simulate too zonal a storm track such that cyclones travel too far south and penetrate too far into central Europe.

To represent uncertainties due to model imperfections and internal climate variability, one should always consider ensembles of GCMs. When selecting GCMs, one should carefully balance the need to comprehensively sample uncertainties against the need to discard ill-performing models (see also Chapter 9). Many impact modellers will not be able to use the full CMIP ensemble, possibly combined with an ensemble of RCMs, as input. Thus one may consider a subset, spanning the range of credible future evolutions of the relevant phenomena. Some downscaling approaches, namely the delta-change approach (Section 12.4.1) and change-factor weather generators (Section 13.4.1), require only long-term climate change signals from climate models. These approaches then provide the possibility to derive a few idealised scenarios summarising the full model spread or even to define scenarios that go beyond the observed model spread to "stress test" the impact system (e.g. Wilby et al. 2011). Note that in this case it will often not be sufficient to define change factors only for some mean aspects. For instance, changes in flood risk will likely also depend on changes in extreme precipitation (see the discussion in Section 17.2).

18.4   Choosing the Downscaling and Bias Correction Options

The choice of downscaling options comprises the following questions: is downscaling and/or bias correction required at all? Is statistical downscaling sufficient, or is prior dynamical downscaling required? Which statistical downscaling approach is suitable?

18.4.1   Is Downscaling Required?

The first question to ask is whether the driving GCMs represent the relevant climate phenomena at the scale of interest. For instance, if temperatures in flat terrain are relevant, a GCM, simply interpolated to the impact-model resolution, may fully suffice as input. But conversely, if local precipitation variability is of interest, a bias corrected GCM will not suffice.

18.4.2   Is Dynamical Downscaling Required?

As discussed in Chapter 12, current bias correction approaches can be used to adjust observed climatologies, but they cannot create non-explained day-to-day sub-grid variability (such as local precipitation or temperature inversions), and more importantly, they cannot represent a local climate change signal that differs from the large-scale signal because of local feedbacks. If substantial sub-grid variability and sub-grid variations of the climate change signal are expected, it makes no sense to bias correct a GCM against high-resolution observations. In regions of complex topography, this is likely the case. Here, dynamical downscaling is required. Note that this argument holds equally well for change-factor weather generators.

In complex terrain or when localised extreme convective precipitation is important, a standard RCM with a horizontal resolution of 10km or coarser may not be able to credibly simulate the local climate change signal. Whether the added value of-change-factor simulations of the order of 3km or finer is required – and sufficient – is an important question of current research.

Depending on the user requirements, PP statistical downscaling might be an alternative to dynamical downscaling. Temperature is a relatively easy case; precipitation at the daily scale, in particular if extremes and spatial fields are relevant, is much more difficult to model statistically – recall also that the choice of predictors to represent climate change is far from trivial (Section 11.6.1). Often the argument is brought forward that statistical downscaling is computationally cheap compared to dynamical downscaling. Whether this statement is true depends again on the user requirements. A regression model for mean temperature is indeed cheap. A generalised linear model simulating spatial fields of extreme precipitation might be computationally very demanding and

---

> **Box 18.1** Statistical Downscaling Ensembles
>
> Statistical downscaling methods should not be combined to multimethod ensembles in the same way as dynamical models. Statistical downscaling methods differ fundamentally from dynamical climate models in that they (1) do not represent the fundamental laws of thermodynamics and fluid dynamics, and (2) in most cases do not simulate full fields of a wide range of climatic variables. Instead, statistical downscaling methods are structurally limited to represent a small selection of statistical climate aspects. Thus, it is rather easy to a priori discard many statistical downscaling methods in a given context based on their structure: a deterministic linear regression model will not be suitable to represent daily extreme events; an unconditional Richardson-type change factor weather generator will not realistically simulate drought; and a single-station weather generator cannot be used if spatial fields are required. Similarly, it is of limited use to consider ensembles of different predictor choices: in case of precipitation amounts, it might be sensible to compare different measures of specific humidity, but it makes no sense to choose models without such predictors. Thus, ensembles of statistical downscaling methods have to be tailored for the specific application. Ensembles should be limited to suitable and well-performing statistical methods to sample model uncertainty.

may crucially miss representing the changing character of extreme precipitation under climate change.

18.4.3 Choosing the Statistical Downscaling Method

The choice of the actual downscaling method is very much determined by the relevant statistical climate aspects (Section 4.2): is mean climate relevant, or is day-to-day variability including extremes? Is long-term temporal variability such as drought or heat waves important? Is a realistic spatial dependence required? A starting point for the selection is given in Section 16.7: there we synthesise the performance of different downscaling methods in different contexts. More experimental approaches will briefly be presented in Chapter 19. A key message is that no method is available that performs well for all aspects.

With the growing database of dynamically downscaled GCMs, bias correction (Chapter 12) is often an obvious choice. Apart from its rather simple implementation, it also has the key advantage that it does not attempt to bridge meso-scale atmospheric processes by simple statistical models and is thus – if carefully applied – potentially more defensible under climate change applications.

The ease of its application, however, often hides the potentially severe problems and artefacts. In particular, these are: first, the difficulties of bias correction to downscale to finer scales; second, the modification of the climate change signal by bias correction; and third, the challenging task of properly evaluating bias corrected climate model output. For a discussion of MOS methods, see Chapter 12, in particular the Sections 12.6

to 12.9 on MOS assumptions and problems. Why it is difficult to sensibly evaluate bias correction methods is discussed in depth in Section 15.7.

As discussed in the previous section, PP might be a cheap alternative to dynamical downscaling. PP is particularly useful when no dynamical downscaling is available or when the climate model (either GCM or RCM) does not simulate input which is representative of the target variable and thus not suitable for bias correction (Chapter 12.6.2).

If local day-to-day variability is important, weather generators might be the method of choice, either conditioned on the large-scale meteorological predictors (i.e. as a stochastic PP method) or modified by change factors. Change-factor weather generators might in many cases be a useful strategy. If long-term variability from weeks to years to decades is important (e.g. when drought is relevant or in ENSO-controlled climates), however, they may not be suitable.

18.5   Interpretation of the Results

18.5.1   Role of Models

All the points discussed should be considered when interpreting regional climate projections and subsequent impact assessments. A useful stance might be a healthy and humble model scepticism: healthy scepticism in the sense that models are models and far away from being "observations of the future". And humble scepticism in the sense that one should not dismiss the value of models. Model ensembles do provide useful information, which is likely more defensible than an ad-hoc expert opinion or an extrapolation of observed trends – in particular because model ensembles can at least partly sample associated uncertainties. Model simulations become a prediction only together with expert knowledge (Stainforth 2016).

18.5.2   Remaining Gaps

As discussed in Chapter 5 and Section 18.3, the simulated output may not satisfy user needs, mainly because of two reasons. First, the large-scale climate and climate change may not be credibly simulated. In such a situation, one should discuss possible ways ahead with global climate modellers. In many cases, an ensemble of noncredible projections might still, in combination with process understanding, provide a plausible – and physically constrained – range future climates.

Second, small-scale processes may not, or not credibly, be represented. This might be the case if convective precipitation, especially at sub-daily scales, is relevant, but also when the model output is not fully representative of the local micro-climate: in complex coastal regions, along lakes and in valleys, where orography and local feedbacks may shape the response to climate change; and finally, if a specific variable is not, or only simulated in a very simplistic way, such as snow cover and depth in typical RCMs. Here one should not take model output at face value and consider alternative approaches: one possible solution might be a downscaling "in one's head": expert knowledge of the local climate might help one understand how the local climate may respond to the simulated large-scale climate. A possible modelling approach might be to dynamically simulate individual relevant events in a future climate at a very high resolution (see Chapter 19). Such solutions, of course would not produce time series that could be directly fed into an impact model. Thus, also impact modellers would have to consider how to make use of such abstract or event-based information.

Even in situations where no credible model simulations exist, one may still be able to provide useful information about a plausible range of changes. Such expert advice might refer to process understanding as well as to comparisons with warmer, wetter or drier climates of other regions. Note again that also a constrained range that spans both positive and negative changes is more valuable than no information.

18.5.3 Interpretation of Uncertainties

The relevance of different sources of uncertainty depends on the user context and affects the interpretation of climate projections in different ways. The usefulness of a certain adaptation measure may depend on the sources of uncertainty in a given context.

Forcing Uncertainty

Forcing uncertainty is intrinsically irreducible because of human reflexive uncertainty (see Section 9.1.1) and should always be considered separately. Which scenarios to consider depends on the user. If their interest is climate change mitigation, the sensitivity of the projected changes to forcing is relevant: will the 1.5°C target substantially reduce the impacts of climate change compared to stronger changes? What would be the cost of inaction? If their interest is adaptation, the choice of scenario depends on the own risk perception and the specific adaptation measure: on the one hand, adaptation to a worst-case scenario could be considered to be on the safe side. On the other hand, if possible, one could assume an optimistic scenario and design the adaptation measures in a way that it could be easily augmented in case changes should become worse than expected.

Model Imperfections versus Internal Variability

At first sight, it might be irrelevant for a user whether model spread stems from model imperfections or internal climate variability. Still, a separation might be very useful for a user. First, on centennial timescales, internal climate variability is essentially irreducible. Model uncertainty, however, could in principle be reduced by improved climate models. Knowing that model spread might (or might not) decrease within the next years of research might affect the adaptation strategy.

Second, the system behaviour will be different for spread related to internal variability than for spread caused by model uncertainty. In the latter case, that is if internal variability is low but models have large uncertainties, one will experience a rather smooth future change, without strong fluctuations on multidecadal scales. Even more, observed trends might provide a good idea of coming future changes. In the former case, however, one will face strong random fluctuations and experience the projected spread.

Past trends will not provide much information about future trends. Likewise, one has to expect decades or even longer periods in which the actual trends will be much stronger than projected or weaker or even reversed. One should therefore not consider the multimodel mean plus or minus some error margin. Instead, one should expect substantial fluctuations.

Note, also, that even though unpredictable internal variability fundamentally limits the skill of weather and seasonal forecasts, it does not limit the skill of climate projections. A user of a seasonal forecast requires a predicted distribution; the skill of the forecast should be better than the climatological forecast. A prediction that could not discriminate between a warmer- or colder-than-average summer would be useless. Such forecast skill can only be provided if the internal climate variability at the seasonal scale is predictable a season ahead.

Also a user of climate projections requires a future climate distribution. Here, skill arises not from predicting internal variability but from credibly representing the response to external forcing. If the outcome of a precipitation projection is plus or minus 10 percent because of internal variability, this information is valuable for the user: they would learn that the response to forcing is negligible, but large positive and negative fluctuations are to be expected. Adaptation to internal climate variability would be required. If, however, the same model spread would arise from model uncertainty, a user may – if possible – want to postpone an adaptation decision until a signal emerges and it becomes clear whether one has to prepare for drier or wetter conditions.

In any case, it is important to acknowledge the presence of uncertainties that are not sampled by the chosen model ensemble. It is likely that the real uncertainty range is broader than the spread derived from a model ensemble. This holds in particular for aspects different from mean climate and at the regional scale. Here, the climate models may actually simulate an uncertainty range that does not encompass the unknown truth.

18.6   Need for Interdisciplinary Co-Operation

In this chapter and previously in Chapter 5 we have discussed the importance of interdisciplinary collaboration. The production and interpretation of statistically downscaled climate projections require expert knowledge from

- the statistical downscalers themselves on the performance of the chosen downscaling method, but also from
- climatologists and meteorologists on the relevant climatic phenomena at all involved scales;
- GCM and RCM modellers on the inherent biases and the credibility of future projections in a given context;
- observational experts on the limitations of the reference data.
- As soon as the model output is interpreted locally, also experts familiar with the actual local climate are required.

Impact modellers and decision makers require this expert knowledge to adequately interpret climate model projections and subsequent impact analyses. In fact, users have to understand the limitations and uncertainties of a given regional climate projection in a given context at least to some extent in order to optimally use the information.

As importantly, downscalers are required to understand user needs to sensibly tailor climate model output for a given application and to provide credible and salient information for a given application. But for downscalers it is also important to recognise that also they are users[1] of information: they have to adequately employ observations and climate model output and may require guidance from local weather experts to adequately interpret model results for local applications.

Of course, both downscalers and users can be experts in a range of fields. Still, it is unlikely that they have sufficient expertise in all the relevant fields. Thus, cooperation is essential and should comprise co-design, co-production and co-exploration. The involvement of relevant experts increases credibility, salience and also legitimacy of the results.[2]

Such in-depth collaboration is time consuming. In practice it will only be possible within funded projects. In particular, developers of generic model output (e.g. via data portals; see Section 18.7) will only have limited time to communicate with all users of their product. Global climate modellers will have neither the capacity nor the interest to directly collaborate with a local administration far away. Thus, in particular the downscaling community – resting at the core of regional climate research – should ensure a sustained knowledge transfer between users, their own discipline and climatology, meteorology and climate modelling into their own field – by workshops, personal collaborations, or at least by reading the key publications.

18.6.1   Role of Climate Services

Boundary organisations play a key role in establishing and facilitating the dialog between downscalers (or regional climate modellers in general) and users. As discussed in the previous section, individual downscalers can directly collaborate with users within specific projects, but their capacity to interact with a wide range of stakeholders is very limited. Here, well-functioning climate services are in demand. One of their key roles could be to identify user needs and to guide users in selecting suitable generic products, appropriately using the model data, and interpreting the results (Sections 18.2 to 18.5). Climate service providers thus need to have working knowledge in climate and meteorology, climate modelling and downscaling – as a consequence, they need to be familiar with state-of-the-art research in these fields. Climate services are crucial in developing concepts for translating the complex information of ensemble projections and their uncertainties to users: as simple as possible, but as precise and detailed as necessary.

---

[1] In fact, everyone should consider themselves "next users" of other information (Hewitson 2016).

[2] Uncertainties are often brought forward as an argument why climate model output so far has had little tangible influence on adaptation decisions (Prudhomme et al. 2010). Another possible reason is a lack of collaboration between users and climate experts.

A suitable public infrastructure needs thus to be provided to meet these requirements and thereby to ensure legitimacy and authority of climate services.

18.7   Preparing Data for Downscaling Portals

Many internet portals make downscaled data available in user-friendly formats, typically based on bias correction or change-factor weather generators (Worldbank n.d., prepdata n.d., Jones et al. 2009*b*, CORDEX 2016). But as discussed in the previous section, statistical downscaling methods are structurally limited to work in specific situations, they typically do not provide broad generic output. Also, as demonstrated in Chapter 12, regional climate projections are uncertain, and statistical downscaling methods may introduce artefacts, in the worst case producing implausible climate change signals. Thus, developers of such products should carefully consider the issues discussed throughout this section. In particular, they should provide detailed user guidance on the limitations of the product that are up to the providers responsibility (see Section 18.9). These guideline documents should include the following information:

- examples of possible applications;
- a brief discussion of the associated uncertainties (including evidence for unsampled uncertainties);
- a summary of the chosen statistical downscaling or bias correction method, including a discussion of structural limitations;
- a list of known problems, such as weak performance in specific contexts;
- metadata, including links to published literature about the downscaling method, and its performance in evaluation and intercomparison studies.

These guidelines should continually be updated.

18.8   Climate Information Distillation

We have amply discussed uncertainties in climate projections and structural limitations of the statistical downscaling approaches and methods. But users are ultimately interested in the question of how climate is expected to change in a given region under a given emission scenario – they are interested in the robust information in all available sources of climate change data, such as ensembles of different generations of global and regional climate models and different statistical downscaling methods, but also observed trends and expert knowledge. The process of extracting this information has been coined climate information distillation, and caught considerable attention over recent years (WCRP WGRC 2014, Stocker et al. 2015). The guidelines discussed throughout this chapter can be thought of as contributing to climate information distillation. Three major questions should be addressed:

- How do different data products compare and potentially disagree?

- What are the causes of these differences?
- Is there evidence of deep unsampled uncertainties?

These questions are all crucial to establish credibility of a climate projection. Even though a wide range of sources should be considered to establish what is robust, these sources should of course be appropriate in the given context: a GCM, also not bias correction, should not be considered if summertime extreme precipitation is of interest. But one should not only rely on the most recent information. For instance, even though the latest EURO-CORDEX simulations may add some value to the older ENSEMBLES simulations (Prein et al. 2013), the latter are by no means proven useless and contain additional information (which is already evident from the fact that both ensembles contradict each other for some regions, variables and seasons; see Section 17.1.2). Distillation does not mean to ignore credible evidence of possible future changes.

18.9 Ethical Responsibility

If the aim of providing regional climate information is to inform real-world adaptation decisions, this provision has an inherently ethical dimension (Hewitson et al. 2014). Climate information is subject to substantial uncertainties, and as we have discussed, its credibility depends critically on the decisions made by the downscaler during its production. Issuing wrong information may prompt maladaptation, but withholding correct information may delay important adaptation decisions. To guide a responsible data provision, Adams et al. (2015) therefore propose an ethical framework for climate services, which partly applies to all providers of climate information. First, they define a list of key values: integrity, transparency, humility and collaboration. Following Adams et al. (2015), regional climate change projections should be credible and defensible; include detailed descriptions of uncertainty, be fit for purpose and; be documented.

These values and principles are key to establish and maintain trust in regional climate modelling. They provide a useful foundation for communicating with decision makers and the broader public. A responsible role is to serve as honest broker, that is to transparently lay out the range of credible evidence about future climate and its potential consequences (Pielke Jr. 2007). But – as laid out in this chapter – also users of climate information have the responsibility to understand the broader concept of global and regional climate change, in particular the relevance of the different sources of uncertainty for their decision context.

18.10 Further Reading

The IPCC Workshop on Regional Climate Projections and their Use in Impacts and Risk Analysis Studies published some tentative guidelines on the use of bias correction in Stocker et al. (2015). These guidelines are complemented by the discussion in Maraun et al. (2017*b*). Contributions calling for good scientific practice are, again, those by Barsugli et al. (2013) and Hewitson et al. (2014). De Elia (2014) discusses the peculiarities of climate modelling, how they may act as barriers when communicating with users and how these barriers may be overcome. A discussion of the ethical responsibility of climate information providers can be found in Adams et al. (2015). The communication between scientists and decision makers in different contexts and the different roles scientists could assume are discussed by Pielke Jr. (2007).

19 Outlook

> *Now here I am, a fool for sure!*
> *No wiser than I was before:*
> *Master, Doctor's what they call me,*
> *And I've been ten years, already,*
> *Crosswise, arcing, to and fro,*
> *Leading my students by the nose,*
> *And see that we can know – nothing!*
> (Faust I, 1808, Johann Wolfgang von Goethe)

In Part I of the book, Chapter 5, we discussed that climate science is post-normal,"where the stakes are high, uncertainties large and decisions urgent, and where values are embedded in the way science is done and spoken" (Hulme 2007). This arena has lead to a serious imbalance of fundamental science and application. Delivery of actionable products is requested from scientists and often promised by eager scientists, where in many cases not even credibility has been established.

In fact, a recurrent theme of this book is that projecting regional climate change is indeed, as stated by Hewitson et al. (2014), still a matter of fundamental research. An often-repeated commonplace in statistical downscaling research is that no single method is applicable to all user problems. This vague statement might leave a user frustrated, and browsing through this book, one might get a similar impression. But is it nothing that we have learned, as sometimes claimed? Do we have to give ourselves to magic art, as Faust did? We believe that in many cases robust answers may be given already. In those cases, where more research is needed, a precise list of research questions exists; answering them may bring us a leap forward. The main purpose of this chapter is to summarise these questions.

Initially, let us revisit the key messages from the book. Downscaling is not always meaningless; it may add substantial value, and it may be crucial to provide credible information about regional climate change. Downscaling is highly uncertain if local climate change is dominated by changes in the large-scale circulation. In fact, large-scale circulation errors of dynamical models cannot be overcome by any statistical post-processing and strongly limit the use of downscaling. But downscaling has the potential for added value if circulation errors are moderate, in particular in complex terrain. And we know quite precisely which type of downscaling approach and method is suitable in which situation. In fact, where local feedbacks or small-scale processes are important, dynamical downscaling might be required.

When selecting a statistical downscaling method for a specific application, two choices are important: first, choosing an appropriate model structure that is capable of reproducing the desired climatic aspects. And second, choosing predictors that adequately represent the relevant physical processes and at the same time are credibly simulated. This choice has a statistical aspect – the appropriate model structure. But the more relevant aspect in statistical downscaling research – the predictor choice – is physical.

In the following sections, we will present an outlook of open research questions. In Section 19.1, we will discuss research that is required to complement ongoing statistical downscaling activities. Beyond that, progress in dynamical and statistical modelling will help overcome some downscaling limitations. We will discuss such developments in Sections 19.2 and 19.3. In some cases, however, one may have to explore alternative approaches to further our understanding of regional climate change. We will sketch such approaches in Section 19.4. We will end with some concluding remarks.

19.1   Research to Complement Ongoing Downscaling Applications

In Chapter 18, we discussed in detail how one may best – against the background sketched earlier – proceed to provide defensible information for practical applications. Such practical downscaling applications rely on several assumptions we have presented in Chapters 11 and 12. For PP, these are the perfect prognosis and informativeness assumption, for MOS the – similar – credibility and representativeness assumption. Both approaches require model structures that are valid in a future climate. Yet little systematic research has been conducted to assess to what extent these assumptions are fulfilled. Many evaluation studies are based on perfect predictors – these are required to assess the informativeness, representativeness and structure assumption for present-day climate. Yet such studies are by design unable to assess the perfect prognosis and credibility assumption. Also most evaluation studies focused on long-term average statistics – only a few studies exist that address the reproduction of observed regional climate change. Finally, almost nothing is known about the likely performance of statistical downscaling methods in future climate. Here a range of studies is urgently required.

Assessing the PP Assumption
For PP, which predictors fulfil the PP assumption, that is which predictors are realistically and bias free simulated by GCMs? Can some smoothly varying variables such as geopotential heights or temperature be taken confidently from a GCM at the grid-box scale? Under which conditions may surface variables be sensible predictors? Do GCM-simulated predictors describing atmospheric stability such as CAPE fulfil the PP assumption? Issues regarding the simulation of future changes will be discussed below.

Assessing the Informativeness Assumption
For PP, which predictors fulfil the informativeness assumption, that is which predictors carry information about variability at all required scales including climate change? Most

assessments are based on daily correlation analyses between predictors and predictands, yet we have discussed amply that such analyses are essentially useless to learn anything about longer timescales: many predictors may be correlated with the predictand on short timescales, but may not carry information about long-term changes. Here evaluation studies of long-term variability along with a precise physics-based interpretation are required. These should address questions such as: which processes are physically relevant for the phenomenon of interest (e.g. one should clearly distinguish between precipitation occurrence and precipitation amount)? Which predictors are physically sensible proxies of these processes (and, importantly, at the same time fulfil the PP assumption)? An important question is whether local feedbacks are involved that are not captured by large-scale predictors. Systematic inter-comparisons of different predictor variables, transformations and domain sizes at long timescales should be carried out. Such analyses should involve the assessment of long-term trends. For precipitation, the signal-to-noise ratio will be low in the historical record such that these analyses should be extended to pseudo-reality studies. A non-trivial question is how predictor–predictand relationships for long timescales can be constrained based on short observational records.

Assessing the Credibility Assumption

For MOS and change-factor weather generators, which predictors fulfil the credibility assumption? Here, we believe, a paradigm shift is required: a common belief in the community is that bias correction is the more sensible if there are larger biases in the driving model. The discussions throughout the book should have illustrated that the opposite is true: for a bias correction to make sense, biases have to be sufficiently small. Biases are an expression of model errors. The larger the biases, the more fundamental the errors, and the lower the credibility of any future projection (see also discussion in Section 19.2).

Assessing the Representativeness Assumption

The representativeness assumption of MOS has received little attention so far. In fact, it has sometimes been argued that one should directly bias correct GCMs to avoid costly dynamical downscaling. In many cases, however, GCM grid box values are not representative of local climate, and RCMs may crucially add value. But in some cases, where small-scale orography or local feedbacks are important, or where local extreme events are of interest, even operational RCMs at horizontal resolutions of some 10km may not credibly represent local climate and climate change. Here, studies of RCM added value and remaining limitations for bias correction are required.

Assessing the Model Structure

For both approaches, it should be assessed whether a given model structure allows for sensible extrapolations. In a 4-degrees-warmer world, predictor distributions will have shifted considerably. For some variables they may have little overlap with present distributions, in particular in the extreme tails. Statistical downscaling methods are rather simple statistical models that are in general not designed for extrapolation. To what extent such an extrapolation is sensible needs to be established by, again, studying the suitability to capture long-term changes either in long historical records or in pseudo-realities.

Assessing GCM and RCM Trends

As mentioned, both PP and MOS – in fact any downscaling – rely on predictors that are credibly simulated in a future climate. For PP downscaling, this concerns mainly the representation of large-scale climate change in GCMs. For MOS, this also concerns regional climate change in GCMs or RCMs. Further studies are required that assess the ability of climate models to reproduce observed trends: regional climate sensitivity of RCMs can be tested with perfect boundary conditions; for free-running climate models, only the compatibility of observed trends with the range of trends simulated by a model ensemble can be assessed. Causes of discrepancies should be analysed and attributed to the misrepresentation of relevant processes or internal variability (or potentially data inhomogeneities). The representation of future climate can, of course, not be validated. Still, credibility may be established: by understanding the mechanisms responsible for regional climate change by tracing uncertainties in regional climate projections to uncertainties in these mechanisms and by linking the climate change signal to present climate biases. Additionally, in particular for extreme events and local feedbacks, we need to understand deep uncertainties: where do we miss or misrepresent crucial processes such that our simulated model spread does not represent model uncertainties?

Modification of Trends by Quantile Mapping

Bias correction modifies the climate change signal as soon as it adjusts the variance of a distribution. In such a situation, it is important to understand the context of the modification: if short-term variability is misrepresented but otherwise the climate change signal is credible, the trend of the driving model should be preserved. In all other situations, it should be carefully assessed by referring to process understanding whether the trend modifications are sensible.

Designing Statistical Model Ensembles

There is a call for ensembles of statistical downscaling methods or even ensembles combining GCMs, RCMs and a range of statistical methods. Research is required about how such ensembles should be designed. There are at least two issues involved. The first is representativeness (and added value): different models in the downscaling chain obviously represent climate at different spatial and temporal scales. In some situations, for example for temperature in flat terrain, a comparison – and combination – may be sensible. But in particular for highly variable processes and extreme events, these simulations may have substantially different characteristics and may be incomparable. Here, one may have to dismiss the GCM and fully rely on the RCMs or the statistical downscaling. The second issue is related to the ontological difference between dynamical and statistical models. Dynamical models are based on first principles. All GCMs (or RCMs) produce qualitatively similar generic output. And even though these models involve heuristic sub-grid parameterisations, it will be difficult to exclude individual models from an ensemble based on a priori considerations. Statistical models, however, have a structure that is designed to represent a limited set of climatic aspects. Thus, a statistical downscaling model cannot be used generically. A deterministic regression model will not realistically represent day-to-day variability, and a single-site model will not be applicable in a spatial context. Thus, in a given context, tailored ensembles have to be designed that rest on our understanding of the model structure as well as on an evaluation of actual skill.

19.2   Progress in Dynamical Modelling

Will downscaling, and in particular statistical downscaling, become obsolete with the ever-increasing resolution of global climate models? A key factor limiting the quality of regional climate projections is errors in the driving models. Errors in the large-scale atmospheric circulation cause uncorrectable biases in surface climate and may induce a qualitatively wrong global warming response. Errors in parameterised processes and local feedbacks may result in implausible local climate change trends. Statistical downscaling will inherit these errors. Thus, reducing climate model errors is crucial in improving the quality of regional climate projections.

In fact, improving climate models is one of the unifying themes of the World Climate Research Programme. Recent contributions have highlighted the need to improve the representation of atmosphere and ocean processes rather than including new model components (Stevens and Bony 2013, Jakob 2014). Improvements are in particular required in the representation of organised convection in the tropics and of the link between clouds, circulation and climate sensitivity (Williams et al. 2013, Bony et al. 2015); the representation of mid-latitude Rossby waves and circulation regimes (Dawson et al. 2012, Palmer 2013, Shepherd 2014); and in the representation of ocean boundary currents and eastern tropical oceans (Richter 2015, Richter et al. 2016, Zuidema et al. 2016).

Key to improving these processes are increased model resolution and improved parameterisations. The Athena project has demonstrated that biases of the ECMWF atmosphere only GCM in simulating mid-latitude climate decrease substantially at a horizontal resolution of approximately 40km (T511; Jung et al. 2012). A horizontal resolution of 16km is required to realistically represent the spatial and temporal patterns of mid-latitude weather regimes such as the North Atlantic Oscillation or blocking (Dawson et al. 2012, Davini et al. 2017). Global cloud-system resolving simulations (Satoh et al. 2008) show promising results in simulating tropical climate variability such as the Madden Julian Oscillation (Liu et al. 2009).

Still, these simulations are cutting edge, and they are limited to atmosphere-only GCMs. Doubling climate model resolution requires about 16 times higher computational resources. Thus, in the foreseeable future, there will be no global transient ensemble projections available that explicitly resolve convection. Coupled simulations at an atmospheric resolution of T511, however, may be realistic in the coming decade. These climate models will not resolve all impact-relevant local-scale climate phenomena, but their improved mean state will remove many of the persistent circulation biases that often limit the quality of downscaling.

Some regional processes will not be well resolved by standard RCMs. Indeed, convection-permitting simulations have been shown to improve the representation of regional climate (Prein et al. 2015) and may prove crucial in simulating a credible response of precipitation extremes to global warming (Kendon et al. 2014, Meredith et al. 2015a, Prein et al. 2017). Recently, a CORDEX initiative has been launched to create an ensemble of convection-permitting simulations for better understanding the response of convection to climate change. But such simulations will be limited to a few case-study domains, and they will still be biased.

Thus, downscaling will still be required and will even be more defensible. Moreover, there will be scope for statistical models to bridge the gap between GCMs or standard RCMs and impact-relevant scales.

19.3   Progress in Statistical Methods

In Section 19.1, we laid out that many of the problems in statistical downscaling concern process understanding rather than statistics. Still, also improvements in statistical methodology are required. A major improvement in PP methods may be achieved by moving towards a combination with weather generators. On the one hand, simple deterministic regression models are not suitable for daily precipitation and are also limited to describing temperature variability. On the other hand, also the limitations of unconditional weather generators in capturing long-term variability are evident. For precipitation, stochastic generalised linear models without explicit temporal dependence already show promising results, but for temperature the development of conditional weather generators will be required.

A major limitation of MOS approaches is their inability to produce sub-grid variability when used for downscaling. Here we expect a separation of bias correction and downscaling to substantially improve performance: bias correction could be carried out using sensible variants of quantile mapping, and the subsequent downscaling step should be carried out with a stochastic model. In climate change research, current bias correction methods do not include any physical information. Since biases are weather dependent, further research should explore the possibilities to exploit this dependence. Also, subsequent stochastic downscaling may be improved by incorporating information on the atmospheric circulation.

The weakest point of statistical downscaling is the simulation of realistic spatial dependence. The analog method, of course, simulates spatially coherent fields but may by construction fail under strong climate change. Multisite weather generators of the Richardson type or based on non-homogeneous hidden Markov models have shown potential but assume rather simplistic spatial models compared to reality. Bias correction inherits spatial dependence above the grid scale but cannot produce sub-grid dependence. It will thus fail to represent sub-grid temperature fields in complex terrain, and sub-grid precipitation fields in general. Here, copula-based approaches may proof useful. Standard copulas will fail to represent the complex dependence in particular in complex terrain. But so-called pair copula constructions may provide a powerful framework to flexibly simulate dependence of a large number of sites. Also semi-parametric kernel-based copulas may be explored.

In any case, the previous discussion highlights a key issue: improvements in current methodology are possible and often possible at only marginal expense. In fact, for single-site stochastic models, the whole methodology is available, and only incremental improvements are required. In principle this holds also for spatial models. The problem, however, is that calibrating spatial statistical models over large domains will be computationally expensive. Still, it is unclear whether a statistical model describing, for example, spatial precipitation fields will be valid in a future climate. Thus, if the target is a continental domain, one may end up with models which come at similar costs as RCMs but without a clear benefit. We believe, thus, that a healthy balance between statistical and dynamical modelling has to be found. The strength of spatial statistical approaches will be in particular in providing tailored solutions for relatively small regions.

19.4   Thinking Out of the Box

We have laid out potential improvements of dynamical models over the next decade. Still, over the coming years, GCMs will suffer from large-scale circulation biases. And even if convection-permitting RCMs may simulate climate change realistically at their resolution, they will be available only for some regions, and they may still fail to simulate local climate change in complex terrain such as narrow Alpine valleys. Thus, the representation of convective precipitation, or elevation-dependent warming will remain a problem. Here, unorthodox approaches are in demand.

Statistical emulators (Walton et al. 2015) may be a promising approach. A limited number of very-high-resolution RCM simulations may be conducted and used to calibrate a statistical emulator, that captures the local response to climate change. This emulator may then be driven with coarser-resolution climate models.

In cases, where large-scale circulation bias is a major issue, surrogate global warming studies (Schär et al. 1996) may be considered in which the large-scale circulation is kept fixed to observed present climate conditions and only the thermodynamic response to climate change is simulated and local dynamics may adjust. Alternatively, atmosphere-only simulations (Section 8.1.1) may be used: sea-surface temperatures from standard coupled atmosphere–ocean GCM projections may be taken as boundary conditions for very-high-resolution atmosphere-only simulations (Hazeleger et al. 2015). Where additionally spatial SST fields are very relevant, a variant of surrogate global warming simulations may be a solution: observed SST fields might be modified by long-term changes derived from standard coupled atmosphere–ocean GCMs and then serve as boundary conditions for high-resolution atmosphere-only GCMs. Of course, these approaches come with a number of limitations and are poor man's solutions rather than an elegant and consistent framework. In particular, all these approaches make strong assumptions about circulation changes: surrogate global warming studies assume that the circulation does not change. Atmosphere-only simulations ignore the influence of errors in the ocean circulation on the atmosphere. And surrogate ocean warming studies assume that the ocean circulation will not change. Still, these approaches might be more defensible than using standard coupled GCM simulations.

If a very high resolution is required, for example to represent local fields of extreme precipitation in alpine terrain, storylines of individual events may provide a way forward (Lackmann 2015, Meredith et al. 2015*b*). Instead of simulating long time slices, one could simulate selected, impact-relevant events over a small domain. The boundary conditions may be taken directly from GCMs or based on observed fields, modified by the climate change signal from a GCM ensemble. The latter would again be a surrogate global warming approach. The short simulation lengths and small domains will allow for much higher resolutions. This approach is also ideal for stress testing, for example by using changes stronger than those simulated by the GCM ensemble to test the robustness of a system against unknown unknowns.

19.5   Concluding Remarks

So how will downscaling look in the foreseeable future? A defensible and feasible default approach for obtaining regional climate change scenarios may be to downscale high-resolution GCMs by standard CORDEX RCMs, followed by a bias correction with improved statistical techniques. If large ensembles of GCMs are to be downscaled, if no RCM is available for the specific region, or if RCMs simulate noncredible or unrepresentative regional output, PP statistical downscaling will be a useful alternative, in particular as conditional weather generators. Unconditional weather generators will be useful to explore the sensitivity of impact models to broad ranges of climate change. These ranges may well exceed simulated climate change uncertainties to test the robustness of a given system.

The strength of statistical downscaling will be over small domains at small scales, which may not be represented by RCMs. In some cases, thinking out of the box may be required to combine different approaches in an unorthodox way. All these cases require an open mind in choosing a suitable approach and method. Such open-mindedness will automatically overcome the superficial antagonism between statistical and dynamical downscaling. In fact, both approaches can also be used to mutually evaluate each other. RCMs provide a reference to test statistical downscaling approaches in a pseudo-reality, and PP statistical downscaling is a powerful diagnostic tool to evaluate the simulated relationship between large- and local-scale climate. At any rate, producing tailored solutions of high quality will involve process understanding: of the controls by large-scale processes, of local factors influencing climate and its response to global warming.

We have amply discussed real and perceived weaknesses of dynamical models and statistical post-processing. In an ideal world, we would first wait until large-scale circulation biases in dynamical models have substantially been reduced and better statistical downscaling approaches have been developed. Still, many adaptation decisions will have to be drawn soon. Completely ignoring the information from imperfect models is not a sensible option, because the alternative of making ad-hoc decision without a scientific basis or just reacting to events that have already occurred is clearly worse, and counting merely on low-regret adaptation measures may defer important decisions. Thus, there is an urgent need to distil the credible information from all available sources of knowledge. We hope this book has made a contribution to synthesise the knowledge in statistical downscaling, to inspire discussions and to develop best practice in downscaling in a real world.

Appendix A   Methods Used in This Book

Several plots in this book are based on evaluation results from the VALUE perfect predictor experiment. The following method descriptions are extracted from the compilation in Gutiérrez et al. (2017).

CFE

Bias correction at station Clausthal-Zellerfeld-Erbprinzentanne. Scaling: simple rescaling of daily intensities. Non-parametric quantile mapping: linear interpolation between neighbouring empirical quantiles. The parametric quantile mapping is based on a two-parameter gamma distribution.

RaiRat-M6/M7/9

Deterministic MOS method. Temperature bias correction methods used in Räisänen and Räty (2013). M6 additively corrects means, M7 additionally rescales the standard deviation. M9 is a semi-empirical quantile mapping, where the empirical transfer function is smoothed with a running mean. The transfer functions are calibrated for each calender month, using a two-month window centred on the month of interest.

Ratyetal-M6-M8

Deterministic MOS method. Precipitation bias correction methods used in Räty et al. (2014). M6 adjusts daily precipitation values by rescaling mean precipitation and separately rescaling anomalies about the mean. M7 adjusts daily precipitation by a power-law scaling. M9 is a parametric quantile mapping based on two different gamma distributions, fitted separately below and above the 95th percentile of daily precipitation on wet days. A 0.1mm threshold was used to define wet days. The transfer functions are calibrated for each calender month, using a three-month time window centred on the month of interest.

ANALOG

PP method. Standard analog technique using Euclidean distance considering the complete fields to compute distances (Gutiérrez et al. 2013, San-Martín et al. 2017). Candidate predictors are sea-level pressure, 2m temperature, temperature at 500hPa, 700hPa and 850hPa, specific humidity at 500hPa and 850hPa, and 500hPa geopotential height. The method has been trained across different zones covering Europe (similar to the Prudence regions) and has no seasonal component.

MLR-AAN/AAI/AAW/RSN/ASW/ASI

PP method. Pointwise multiple linear regression for temperature using gridpoint raw data (or standardised anomalies) as predictors (Huth 2002, Huth et al. 2015). The first letter of the code refers to the raw (R) or anomaly (A) data used as predictors, the second letter refers to the annual (A) or seasonal (S) training, and the third letter refers to infla- tion (I) or white noise (W) variance correction (N for no correction). For comparison, the method has also been applied to precipitation in VALUE. Predictors are sea-level pressure and temperature at 850hPa.

MLR-T/GLM-P

Semi-deterministic PP method. Adapted from Wilby et al. (2002). Temperature is mod- elled as a deterministic regression. Precipitation is modelled in a two-step approach. Occurrence is modelled stochastically via a logistic regression; precipitation amounts given a wet day are simulated based on the variance inflated, predicted mean of an expo- nential regression. The selection of predictors changes from one site to another and from one variable to another and is based on a step-wise approach building on the adjusted determination coefficient. Candidate predictors are sea-level pressure, 2m temperature, 10m horizontal wind components, temperature, specific humidity and horizontal wind components at 500hPa, 700hPa and 850hPa. The method is not trained on individual seasons.

GLM

Stochastic PP method. Standard two-stage implementation of a generalized linear model (GLM) for precipitation (San-Martín et al. 2017). Precipitation occurrence is simulated by a logistic regression (threshold of 0.1mm), and a GLM with gamma error distribution and log canonical link-function is applied to downscale daily precipitation amounts. The method is trained across different zones covering Europe (similar to the Prudence regions) with no seasonal component. The predictors are the 20 leading PCs of the joined predictor fields. Candidate predictors are sea-level pressure, 2m temperature, temperature at 500hPa, 700hPa and 850hPa, specific humidity at 500hPa and 850hPa and geopotential height at 500hPa. The method is not trained on individual seasons.

WT-WG

Stochastic PP method. Gaussian/binomial-gamma distributions are fitted to the observed temperature/precipitation values within each of the 100 weather types obtained by applying k-means to sea-level pressure fields. These distributions are obtained to sim- ulate downscaled values. More details in Gutiérrez et al. (2013) and San-Martín et al. (2017).

SS-WG

Multivariable (sic!) Richardson-type weather generator simulating daily time-series of precipitation, minimum and maximum temperature (Keller et al. 2015, 2016). First, daily precipitation occurrence is modelled based on a first-order two-state Markov chain using 1mm/day as a wet threshold. Precipitation intensities are simulated from a mixture model of two exponential distributions. To ensure inter-variable consistency, the parameters of the temperature statistics are conditioned on the precipitation state. Synthetic temperature time-series are simulated using a first-order autoregressive model (AR1). All WG parameters are determined for each station and each month separately.

MARFI-BASIC and MARFI-M3

Multivariate stochastic Richardson-type weather generator building upon earlier variants (Dubrovsky 1997, Dubrovský et al. 2004). Precipitation occurrence is modelled by Markov chain and precipitation amounts on wet days are sampled from a gamma distribution. Standardised values of the temperature variables are modelled by a first-order bi-variate auto-regressive model, in which the means and standard deviations of the two variables are conditioned on the state (wet or dry) of the day. In BASIC, the two temperature variables are TMAX and TMIN and the order of the Markov chain is one. In M3 a third-order Markov chain is used to model wet day occurrence.

Appendix B   Useful Resources

A dynamic version of these pages will be available at www.cambridge.org/climate research. The authors intend to regularly update these resources for the coming years.

B.1   Statistical Downscaling Software Packages and Portals

Here we present (in alphabetical order) a selection of open-source software packages and online portals to perform statistical downscaling. This list is not comprehensive. We may have missed useful and important resources. Moreover, the methods listed here are affected by the limitations discussed throughout the book and should thus be selected and applied carefully in a given context.

downscaleR
R package by the Santander Meteorology group containing functions for MOS such as scaling, and different quantile mapping versions, and PP methods such as linear regression, generalised linear models, weather-type-based downscaling and the analog method.

https://github.com/SantanderMetGroup/downscaleR/

ENSEMBLES Downscaling Portal
Online portal of the Santander Meteorology Group to carry out downscaling. The webpage provides different sets of predictor and predictand data (with upload options) and a range of MOS and PP methods.

https://www.meteo.unican.es/downscaling/ensembles

ESD
R package by Rasmus Benestad, Abdelkader Mezghani and Kajsa Parding. The package contains a range of functions to post-process and analyse large data sets (e.g. in NetCDF format) including PP statistical downscaling based on linear regression and temporal disaggregation.

https://github.com/metno/esd

Rglimclim

R package for a conditional multisite, multivariate weather generator based on generalised linear models developed by Richard Chandler.

http://www.ucl.ac.uk/~ucakarc/work/glimclim.html

MeteoLab

Matlab toolbox of the Santander Meteorology group for statistical analysis and data mining in meteorology, focusing on statistical downscaling methods.

https://meteo.unican.es/trac/MLToolbox

SDSM

Online statistical downscaling tool by Rob Wilby for perfect prognosis and change-factor weather generators.

http://co-public.lboro.ac.uk/cocwd/SDSM/

qmap

R package by Lukas Gudmundsson for quantile mapping.

https://cran.r-project.org/web/packages/qmap/

B.2   Programmes and Initiatives

CMIP

The Coupled Model Intercomparison Project (CMIP) defined a framework to conduct and intercompare simulations with coupled atmosphere–ocean GCMs. The most recent phase is CMIP5 (Taylor et al. 2012).

http://cmip-pcmdi.llnl.gov/

CORDEX

Initiative of the World Climate Research Programme (WCRP) to advance and coordinate the science and application of climate downscaling through global partnerships, in particular to generate large high-resolution multimodel ensembles for all regions of the Earth.

http://www.cordex.org

Climate Data Guide

Portal of the National Center for Atmospheric Research (NCAR) and the University of Colorado, Boulder. The Climate Data Guide provides concise information on the strengths and limitations of the key observational data sets, tools and methods used to evaluate Earth system models and to understand the climate system.

https://climatedataguide.ucar.edu/

Climate Explorer
An online portal of the dutch KNMI to analyse and plot a wide range of observational and simulated climate data. Several of the figures in this book have been created with the Climate Explorer.

https://climexp.knmi.nl

ENSEMBLES
A major European downscaling project funded by the European Union. A major output was a large ensemble projection based on dynamically downscaled GCMs.

http://ensembles-eu.metoffice.com/ (Project webpage)
http://ensemblesrt3.dmi.dk/ (RCM data portal)

ESGF
The Earth System Grid Federation is an open-source effort providing a robust, distributed data and computation platform, enabling worldwide access to Peta/Exa-scale scientific data. ESGF hosts, for example, CMIP and CORDEX data.

https://esgf.llnl.gov/

Future Earth
Major international research platform providing the knowledge and support to accelerate transformations to a sustainable world.

http://www.futureearth.org/

Global Framework for Climate Services
The WMO-led Global Framework for Climate Services (GFCS) guides the development of science-based climate information and services to support decision-making in climate-sensitive sectors.

http://www.gfcs-climate.org/

NARCCAP
The North American Regional Climate Change Assessment Program (NARCCAP) is an international program to produce high resolution climate change simulations.

http://narccap.ucar.edu/

VALUE
European network that conducted the most comprehensive evaluation of statistical downscaling and bias correction methods.

www.value-cost.eu

World Climate Research Programme
Programme founded by the World Meteorological Organisation (WMO) and the International Council of Scientific Unions (ICSU) to facilitate the analysis and prediction of

Earth system variability and change for use in an increasing range of practical applications of direct relevance, benefit and value to society.

https://www.wcrp-climate.org/

Working Group on Regional Climate
Working group of the WCRP to coordinate regional climate research within WCRP and serve as the conduit for two-way information exchange between WCRP and the various institutions and coordinating bodies that provide climate services.

https://www.wcrp-climate.org/regional-climate

B.3   Observational Data Sets

Next we present a selection of large-domain gridded observational data sets with daily resolution. Again, this list is not comprehensive, and important resources may be missing. National weather services often have access to much denser station networks and may therefore provide gridded data sets of higher quality. A comprehensive overview of data products (including reanalyses) with detailed information can be found on https://climatedataguide.ucar.edu/climate-data. In many situations, statistical downscaling is applied to station data. These are often distributed via national weather services and not listed here. A valuable exception is the European Climate Assessment and Dataset (ECA&D; Klein Tank et al. 2002) http://www.ecad.eu/.

B.3.1   Global

CHIRPS

name	Climate Hazards Group InfraRed Precipitation with Station data (CHIRPS)
project/institution	University of California, Santa Barbara
reference	Funk et al. (2015)
region	global
period	1981–present
variables	precipitation
data base	infrared global cold cloud duration data, calibrated with multisatellite precipitation analysis and bias corrected against rain gauges
nominal resolution	0.05°
homogenisation	not known
link	http://chg.ucsb.edu/data/chirps/

CMORPH

name	Climate Prediction Center MORPHing technique
project/institution	National Oceanic and Atmospheric Administration (NOAA), Climate Prediction Cente (CPC)
reference	Joyce et al. (2004)
region	global
period	1998–2014
variables	precipitation
data base	passive microwave satellite scans and satellite infrared imagery
nominal resolution	0.07277° (8 km at equator)
homogenisation	not known
link	https://www.dwd.de/EN/ourservices/gpcc/gpcc.html

CPC-UNI

name	CPC Unified Gauge-Based Analysis of Global Daily Precipitation (CPC-UNI)
project/institution	National Oceanic and Atmospheric Administration (NOAA), Climate Prediction Cente (CPC)
reference	Xie et al. (2010)
region	global
period	1979–present
variables	precipitation
data base	gauge data from more than 30,000 stations from multiple sources (GTS, COOP, other national and international agencies)
nominal resolution	0.5°
station density	approx. 0.2 stations per grid box, with low density over Africa and Antarctica (1979-2005, approx. 0.12 since 2006)
homogenisation	not homogenised
link	https://climatedataguide.ucar.edu/climate-data/cpc-unified-gauge-based-analysis-global-daily-precipitation

GPCC

name	GPCC Full Data Daily
project/institution	DWD, Global Precipitation Climatology Centre
reference	Schamm et al. (2015)

region	global
period	1988–2013
variables	precipitation
data base	approx. 27,000 rain gauges provided by national meteorological and hydrological services, global and regional data collections as well as WMO GTS-data
nominal resolution	1°
station density	1.8 stations per grid box
homogenisation	not homogenised

link	https://www.dwd.de/EN/ourservices/gpcc/gpcc.html

GPCP

name	Global Precipitation Climatology Project (GPCP)
project/institution	Global Precipitation Climatology Project, GEWEX, WCRP
reference	Adler et al. (2003)

region	global
period	1997–present
variables	precipitation
data base	low-orbit satellite microwave data, geosynchronous-orbit satellite infrared data, and GPCC rain gauge observations
nominal resolution	1°
homogenisation	not known

link	https://rda.ucar.edu/datasets/ds728.3/

WATCH

name	WATCH Forcing Data 20th Century
project/institution	Water and Global Change (WATCH)
reference	Weedon et al. (2011)

region	global
period	1901–2001
variables	temperature, precipitation, short-/longwave downward radiation, wind, surface pressure, specific humidity
data base	ERA-40 reanalysis bias corrected against monthly gridded observations; before 1958 years from 1958–2001 have been resampled
nominal resolution	0.5° (effectively the 1° resolution of ERA-40)
homogenisation	as ERA-40, i.e. inhomogeneous mainly due to the varying data sources

link	http://www.eu-watch.org/data_availability

name	WATCH Forcing Data ERA-Interim
project/institution	Water and Global Change (WATCH)
reference	Weedon et al. (2014)

region	global
period	1979–2014
variables	temperature, precipitation, short-/longwave downward radiation, wind, surface pressure, specific humidity
data base	ERA-Interim reanalysis bias corrected against monthly gridded observations, available at 3h temporal resolution
nominal resolution	0.5° (effectively the 0.75° resolution of ERA-Interim)
homogenisation	as ERA-Interim

link	http://www.eu-watch.org/data_availability

B.3.2   Africa

ARC2

name	African Rainfall Climatology 2 (ARC2)
project/institution	National Oceanic and Atmospheric Administration (NOAA), Climate Prediction Cente (CPC)
reference	Novella and Thiaw (2013)

region	Africa
period	1983–present
variables	precipitation
data base	geostationary infrared (IR) data (EUMETSAT) and Global Telecommunication System (GTS) gauge observations
nominal resolution	0.1°
homogenisation	not known

link	ftp://ftp.cpc.ncep.noaa.gov/fews/fewsdata/africa/arc2/

B.3.3   Americas

Livneh et al.

name	Daily Obserational Hydrometeorology data set: North American Extent
project/institution	University of Colorado, Boulder, USA
reference	Livneh et al. (2015)

region	USA, southern Canada, Mexico
period	1950–2013
variables	temperature (minimum, maximum), precipitation, 10-m wind speed
data base	US National Climatic Data Centre, Environment Canada, Mexican Meteorological Service and regional offices (temperature, precipitation) NCEP/NCAR reanalysis (temperature, wind speed)
nominal resolution	1/16°
station density	around 0.1 stations per grid box
homogenisation	not known

link	http://www.colorado.edu/lab/livneh/data

CLARIS

name	CLARIS La Plata Basin
project/institution	CLARIS (EU FP7 project), University of Buenos Aires, Argentina
reference	Penalba et al. (2014), Boulanger et al. (2010)

region	La Plata Basin
period	1961–2000
variables	temperature (minimum, maximum), precipitation
data base	station data from more than 60 institutions
nominal resolution	0.5°
station density	between aprox. 0.2 and more than 0.8 stations per grid box, varying strongly in time
homogenisation	not homogenised

link	http://www.cima.fcen.uba.ar/ClarisLPB/

B.3.4   Asia

APHRODITE

name	APHRODITE
project/institution	Research Institute for Humanity and Nature
reference	Yatagai et al. (2012)

region	Asia
period	1950–2007
variables	precipitation
data base	purely based on station data
nominal resolution	0.25°
station density	approx. 0.17 stations per grid box, but strongly varying from region to region.
homogenisation	not homogenised

link	http://www.chikyu.ac.jp/precip/english/products.html

B.3.5   Australia

AWAP

name	Australian Water Availability Project (AWAP) Data
project/institution	Australian Bureau of Meteorology
reference	Jones et al. (2009*a*)

region	Australia
period	1900 (precipitation); 1910 (temperature); 1971 (humidity) - present
variables	temperature (minimum, maximum), precipitation, humidity
data base	Australian Data Archive for Meteorology (ADAM), precipitation: 3000 ($\sim$ 1900) to 7000 ($\sim$ 1970) gauges; temperature: > 600 stations (since 1960s); humidity: 300 ($\sim$ 1970) - 800 ($\sim$ 2010) stations.
nominal resolution	0.05°
station density	around 0.02 rain gauges per grid box (since 1970s)
homogenisation	available for temperature (0.25°, 109 stations)

link	http://www.bom.gov.au/climate/maps/ and http://www.bom.gov.au/climate/data-services/ moderate data charges apply.

B.3.6   Europe

E-OBS

name	E-OBS
project/institution	ENSEMBLES/KNMI
reference	Haylock et al. (2008)

region	Europe, Turkey, Mediterranean Coast
period	1950 – present
variables	temperature (min, max, mean), precipitation, sea level pressure
data base	purely based on station data
resolution	0.25°
station density	approx. 0.14 stations per grid box, varying across region (original version, has much improved since 2008).
homogenisation	not homogenised

link	http://www.ecad.eu/download/ensembles/ensembles.php

References

Adams, P., Eitland, E., Hewitson, B., Vaughan, C., Wilby, R. and Zebiak, S. (2015), 'Toward an ethical framework for climate services. A white paper of the climate services partnership working group on climate services ethics'.

Addor, N., Rohrer, M., Furrer, R. and Seibert, J. (2016), 'Propagation of biases in climate models from the synoptic to the regional scale: Implications for bias adjustment', *J. Geophys. Res.* **121**(5), 2075–2089.

Adler, R. F., Huffman, G. J., Chang, A., Ferraro, R., Xie, P.-P., Janowiak, J., Rudolf, B., Schneider, U., Curtis, S., Bolvin, D., Gruber, A., Suesskind, J., Arkin, P. and Nelkin, E. (2003), 'The version-2 global precipitation climatology project (GPCP) monthly precipitation analysis (1979–present)', *J. Hydrometeorol.* **4**(6), 1147–1167.

Aguilar, E., Auer, I., Brunet, M., Peterson, T. C. and Wieringa, J. (2003), 'Guidelines on climate metadata and homogenization'. WMO-TD No. 1186. World Meteorological Organization, Geneva, Switzerland, 2003.

Akaike, H. (1973), "Information theory and an extension of the maximum likelihood principle", *in* B. N. Petrov and F. Csaki, eds., *2nd International Symposium on Information Theory*, Budapest, pp. 267–281.

Allcroft, D. J. and Glasbey, C. A. (2003), 'A latent Gaussian Markov random-field model for spatiotemporal rainfall disaggregation', *J. Roy. Stat. Soc. C.* **52**(4), 487–498.

Anagnostopoulos, G. G., Koutsoyiannis, D., Christofides, A., Efstratiadis, A. and Mamassis, N. (2010), 'A comparison of local and aggregated climate model outputs with observed data', *Hydrol. Sci. J.* **55**(7), 1094–1110.

Arakawa, A. and Schubert, W. H. (1974), 'Interaction of a cumulus cloud ensemble with the large-scale environment, part i', *J. Atmos. Sci.* **31**(3), 674–701.

Auer, I., Böhm, R., Jurković, A., Orlik, A., Potzmann, R., Schöner, W., Ungersböck, M., Brunetti, M., Nanni, T., Maugeri, M., Briffa, K., Jones, P., Efthymiadis, D., Mestre, O., Moisselin, J.-M., Begert, M., Brazdil, R., Bochnicek, O., Cegnar, T., Gajic-Capka, M., Zaninovic, K., Mahstorovic, Z., Szalai, S., Szentimrey, T. and Mercalli, L. (2005), 'A new instrumental precipitation dataset for the greater alpine region for the period 1800–2002', *Int. J. Climatol.* **25**(2), 139–166.

Austin, P. M. (1987), 'Relation between measured radar reflectivity and surface rainfall', *Mon. Wea. Rev.* **115**(5), 1053–1070.

Bagrov, N. A. (1959), 'Analytic representation of a sequence of meteorological fields via natural orthogonal components', *Trudy Tsentr. Inst. Progn* **74**, 3–24.

Baldwin, M. P. and Dunkerton, T. J. (2001), 'Stratospheric harbingers of anomalous weather regimes', *Science* **294**(5542), 581–584.

Bardossy, A., Bogardi, I. and Matyasovszky, I. (2005), 'Fuzzy rule-based downscaling of precipitation', *Theor. Appl. Climatol.* **82**(1), 119–129.

Bárdossy, A. and Pegram, G. G. S. (2009), 'Copula based multisite model for daily precipitation simulation', *Hydrol. Earth Syst. Sci.* **13**(12), 2299.

Bárdossy, A. and Plate, E. J. (1991), 'Modeling daily rainfall using a semi-Markov representation of circulation pattern occurrence', *J. Hydrol.* **122**, 33–47.

Bárdossy, A. and Plate, E. J. (1992), 'Space-time model for daily rainfall using atmospheric circulation patterns', *Wat. Resour. Res.* **28**, 1247–1259.

Barkhordarian, A., von Storch, H. and Bhend, J. (2013), 'The expectation of future precipitation change over the Mediterranean region is different from what we observe', *Clim. Dynam.* **40**(1–2), 225–244.

Barnett, T. P. and Preisendorfer, R. W. (1978), 'Multifield analog prediction of short-term climate fluctuations using a climate state vector', *J. Atmos. Sci.* **35**(10), 1771–1787.

Barry, R. G. (2008), *Mountain weather and climate*, Cambridge University Press.

Barry, R. G. and Blanken, P. D. (2016), *Microclimate and local climate*, Cambridge University Press.

Barry, R. G. and Chorley, R. J. (2009), *Atmosphere, weather and climate*, Routledge.

Barsugli, J. J., Guentchev, G., Horton, R. M., Wood, A., Mearns, L. O., Liang, X.-Z., Winkler, J. A., Dixon, K., Hayhoe, K., Rood, R. B., Goddard, L., Ray, A., Buja, L. and Ammann, C. (2013), 'The practitioner's dilemma: How to assess the credibility of downscaled climate projections', *EOS* **94**(46), 424–425.

Bartholy, J., Bogardi, I. and Matyasovszky, I. (1995), 'Effect of climate change on regional precipitation in Lake Balaton watershed', *Theor. Appl. Climatol.* **51**(4), 237–250.

Bates, B. C., Charles, S. P. and Hughes, J. P. (1998), 'Stochastic downscaling of numerical climate model simulations', *Env. Mod. Soft.* **13**(3), 325–331.

Bechtold, P. (2015), 'Atmospheric moist convection. ECMWF Lecture Notes'.

Becker, N., Ulbrich, U. and Klein, R. (2015), 'Systematic large-scale secondary circulations in a regional climate model', *Geophys. Res. Lett.* **42**(10), 4142–4149.

Beckmann, B.-R. and Buishand, T. A. (2002), 'Statistical downscaling relationships for precipitation in the Netherlands and North Germany', *Int. J. Climatol.* **22**(1), 15–32.

Beersma, J. J. and Buishand, T. A. (2003), 'Multi-site simulation of daily precipitation and temperature conditional on the atmospheric circulation', *Clim. Res.* **25**, 121–133.

Befort, D. J., Wild, S., Kruschke, T., Ulbrich, U. and Leckebusch, G. C. (2016), 'Different long-term trends of extra-tropical cyclones and windstorms in ERA-20C and NOAA-20CR reanalyses', *Atmos. Sci. Lett.* **17**(11), 586–595.

Bellenger, H., Guilyardi, E., Leloup, J., Lengaigne, M. and Vialard, J. (2014), 'ENSO representation in climate models: from CMIP3 to CMIP5', *Clim. Dynam.* **42**, 1999–2018.

Bellone, E., Hughes, J. P. and Guttorp, P. (2000), 'A hidden Markov model for downscaling synoptic atmospheric patterns to precipitation amounts', *Clim. Res.* **15**(1), 1–12.

Bellprat, O., Kotlarski, S., Lüthi, D. and Schär, C. (2013), 'Physical constraints for temperature biases in climate models', *Geophys. Res. Lett.* **40**, 4042–4047.

Beltrami, E. (1873), 'Sulle funzioni bilineari', *Giornale di Matematiche ad Uso degli Studenti Delle Universita* **11**(2), 98–106.

Benestad, R. E. (2001), 'A comparison between two empirical downscaling strategies', *Int. J. Climatol.* **21**(13), 1645–1668.

Benestad, R. E. (2002), 'Empirically downscaled temperature scenarios for northern Europe based on a multi-model ensemble', *Clim. Res.* **21**(2), 105–125.

Benestad, R. E. (2005), 'Climate change scenarios for northern Europe from multi-model IPCC AR4 climate simulations', *Geophys. Res. Lett.* **32**(17).

Benestad, R. E. (2011), 'A new global set of downscaled temperature scenarios', *J. Climate* **24**(8), 2080–2098.

Benestad, R. E., Chen, D., Mezghani, A., Fan, L. and Parding, K. (2015), 'On using principal components to represent stations in empirical-statistical downscaling', *Tellus A* **67**, 28326.

Benestad, R. E., Hanssen-Bauer, I. and Chen, D. (2008), *Empirical-statistical downscaling*, World Scientific Publishing Co Inc.

Benestad, R. E., Hanssen-Bauer, I. and Førland, E. J. (2007), 'An evaluation of statistical models for downscaling precipitation and their ability to capture long-term trends', *Int. J. Climatol.* **27**(5), 649–665.

Bengtsson, L., Arkin, P., Berrisford, P., Bougeault, P., Folland, C. K., Gordon, C., Haines, K., Hodges, K. I., Jones, P., Kallberg, P., Rayner, N., Simmons, A. J., Stammer, D., Thorne, P. W., Uppala, S. and Vose, R. S. (2007), 'The need for a dynamical climate reanalysis', *Bull. Amer. Meteorol. Soc.* **88**(4), 495–501.

Bengtsson, L., Hagemann, S. and Hodges, K. I. (2004), 'Can climate trends be calculated from reanalysis data?', *J. Geophys. Res.* **109**, D11111.

Berger, A. (1988), 'Milankovitch theory and climate', *Rev. Geophys.* **26**(4), 624–657.

Betts, A. K. and Miller, M. J. (1986), 'A new convective adjustment scheme. Part II: Single column tests using GATE wave, BOMEX, ATEX and arctic air-mass data sets', *Quart. J. Roy. Meteorol. Soc.* **112**(473), 693–709.

Bevacqua, E., Maraun, D., Hobæk Haff, I., Widmann, M. and Vrac, M. (2017), 'Multivariate Statistical Modelling of Compound Events via Pair-Copula Constructions: Analysis of Floods in Ravenna', *Hydrol. Earth Syst. Sci.* **21**(6), 2701–2723.

Bhend, J. and Whetton, P. (2013), 'Consistency of simulated and observed regional changes in temperature, sea level pressure and precipitation', *Clim. Change* **118**(3–4), 799–810.

Boberg, F. and Christensen, J. H. (2012), 'Overestimation of Mediterranean summer temperature projections due to model deficiencies', *Nat. Clim. Change* **2**(6), 433–436.

Boé, J., Terray, L., Habets, F. and Martin, E. (2007), 'Statistical and dynamical downscaling of the Seine basin climate for hydro-meteorological studies', *Int. J. Climatol.* **27**, 1643–1655.

Böhm, R., Auer, I., Brunetti, M., Maugeri, M., Nanni, T. and Schöner, W. (2001), 'Regional temperature variability in the European Alps: 1760–1998 from homogenized instrumental time series', *Int. J. Climatol.* **21**(14), 1779–1801.

Bony, S., Colman, R., Kattsov, V. M., Allan, R. P., Bretherton, C. S., Dufresne, J.-L., Hall, A., Hallegatte, S., Holland, M. M., Ingram, W., Randall, D. A., Soden, B. J., Tselioudis, G. and Webb, M. J. (2006), 'How well do we understand and evaluate climate change feedback processes?', *J. Climate* **19**(15), 3445–3482.

Bony, S., Stevens, B., Frierson, D. M. W., Jakob, C., Kageyama, M., Pincus, R., Shepherd, T. G., Sherwood, S. C., Siebesma, A. P., Sobel, A. H., Watanabe, M. and Webb, M. J. (2015), 'Clouds, circulation and climate sensitivity', *Nat. Geosci.* **8**(4), 261–268.

Bothe, O., Evans, M., Fernández Donado, L., Garcia Bustamante, E., Gergis, J., Gonzalez-Rouco, J. F., Goosse, H., Hegerl, G., Hind, A., Jungclaus, J. H. et al. (2015), 'Continental-scale temperature variability in PMIP3 simulations and PAGES 2k regional temperature reconstructions over the past millennium', *Clim. Past* **11**, 1673–1699.

Boucher, O., Randall, D., Artaxo, P., Bretherton, C., Feingold, G., Forster, P., Kerminen, V.-M., Kondo, Y., Liao, H., Lohmann, U., Rasch, P., Satheesh, S. K., Sherwood, S., Stevens, B. and Zhang, X. Y. (2013), *Climate Change 2013: The Physical Science Basis. Contribution of*

Working Group I to the Fifth Assessment Report of the Intergovernmental Panel on Climate Change, Cambridge University Press, Cambridge, United Kingdom, and New York, NY, USA, chapter 'Clouds and Aerosols'.

Boulanger, J., Brasseur, G., Carril, A. F., de Castro, M., Degallier, N., Ereño, C., Le Treut, H., Marengo, J. A., Menendez, C. G., Nuñez, M. N., Penalba, O. C., Rolla, A. L., Rusticucci, M. and Terra, R. (2010), 'A Europe–South America network for climate change assessment and impact studies', *Clim. Change* **98**(3), 307–329.

Box, G. E. P. and Draper, N. R. (1987), *Empirical model-building and response surfaces*, Vol. 424, Wiley New York.

Brands, S., Gutiérrez, J. M. and Herrera, S. (2012), 'On the use of reanalysis data for downscaling', *J. Climate* **25**, 2517–2526.

Brandsma, T. and Van der Meulen, J. P. (2008), 'Thermometer screen intercomparison in De Bilt (the Netherlands) Part II: Description and modeling of mean temperature differences and extremes', *Int. J. Climatol.* **28**(3), 389–400.

Brayshaw, D. J., Hoskins, B. and Blackburn, M. (2015), 'The basic ingredients of the North Atlantic storm track. part I: Land–sea contrast and orography', *J. Atmos. Sci.* **72**(9).

Bretherton, C. S., Smith, C. and Wallace, J. M. (1992), 'An intercomparison of methods for finding coupled patterns in climate data', *J. Climate* **5**(6), 541–560.

Brohan, P., Kennedy, J. J., Harris, I., Tett, S. F. B. and Jones, P. D. (2006), 'Uncertainty estimates in regional and global observed temperature changes: A new dataset from 1850', *J. Geophys. Res.* **111**, D12106.

Brown, C., Ghile, Y., Laverty, M. and Li, K. (2012), 'Decision scaling: Linking bottom-up vulnerability analysis with climate projections in the water sector', *Wat. Resour. Res.* **48**, W09537.

Brown, C. and Wilby, R. L. (2012), 'An alternate approach to assessing climate risks', *EOS* **93**(41), 401–402.

Buell, C. E. (1975), The topography of empirical orthogonal functions, *in* 'Preprints Fourth Conference on Probability and Statistics in Atmospheric Science', p. 188.

Buell, C. E. (1979), On the physical interpretation of empirical orthogonal functions, *in* 'Preprints Sixth Conference on Probability and Statistics in Atmospheric Science', p. 112.

Buishand, T. A. (1977), Stochastic modeling of daily rainfall sequences., Technical Report 77-3.

Buishand, T. A. and Brandsma, T. (2001), 'Multisite simulation of daily precipitation and temperature in the Rhine basin by nearest-neighbor resampling', *Wat. Resour. Res.* **37**(11), 2761–2776.

Bukovsky, M. S. (2012), 'Temperature trends in the NARCCAP regional climate models', *J. Climate* **24**, 3985–3991.

Bukovsky, M. S. and Karoly, D. J. (2009), 'Precipitation simulations using WRF as a nested regional climate model', *J. Appl. Meteorol. Climatol.* **48**(10), 2152–2159.

Bürger, G., Murdock, T. Q., Werner, A. T., Sobie, S. R. and Cannon, A. J. (2012), 'Downscaling extremes - an intercomparison of multiple statistical methods for present climate', *J. Climate* **25**, 4366–4388.

Buser, C. M., Künsch, H. R., Lüthi, D., Wild, M. and Schär, C. (2009), 'Bayesian multi-model projection of climate: bias assumptions and interannual variability', *Clim. Dynam.* **33**, 849–868.

Busuioc, A., von Storch, H. and Schnur, R. (1999), 'Verification of GCM-generated regional seasonal precipitation for current climate and of statistical downscaling estimates under changing climate conditions', *J. Climate* **12**, 258–272.

Butler, A. H., Thompson, D. W. J. and Heikes, R. (2010), 'The steady-state atmospheric circulation response to climate change - like thermal forcings in a simple general circulation model', *J. Climate* **23**(13), 3474–3496.

Cabré, M. F., Solman, S. A. and Nuñez, M. N. (2010), 'Creating regional climate change scenarios over southern South America for the 2020's and 2050's using the pattern scaling technique: validity and limitations', *Clim. Change* **98**(3–4), 449–469.

Caldwell, P., Chin, H.-N. S., Bader, D. C. and Bala, G. (2009), 'Evaluation of a WRF dynamical downscaling simulation over California', *Clim. Change* **95**, 499–521.

Cannon, A. J. (2011), 'Quantile regression neural networks: Implementation in R and application to precipitation downscaling', *Comp. Geosci.* **37**(9), 1277–1284.

Cannon, A. J. (2016), 'Multivariate Bias Correction of Climate Model Output: Matching Marginal Distributions and Intervariable Dependence Structure', *J. Climate* **29**(19), 7045–7064.

Cannon, A. J. (2017), 'Multivariate quantile mapping bias correction: an n-dimensional probability density function transform for climate model simulations of multiple variables', *Clim. Dynam.* pp. 1–19, DOI: 10.1007/s00382-017-3580-6.

Casanueva, A., Frías, M. D., Herrera, S., San-Martín, D., Zaninovic, K. and Gutiérrez, J. M. (2014), 'Statistical downscaling of climate impact indices: testing the direct approach', *Clim. Change* **127**(3–4), 547–560.

Cash, D., Clark, W. C., Alcock, F., Dickson, N. M., Eckley, N. and Jäger, J. (2002), 'Salience, credibility, legitimacy and boundaries: linking research, assessment and decision making', John F. Kennedy School of Government, Harvard University, Faculty Research Working Papers Series.

Casola, J. H. and Wallace, J. M. (2007), 'Identifying weather regimes in the wintertime 500-hPa geopotential height field for the Pacific–North American sector using a limited-contour clustering technique', *J. Appl. Meteorol. Climatol.* **46**(10), 1619–1630.

Castro, C. L., Pielke, R. A. and Leoncini, G. (2005), 'Dynamical downscaling: Assessment of value retained and added using the Regional Atmospheric Modeling System (RAMS)', *J. Geophys. Res.* **110**(D5), D05108.

Caussinus, H. and Mestre, O. (2004), 'Detection and correction of artificial shifts in climate series', *J. Roy. Stat. Soc. C* **53**(3), 405–425.

Cavazos, T. and Hewitson, B. C. (2005), 'Performance of NCEP-NCAR reanalysis variables in statistical downscaling of daily precipitation', *Clim. Res.* **28**, 95–107.

Cayan, D., Kunkel, K., Castro, C., Gershunov, A., Barsugli, J., Ray, A., Overpeck, J., Anderson, M., Russell, J., Rajagopalan, B., Rangwala, I. and Duffy, P. (2013), *Assessment of Climate Change in the Southwest United States: A Report Prepared for the National Climate Assessment*, Island Press, chapter 'Future climate: Projected average', pp. 101–125.

Ceppi, P., Scherrer, S. C., Fischer, A. M. and Appenzeller, C. (2012), 'Revisiting Swiss temperature trends 1959–2008', *Int. J. Climatol.* **32**, 203–213.

Chandler, R. E. (2005), 'On the use of generalized linear models for interpreting climate variability', *Environmetrics* **16**, 699–715.

Chandler, R. E. (2013), 'Exploiting strength, discounting weakness: combining information from multiple climate simulators', *Phil. Trans. R. Soc. A* **371**(1991), 20120388.

Chandler, R. E. and Bate, S. (2007), 'Inference for clustered data using the independence loglikelihood', *Biometrika* **94**, 167–183.

Chandler, R. E. and Wheater, H. S. (2002), 'Analysis of rainfall variability using generalized linear models: A case study from the west of Ireland', *Wat. Resour. Res.* **38**(10), 1192.

Charles, S. P., Bari, M. A., Kitsios, A. and Bates, B. C. (2007), 'Effect of GCM bias on downscaled precipitation and runoff projections for the Serpentine catchment, Western Australia', *Int. J. Climatol.* **27**(12), 1673–1690.

Charles, S. P., Bates, B. C., Whetton, P. H. and Hughes, J. P. (1999), 'Validation of downscaling models for changed climate conditions: case study of southwestern Australia', *Clim. Res.* **12**, 1–14.

Chen, L., Li, T. and Yu, Y. (2015), 'Causes of strengthening and weakening of ENSO amplitude under global warming in four CMIP5 models', *J. Climate* **28**, 3250–3274.

Christensen, J. H., Boberg, F., Christensen, O. B. and Lucas-Picher, P. (2008), 'On the need for bias correction of regional climate change projections of temperature and precipitation', *Geophys. Res. Lett.* **35**, L20709.

Christensen, J. H. and Christensen, O. B. (2007), 'A summary of the PRUDENCE model projections of changes in European climate by the end of this century', *Clim. Change* **81**, 7–30.

Christensen, J. H., Hewitson, B., Busuioc, A., Chen, A., Gao, X., Held, I., Jones, R., Kolli, R. K., Kwon, W.-T., Laprise, R., Rueda, V. M., Mearns, L., Menéndez, C. G., Räisänen, J., Rinke, A., Sarr, A. and Whetton, P. (2007), *Climate Change 2007: The Physical Science Basis. Contribution of Working Group I to the Fourth Assessment Report of the Intergovernmental Panel on Climate Change*, Cambridge University Press, Cambridge, United Kingdom and New York, NY, USA, chapter 'Regional Climate Projections'.

Christensen, J. H., Machenhauer, B., Jones, R. G., Schär, C., Ruti, P. M., Castro, M. and Visconti, G. (1997), 'Validation of present-day regional climate simulations over Europe: LAM simulations with observed boundary conditions', *Clim. Dynam.* **13**, 489–506.

Chu, J.-L., Kang, H., Tam, C.-Y., Park, C.-K. and Chen, C.-T. (2008), 'Seasonal forecast for local precipitation over northern taiwan using statistical downscaling', *J. Geophys. Res.* **113**(D12).

Chu, J.-L. and Yu, P.-S. (2010), 'A study of the impact of climate change on local precipitation using statistical downscaling', *J. Geophys. Res.* **115**, D10105.

Clark, P. (2009), 'Issues with high-resolution NWP', Technical report, United Kingdom Met Office.

Coe, R. and Stern, R. D. (1982), 'Fitting models to daily rainfall data', *J. Appl. Meteorol.* **21**(7), 1024–1031.

Cohen, S. J. and Allsopp, T. R. (1988), 'The potential impacts of a scenario of CO_2-induced climatic change on Ontario, Canada', *J. Climate* **1**, 669–681.

Coles, S. (2001), *An introduction to statistical modeling of extreme values*, Springer Series in Statistics, Springer.

Colette, A., Vautard, R. and Vrac, M. (2012), 'Regional climate downscaling with prior statistical correction of the global climate forcing', *Geophys. Res. Lett.* **39**(13), L13707.

Collins, M., Booth, B. B. B., Harris, G. R., Murphy, J. M., Sexton, D. M. H. and Webb, M. J. (2006), 'Towards quantifying uncertainty in transient climate change', *Clim. Dynam.* **27**, 127–147.

Collins, M., Knutti, R., Arblaster, J., Dufresne, J.-L., Fichefet, T., Friedlingstein, P., Gao, X., Gutowski, W. J., Johns, T., Krinner, G., Shongwe, M., Tebaldi, C., Weaver, A. J. and Wehner, M. (2013), *Climate Change 2013: The Physical Science Basis. Contribution of Working Group I to the Fifth Assessment Report of the Intergovernmental Panel on Climate Change*, Cambridge University Press, Cambridge, United Kingdom, and New York, NY, USA, chapter 'Long-Term Climate Change: Projections, Committments and Irreversibility'.

Compo, G. P., Whitaker, J. S., Sardeshmukh, P. D., Matsui, N., Allan, R. J., Yin, X., Gleason, B. E., Vose, R. S., Rutledge, G., Bessemoulin, P., Brönnimann, S., Brunet, M., Crouthamel, R. I., Grant, A. N., Groisman, P. Y., Jones, P. D., Kruk, M. C., Kruger, A. C., Marshall, G. J., Maugeri, M., Mok, H. Y., Nordli, O., Ross, T. F., Trigo, R. M., Wang, X. L., Woodruff, S. D. and Worley, S. J. (2011), 'The twentieth century reanalysis project', *Quart. J. Roy. Meteorol. Soc.* **137**(654), 1–28.

Cooley, D., Nychka, D. and Naveau, P. (2007), 'Bayesian spatial modeling of extreme precipitation return levels', *J. Am. Stat. Ass.* **102**(479), 824–840.

CORDEX (2016), 'Bias-adjusted RCM data', http://www.cordex.org/index.php?option=com_content&view=article&id=275&Itemid=785.

Cowpertwait, P., Isham, V. and Onof, C. (2007), 'Point process models of rainfall: developments for fine-scale structure', *Proc. Roy. Soc. A* **463**(2086), 2569–2588.

Cowpertwait, P. S. P. (1994), A generalized point process model for rainfall, *in* 'Proc. Roy. Soc. A', Vol. 447, pp. 23–37.

Cowpertwait, P. S. P., Kilsby, C. G. and O'Connell, P. E. (2002), 'A space-time Neyman-Scott model of rainfall: Empirical analysis of extremes', *Wat. Resour. Res.* **38**(8), 1131.

Cox, D. R. and Hinkley, D. V. (1994), *Theoretical statistics*, Chapman & Hall, London.

Cox, D. R. and Isham, V. (1988), A simple spatial-temporal model of rainfall, *in* 'Proc. Roy. Soc. A', Vol. 415, pp. 317–328.

Cox, D. R. and Isham, V. S. (1994), *Statistics for the environment, 2: Water related issues*, Wiley, chapter 'Stochastic Models of Precipitation', pp. 3–18.

Craddock, J. and Flood, C. (1969), 'Eigenvectors for representing the 500 mb geopotential surface over the Northern Hemisphere', *Quart. J. Roy. Meteorol. Soc.* **95**(405), 576–593.

Crowley, T. (1990), 'Are there any satisfactory geologic analogs for a future greenhouse warming?', *J. Climate* **3**, 1282–1292.

Daly, C., Neilson, R. P. and Phillips, D. L. (1994), 'A statistical-topographic model for mapping climatological precipitation over mountainous terrain', *J. Appl. Meteorol.* **33**(2), 140–158.

Davini, P., von Hardenberg, J., Corti, S., Christensen, H. M., Juricke, S., Subramanian, A., Watson, P. A. G., Weisheimer, A. and Palmer, T. N. (2017), 'Climate SPHINX: evaluating the impact of resolution and stochastic physics parameterisations in the EC-Earth global climate model', *Geosci. Model Dev.* **10**(3), 1383–1402.

Davison, A. C. (2003), *Statistical models*, Cambridge Series in Statistical and Probabilistic Mathematics, Cambridge University Press.

Dawson, A., Palmer, T. N. and Corti, S. (2012), 'Simulating regime structures in weather and climate prediction models', *Geophys. Res. Lett.* **39**(21).

Dayon, G., Boé, J. and Martin, E. (2015), 'Transferability in the future climate of a statistical downscaling method for precipitation in France', *J. Geophys. Res.* **120**(3), 1023–1043.

De Elia, R. (2014), 'Specificities of climate modeling research and the challenges in communicating to users', *Bull. Amer. Meteorol. Soc.* **95**(7), 1003–1010.

Deardorff, J. W. (1972), 'Numerical investigation of neutral and unstable planetary boundary layers', *J. Atmos. Sci.* **29**(1), 91–115.

Dee, D. P., Källén, E., Simmons, A. J. and Haimberger, L. (2011*a*), 'Comments on "Reanalyses suitable for characterizing long-term trends"', *Bull. Amer. Meteorol. Soc.* **92**(1), 65–70.

Dee, D. P., Uppala, S. M., Simmons, A. J., Berrisford, P., Poli, P., Kobayashi, S., Andrae, U., Balmaseda, M. A., Balsamo, G., Bauer, P., Bechtold, P., Beeljars, A. C. M., van den Berg, L., Bidlot, J., Bormann, N., Delsol, C., Dragani, R., Fuentes, M., Geer, A. J., Haimberger, L., Healy, S. B., Hersbach, H., Hólm, E. V., Isaksen, L., Kållberg, P., Köhler, M., Matricardi, M., McNally, A. P., Monge-Sanz, B. M., Morcrette, J.-J., Park, B.-K., Peubey, C., de Rosnay, P., Tavolato, C., Thépaut, J.-N. and Vitart, F. (2011*b*), 'The ERA-Interim reanalysis: configuration and performance of the data assimilation system', *Quart. J. Roy. Meteorol. Soc.* **137**, 553–597.

Deidda, R., Badas, M. G. and Piga, E. (2006), 'Space-time multifractality of remotely sensed rainfall fields', *J. Hydrol.* **322**(1), 2–13.

Della-Marta, P. M. and Wanner, H. (2006), 'A method of homogenizing the extremes and mean of daily temperature measurements', *J. Climate* **19**(17), 4179–4197.

Denis, B., Laprise, R., Caya, D. and Côté, J. (2002), 'Downscaling ability of one-way nested regional climate models: the Big-Brother Experiment', *Clim. Dynam.* **18**(8), 627–646.

Déqué, M., Rowell, D. P., Luthi, D., Giorgi, F., Christensen, J. H., Rockel, B., Jacob, D., Kjellström, E., de Castro, M. and van den Hurk, B. (2007), 'An intercomparison of regional climate simulations for Europe: assessing uncertainties in model projections', *Clim. Change* **81**, 53–70.

Deser, C., Knutti, R., Solomon, S. and Phillips, A. S. (2012), 'Communication of the role of natural variability in future North American climate', *Nat. Clim. Change* **2**, 775–779.

Deser, F., Rockel, B., von Storch, H., Winterfeldt, J. and Zahn, M. (2011), 'Regional climate models add value to global model data. a review and selected examples', *Bull. Amer. Meteorol. Soc.* **92**, 1181–1192.

Dessai, S. (2009), 'Do we need better predictions to adapt to a changing climate?', *EOS* **90**, 111–112.

Dessai, S. and Hulme, M. (2004), 'Does climate adaptation policy need probabilities?', *Climate Policy* **4**(2), 107–128.

Di Luca, A., de Elía, R. and Laprise, R. (2015), 'Challenges in the quest for added value of regional climate dynamical downscaling', *Curr. Clim. Change Rep.* **1**(1), 10–21.

Dickinson, R. E., Errico, R. M., Giorgi, F. and Bates, G. T. (1989), 'A regional climate model for the Western United States', *Clim. Change* **15**, 383–422.

Diggle, P. J. and Ribeiro, P. J. (2007), *Model-based geostatistics*, Springer Series in statistics, Springer.

Ding, H., Keenlyside, N., Latif, M., Park, W. and Wahl, S. (2015), 'The impact of mean state errors on equatorial atlantic interannual variability in a climate model', *J. Geophys. Res.* **120**(2), 1133–1151.

Director, H. and Bornn, L. (2015), 'Connecting point-level and gridded moments in the analysis of climate data', *J. Climate* **28**(9), 3496–3510.

Dobson, A. J. (2001), *An introduction to generalized linear models*, Chapman and Hall, London.

Dommenget, D. and Latif, M. (2002), 'A cautionary note on the interpretation of EOFs', *J. Clim.* **15**(2), 216–225.

Döscher, R., Willén, U., Jones, C., Rutgersson, A., Meier, H. E. M., Hansson, U. and Graham, L. P. (2002), 'The development of the regional coupled ocean-atmosphere model RCAO', *Boreal Env. Res.* **7**(3), 183–192.

Dosio, A. (2016), 'Projections of climate change indices of temperature and precipitation from an ensemble of bias-adjusted high-resolution EURO-CORDEX regional climate models', *J. Geophys. Res.* **121**(10), 5488–5511.

Dosio, A., Paruolo, P. and Rojas, R. (2012), 'Bias correction of the ENSEMBLES high resolution climate change projections for use by impact models: Analysis of the climate change signal', *J. Geophys. Res. Atmos.* **117**, D17110.

Dubrovsky, M. (1997), 'Creating daily weather series with use of the weather generator', *Environmetrics* **8**(5), 409–424.

Dubrovský, M., Buchtele, J. and Žalud, Z. (2004), 'High-frequency and low-frequency variability in stochastic daily weather generator and its effect on agricultural and hydrologic modelling', *Clim. Change* **63**(1), 145–179.

Dufresne, J. L. and Bony, S. (2008), 'An assessment of the primary sources of spread of global warming estimates from coupled atmosphere-ocean models', *J. Climate* **21**, 5135–5144.

Dunn, P. K. (2004), 'Occurrence and quantity of precipitation can be modelled simultaneously', *Int. J. Climatol.* **24**(10), 1231–1239.

Easterling, D. R. (1999), 'Development of regional climate scenarios using a downscaling approach', *Clim. Change* **41**(3–4), 615–634.

Eckart, C. and Young, G. (1936), 'The approximation of one matrix by another of lower rank', *Psychometrika* **1**(3), 211–218.

Eckart, C. and Young, G. (1939), 'A principal axis transformation for non-hermitian matrices', *Bull. Amer. Math. Soc.* **45**(2), 118–121.

ECMWF (2016), *IFS documentation – Cy41r2; operational implementation 8 March 2016*, ECMWF, chapter 'Part IV: Physical Processes'.

Eden, C., Greatbatch, R. J. and Böning, C. W. (2004), 'Adiabatically correcting an eddy-permitting model using large-scale hydrographic data: Application to the Gulf Stream and the North Atlantic Current', *J. Phys. Oceanogr.* **34**(4), 701–719.

Eden, J. M., Widmann, M., Maraun, D. and Vrac, M. (2014), 'Comparison of GCM-and RCM-simulated precipitation following stochastic postprocessing', *J. Geophys. Res.* **119**(19).

Eden, J., Widmann, M., Grawe, D. and Rast, S. (2012), 'Skill, correction, and downscaling of GCM-simulated precipitation', *J. Climate* **25**, 3970–3984.

Edwards, P. N. (2011), 'History of climate modeling', *WIRES Clim. Change* **2**(1), 128–139.

Ehret, U., Zehe, E., Wulfmeyer, V., Warrach-Sagi, K. and Liebert, J. (2012), 'Should we apply bias correction to global and regional climate model data?', *Hydrol. Earth Syst. Sci.* **16**, 3391–3404.

Eitzinger, J. and Thaler, S. (2016), 'STARC-Impact Kick-Off Meeting. Introduction to Workpackage 4'.

Embrechts, P., Klüppelberg, C. and Mikosch, T. (1997), *Modelling extremal events for insurance and finance*, Applications in Mathematics, Springer.

Engen-Skaugen, T. (2007), 'Refinement of dynamically downscaled precipitation and temperature scenarios', *Clim. Change* **84**(3–4), 365–382.

Entekhabi, D., Rodriguez-Iturbe, I. and Eagleson, P. S. (1989), 'Probabilistic representation of the temporal rainfall process by a modified Neyman-Scott Rectangular Pulses Model: Parameter estimation and validation', *Wat. Resour. Res.* **25**(2), 295–302.

European Commission (2013), 'An EU Strategy on adaptation to climate change', http://ec.europa.eu/clima/policies/adaptation.

Farmer, S. (1971), 'An investigation into the results of principal component analysis of data derived from random numbers', *The Statistician* **20**(4), 63–72.

Fawcett, L. and Walshaw, D. (2007), 'Improved estimation for temporally clustered extremes', *Environmetrics* **18**(2), 173–188.

Ferraris, L., Gabellani, S., Rebora, N. and Provenzale, A. (2003), 'A comparison of stochastic models for spatial rainfall downscaling', *Wat. Resour. Res.* **39**(12), 1368.

Fischer, E. M., Seneviratne, S. I., Vidale, P. L., Lüthi, D. and Schär, C. (2007), 'Soil moisture-atmosphere interactions during the 2003 European summer heat wave', *J. Climate* **20**, 5081–5099.

Flato, G., Marotzke, J., Abiodun, B., Braconnot, P., Chou, S. C., Collins, W., Cox, P., Driouech, F., Emori, S., Eyring, V., Forest, C., Gleckler, P., Guilyardi, E., Jakob, C., Kattsov, V., Reason, C. and Rummukainen, M. (2013), *Climate Change 2013: The Physical Science Basis. Contribution of Working Group I to the Fifth Assessment Report of the Intergovernmental Panel on Climate Change*, Cambridge University Press, Cambridge, United Kingdom, and New York, NY, USA, chapter 'Evaluation of Climate Models'.

Foley, A. M. (2010), 'Uncertainty in regional climate modelling: A review', *Prog. Phys. Geogr.* **34**(5), 647–670.

Fowler, H. J., Blenkinsop, S. and Tebaldi, C. (2007), 'Linking climate change modelling to impacts studies: recent advances in downscaling techniques for hydrological modelling', *Int. J. Climatol.* **27**, 1547–1578.

Fowler, H. J., Kilsby, C. G. and O'Connell, P. E. (2000), 'A stochastic rainfall model for the assessment of regionl water resources systems under changed climate conditions', *Hydrol. Earth. Syst. Sci.* **4**(2), 263–282.

Fraley, C. and Raftery, A. E. (2002), 'Model-based clustering, discriminant analysis, and density estimation', *J. Amer. Stat. Assoc.* **97**(458), 611–631.

Fraley, C. and Raftery, A. E. (2007), 'Model-based methods of classification: using the mclust software in chemometrics', *J. Stat. Soft.* **18**(6), 1–13.

Frei, C., Christensen, J. H., Deque, M., Jacob, D., Jones, R. G. and Vidale, P. L. (2003), 'Daily precipitation statistics in regional climate models: Evaluation and intercomparison for the European Alps', *J. Geophys. Res.* **108**(D3), 4124.

Frei, C. and Schär, C. (1998), 'A precipitation climatology of the Alps from high-resolution raingauge observations', *Int. J. Climtol.* **18**(8), 873–900.

Frei, C., Schöll, R., Fukutome, S., Schmidli, J. and Vidale, P. L. (2006), 'Future change of precipitation extremes in Europe: an intercomparison of scenarios from regional climate models', *J. Geophys. Res.* **111**, D06105.

Frey-Buness, F., Heimann, D. and Sausen, R. (1995), 'A statistical-dynamical downscaling procedure for global climate simulations', *Theor. Appl. Climatol.* **50**(3–4), 117–131.

Frías, M. D., Zorita, E., Fernández, J. and Rodríguez-Puebla, C. (2006), 'Testing statistical downscaling methods in simulated climates', *Geophys. Res. Lett.* **33**, L19807.

Friederichs, P. and Hense, A. (2007), 'Statistical downscaling of extreme precipitation events using censored quantile regression', *Mon. Wea. Rev.* **135**(6), 2365–2378.

Fritsch, J. M., Chappell, C. F. and Hoxit, L. R. (1976), 'The use of large-scale budgets for convective parameterization', *Mon. Wea. Rev.* **104**(11), 1408–1418.

Frost, A. J., Charles, S. P., Timbal, B., Chiew, F. H. S., Mehrotra, R., Nguyen, K. C., Chandler, R. E., McGregor, J. L., Fu, G., Kirono, D. G. C. et al. (2011), 'A comparison of multi-site daily rainfall downscaling techniques under Australian conditions', *J. Hydrol.* **408**(1), 1–18.

Fuentes, U. and Heimann, D. (2000), 'An improved statistical-dynamical downscaling scheme and its application to the Alpine precipitation climatology', *Theor. Appl. Climatol.* **65**(3–4), 119–135.

Funk, C., Peterson, P., Landsfeld, M., Pedreros, D., Verdin, J., Shukla, S., Husak, G., Rowland, J., Harrison, L., Hoell, A. and Michaelsen, J. (2015), 'The climate hazards infrared precipitation with stations - a new environmental record for monitoring extremes', *Scientific Data* **2**, 150066.

Gabriel, K. R. and Neumann, J. (1962), 'A Markov chain model for daily rainfall occurrence at Tel Aviv', *Quart. J. Roy. Meteorol. Soc.* **88**, 90–95.

Gangopadhyay, S., Pruitt, T., Brekke, L. and Raff, D. (2011), 'Hydrologic projections for the Western United States', *EOS* **92**(48), 441–442.

García-Morales, M. B. and Dubus, L. (2007), 'Forecasting precipitation for hydroelectric power management: how to exploit GCM's seasonal ensemble forecasts', *Int. J. Climatol.* **27**(12), 1691–1705.

Gates, W. L. (1985), 'The use of general circulation models in the analysis of the ecosystem impacts of climatic change', *Clim. Change* **7**, 267–284.

GCOS (1998), 'Report on the adequacy of the global climate observing system', GCOS-48, Geneva.

Genest, C. and Favre, A.-C. (2007), 'Everything you always wanted to know about copula modeling but were afraid to ask', *J. Hydrol. Eng.* **12**(4), 347–368.

Georgakakos, A., Fleming, P., Dettinger, M., Peters-Lidard, C., Richmond, T. C., Reckhow, K., White, K. and Yates, D. (2014), *Climate Change Impacts in the United States: The Third National Climate Assessment*, U.S. Global Change Research Program, chapter 'Water Resources', pp. 69–112.

Gerrity, J. P. and McPherson, R. D. (1969), 'Development of a limited area fine-mesh prediction model', *Mon. Wea. Rev.* **97**(9), 665–669.

Giannini, A., Saravanan, R. and Chang, P. (2003), 'Oceanic forcing of Sahel rainfall on interannual to interdecadal time scales', *Science* **302**(5647), 1027–1030.

Gilks, W. R., Richardson, S. and Spiegelhalter, D. (1995), *Markov chain Monte Carlo in practice*, CRC Press.

Giorgi, F. (1990), 'Simulation of regional climate using a limited-area model nested in a general circulation model', *J. Climate* **3**, 941–963.

Giorgi, F. and Bates, G. T. (1989), 'The climatological skill of a regional climate model over complex terrain', *Mon. Wea. Rev.* **117**, 2325–2347.

Giorgi, F. and Gutowski, W. J. (2016), 'Coordinated experiments for projections of regional climate change', *Curr. Clim. Change Rep.* **2**(4), 202–210.

Giorgi, F., Hewitson, B., Christensen, J., Hulme, M., von Storch, H., Whetton, P., Jones, R., Mearns, L. and Fu, C. (2001), *Climate Change 2001: The Scientific Basis. Contribution of Working Group I to the Third Assessment Report of the Intergovernmental Panel on Climate Change*, Cambridge University Press, Cambridge, United Kingdom, and New York, NY, USA, chapter 'Regional Climate Information – Evaluation and Projections'.

Giorgi, F., Jones, C. and Asrar, G. R. (2009), 'Addressing climate information needs at the regional level: the CORDEX framework', *WMO Bulletin* **58**(3), 175–183.

Giorgi, F., Marinucci, M. R. and Visconti, G. (1991), 'A 2XCO2 climate change scenario over Europe generated using a limited area model nested in a general circulation model 2. Climate change scenario', *J. Geophys. Res.* **97**, 10011–10028.

Giorgi, F. and Mearns, L. O. (1999), 'Introduction to special section: Regional Climate Modeling Revisited', *J. Geophys. Res.* **D6**, 6335–6352.

Giorgi, F., Torma, C., Coppola, E., Ban, N., Schär, C. and Somot, S. (2016), 'Enhanced summer convective rainfall at Alpine high elevations in response to climate warming', *Nat. Geosci.* **9**(8), 584–589.

Girshick, M. A. (1939), 'On the sampling theory of roots of determinantal equations', *The Annals of Mathematical Statistics* **10**(3), 203–224.

Girvetz, E. H., Maurer, E. P., Duffy, P. B., Ruesch, A., Thrasher, B. and Zganjar, C. (2013), *Making climate data relevant to decision making: the important details of spatial and temporal downscaling*, The World Bank.

Glahn, H. R. (1962), 'An experiment in forecasting rainfall probabilities by objective methods', *Mon. Weather Rev* **90**, 59–67.

Glahn, H. R. and Allen, R. A. (1965), 'A note concerning the "inflation" of regression forecasts', *J. Appl. Meteorol.* **5**, 124–126.

Glahn, H. R. and Lowry, D. A. (1972), 'The use of model output statistics (MOS) in objective weather forecasing', *J. Appl. Meteorol.* **11**, 1203–1211.

Glasbey, C. A. and Nevison, I. M. (1997), Rainfall modelling using a latent Gaussian variable, *in Modelling longitudinal and spatially correlated data*, Springer, pp. 233–242.

Gleick, P. H. (1986), 'Methods for evaluating the regional hydrologic impacts of global climatic changes', *J. Hydrol.* **88**(1), 97–116.

Gneiting, T., Raftery, A. E., Westveld, A. H. and Goldman, T. (2005), 'Calibrated probabilistic forecasting using ensemble model output statistics and minimum CRPS estimation', *Mon. Wea. Rev.* **133**, 1098–1118.

Gobiet, A., Suklitsch, M. and Heinrich, G. (2015), 'The effect of empirical-statistical correction of intensity-dependent model errors on the temperature climate change signal', *Hydrol. Earth Syst. Sci.* **19**, 4055–4066.

Goodess, C. M., Anagnostopoulou, C., Bárdossy, A., Frei, C., Harpham, C., Haylock, M. R., Hundecha, Y., Maheras, P., Ribalaygua, J., Schmidli, J., Schmith, T., Tolika, K., Tomozeiu, R. and Wilby, R. L. (2010), 'An intercomparison of statistical downscaling methods for Europe and European regions – assessing their performance with respect to extreme weather events and the implications for climate change applications', Project report, Climatic Research Unit, University of East Anglia, Norwich, UK.

Goosse, H. (2015), *Climate system dynamics and modeling*, Cambridge University Press.

Goosse, H. (2017), 'Reconstructed and simulated temperature asymmetry between continents in both hemispheres over the last centuries', *Clim. Dynam.* **48**(5–6), 1483–1501.

Goosse, H., Renssen, H., Timmermann, A., Bradley, R. S. and Mann, M. E. (2006), 'Using paleoclimate proxy-data to select optimal realisations in an ensemble of simulations of the climate of the past millennium', *Clim. Dynam.* **27**(2–3), 165–184.

Groisman, P. Y. and Legates, D. R. (1994), 'The accuracy of United States precipitation data', *Bull. Amer. Meteorol. Soc.* **75**(2), 215–227.

Groot, A., Swart, R., Hygen, H., Benestad, R. E., Cauchy, A., Betgen, C. and Dubois, G. (2004), 'ClipC deliverable user requirements, part 1: Strategies for user consultation and engagement and user requirements: Synthesis from past efforts'. www.clipc.eu/media/clipc/org/documents/clipc_deliverable2_1_final_intemplate.pdf, accessed 2 August 2017.

Grotch, S. L. and MacCracken, M. C. (1991), 'The use of general circulation models to predict regional climate change', *J. Climate* **4**, 286–303.

Güntner, A., Olsson, J., Calver, A. and Gannon, B. (2001), 'Cascade-based disaggregation of continuous rainfall time series: the influence of climate', *Hydrol. Earth Syst. Sci.* **5**(2), 145–164.

Gutiérrez, J. M. et al. (2017), 'An intercomparison of a large ensemble of statistical downscaling methods for Europe: overall results from the VALUE perfect predictor cross-validation experiment', *Int. J. Climatol., subm.*

Gutiérrez, J. M., San-Martín, D., Brands, S., Manzanas, R. and Herrera, S. (2013), 'Reassessing statistical downscaling techniques for their robust application under climate change conditions', *J. Climate* **26**(1), 171–188.

Gutmann, E., Pruitt, T., Clark, M. P., Brekke, L., Arnold, J. R., Raff, D. A. and Rasmussen, R. M. (2014), 'An intercomparison of statistical downscaling methods used for water resource assessments in the United States', *Wat. Resour. Res.* **50**(9), 7167–7186.

Gutowski, W. J., Decker, S. G., Donavon, R. A., Pan, Z., Arritt, R. W. and Takle, E. S. (2003), 'Temporal-spatial scales of observed and simulated precipitation in central U.S. climate', *J. Climate* **16**, 3841–3847.

Haas, R. and Pinto, J. G. (2012), 'A combined statistical and dynamical approach for downscaling large-scale footprints of European windstorms', *Geophys. Res. Lett.* **39**(23).

Haas, R., Pinto, J. G. and Born, K. (2014), 'Can dynamically downscaled windstorm footprints be improved by observations through a probabilistic approach?', *J. Geophys. Res.* **119**(2), 713–725.

Haerter, J. O., Eggert, B., Moseley, C., Piani, C. and Berg, P. (2015), 'Statistical precipitation bias correction of gridded model data using point measurements', *Geophys. Res. Lett.* **42**, 1919–1929.

Haerter, J. O., Hagemann, S., Moseley, C. and Piani, C. (2011), 'Climate model bias correction and the role of timescales', *Hydrol. Earth Syst. Sci.* **15**(3), 1065–1079.

Hagemann, S., Chen, C., Clark, D., Folwell, S., Gosling, S. N., Haddeland, I., Hannasaki, N., Heinke, J., Ludwig, F., Voss, F. and Wiltshire, A. (2013), 'Climate change impact on available water resources obtained using multiple global climate and hydrology models', *Earth Syst. Dynam.* **4**, 129–144.

Hagemann, S., Chen, C., Haerter, J. O., Heinke, J., Gerten, D. and Piani, C. (2011), 'Impact of a statistical bias correction on the projected hydrological changes obtained from three GCMs and two hydrology models', *J. Hydrometeorol.* **12**(4), 556–578.

Haiden, T., Kann, A., Wittmann, C., Pistotnik, G., Bica, B. and Gruber, C. (2011), 'The Integrated Nowcasting through Comprehensive Analysis (INCA) system and its validation over the Eastern Alpine region', *Wea. Forecast.* **26**(2), 166–183.

Haines, K. and Hannachi, A. (1995), 'Weather regimes in the Pacific from a GCM', *J. Atmos. Sci.* **52**(13), 2444–2462.

Hall, A. (2014), 'Projecting regional change', *Science* **346**(6216), 1461–1462.

Hall, A., Qu, X. and Neelin, J. D. (2008), 'Improving predictions of summer climate change in the United States', *Geophys. Res. Lett.* **35**, L01702.

Hall, T., Brooks, H. E., Doswell, I. I. I. and Charles, A. (1999), 'Precipitation forecasting using a neural network', *Wea. Forecast* **14**(3), 338–345.

Hannachi, A. (1997), 'Low-frequency variability in a GCM: Three-dimensional flow regimes and their dynamics', *J. Climate* **10**(6), 1357–1379.

Hannachi, A., Jolliffe, I. T. and Stephenson, D. B. (2007), 'Empirical orthogonal functions and related techniques in atmospheric science: A review', *Int. J. Climatol.* **27**(9), 1119–1152.

Hanssen-Bauer, I., Achberger, C., Benestad, R. E., Chen, D. and Forland, E. J. (2005), 'Statistical downscaling of climate scenarios over Scandinavia', *Clim. Res.* **29**(3), 255–268.

Harris, I., Jones, P. D., Osborn, T. J. and Lister, D. H. (2014), 'Updated high-resolution grids of monthly climatic observations - the CRU TS3.10 Dataset', *Int. J. Climatol.* **34**(3), 623–642.

Harvey, B. J., Shaffrey, L. C. and Woollings, T. J. (2015), 'Deconstructing the climate change response of the Northern Hemisphere wintertime storm tracks', *Clim. Dynam.* **45**(9-10), 2847–2860.

Hastie, T. J. and Tibshirani, R. J. (1990), *Generalized additive models*, Chapman & Hall.

Hawkins, E., Smith, R. S., Gregory, J. M. and Stainforth, D. A. (2016), 'Irreducible uncertainty in near-term climate projections', *Clim. Dynam.* **46**(11–12), 3807–3819.

Hawkins, E. and Sutton, R. (2009), 'The potential to narrow uncertainty in regional climate predictions', *Bull. Amer. Meteorol. Soc.* **90**(8), 1095–1107.

Hawkins, E. and Sutton, R. (2011), 'The potential to narrow uncertainty in projections of regional precipitation change', *Clim. Dynam.* DOI:10.1007/s00382-010-0810-6.

Hawkins, E. and Sutton, R. (2012), 'Time of emergence of climate signals', *Geophys. Res. Lett.* **39**, L01702.

Hay, L. E. and Clark, M. P. (2003), 'Use of statistically and dynamically downscaled atmospheric model output for hydrologic simulations in three mountainous basins in the western United States', *J. Hydrol.* **282**, 56–75.

Hay, L. E., Clark, M. P., Wilby, R. L., Gutowski, W. J., Leavesley, G. H., Pan, Z., Arritt, R. W. and Takle, E. S. (2002), 'Use of regional climate model output for hydrologic simulations', *J. Hydrometeorol.* **3**(5), 571–590.

Hay, L. E., McCabe, G. J., Wolock, D. M. and Ayers, M. A. (1991), 'Simulation of precipitation by weather type analysis', *Wat. Resour. Res.* **27**, 493–501.

Haylock, M. R., Gawley, G. C., Harpham, C., Wilby, R. L. and Goodess, C. M. (2006), 'Downscaling heavy precipitation over the United Kingdom: A comparison of dynamical and statistical methods and their future scenarios', *Int. J. Climatol.* **26**(10), 1397–1415.

Haylock, M. R., Hofstra, N., Klein Tank, A. M. G., Klok, E. J., Jones, P. D. and New, M. (2008), 'A European daily high-resolution gridded data set of surface temperature and precipitation for 1950–2006', *J. Geophys. Res.* **113**, 20119.

Hazeleger, W., Severijns, C., Semmler, T., Ştefănescu, S., Yang, S., Wang, X., Wyser, K., Dutra, E., Baldasano, J. M., Bintanja, R., Bougeault, P., Caballero, R., Ekman, A. M. L., Christensen, J. H., van den Hurk, B., Jimenez, P., Jones, C., Kållberg, P., Koenigk, T., Mc Grath, R., Miranda, P., van Noije, T., Palmer, T., Parodi, J. A., Schmith, T., Selten, F., Storelvmo, T., Sterl, A., Tapamo, H., Vancoppenolle, M., Viterbo, P. and Willen, U. (2010), 'EC-Earth: a seamless earth-system prediction approach in action', *Bull. Amer. Meteorol. Soc.* **91**(10), 1357–1363.

Hazeleger, W., van den Hurk, B. J. J. M., Min, E., van Oldenborgh, G. J., Petersen, A. C., Stainforth, D. A., Vasileiadou, E. and Smith, L. A. (2015), 'Tales of future weather', *Nat. Clim. Change* **5**, 107–113.

He, J. and Soden, B. J. (2015), 'Anthropogenic weakening of the tropical circulation: The relative roles of direct CO_2 forcing and sea surface temperature change', *J. Climate* **28**(22), 8728–8742.

Heimann, D. (1986), 'Estimation of regional surface layer wind field characteristics using a three-layer mesoscale model', *Beiträge zur Physik der Atmosphäre* **59**, 518–537.

Held, I. M. and Soden, B. J. (2006), 'Robust responses of the hydrological cycle to global warming', *J. Climate* **19**(21), 5686–5699.

Held, I. M., Ting, M. and Wang, H. (2002), 'Northern winter stationary waves: theory and modeling', *J. Climate* **15**(16), 2125–2144.

Hempel, S., Frieler, K., Warszawski, L., Schewe, J. and Piontek, F. (2013), 'A trend-preserving bias correction - the ISI-MIP approach', *Earth Syst. Dynam.* **4**, 219–236.

Henderson-Sellers, A. (1996), 'Can we integrate climatic modelling and assessment?', *Environ. Mod. Assess.* **1**(1–2), 59–70.

Hertig, E., Beck, C., Wanner, H. and Jacobeit, J. (2015), 'A review of non-stationarities in climate variability of the last century with focus on the North Atlantic–European sector', *Earth Sci. Rev.* **147**, 1–17.

Hertig, E. et al. (2017), 'Validation of extremes from the Perfect-Predictor Experiment of the COST Action VALUE', *Int. J. Climatol., subm.*

Hess, P. and Brezowsky, H. (1977), *Katalog der Großwetterlagen Europas (1861–1976), Selbstverlag des Deutschen Wetterdienstes Bd. 15*, Berichte des Deutschen Wetterdienstes, Offenbach am Main.

Hewitson, B. C. (2011), 'Meeting user needs: climate service limits, ideals, & realities', http://www.wcrp-climate.org/conference2011/orals/A6/Hewitson_A6.pdf.

Hewitson, B. C. (2016), 'CORDEX gaps and the distillation dilemma', http://www.icrc-cordex2016.org/images/pdf/Programme/presentations/plenary_1/Pl1_4_Hewitson_Bruce.pdf.

Hewitson, B. C. and Crane, R. G. (1994), *Neural nets: applications in Geography*, Springer.

Hewitson, B. C. and Crane, R. G. (1996), 'Climate downscaling: techniques and application', *Clim. Res.* **7**, 85–95.

Hewitson, B. C., Daron, J., Crane, R. G., Zermoglio, M. F. and Jack, C. (2014), 'Interrogating empirical-statistical downscaling', *Clim. Change* **122**, 539–554.

Hewitt, C. D. (2005), 'The ENSEMBLES Project: Providing ensemble-based predictions of climate changes and their impacts', *EGGS newsletter* **13**, 22–25.

Hewitt, C., Mason, S. and Walland, D. (2012), 'The global framework for climate services', *Nat. Clim. Change* **2**(12), 831–832.

Hidalgo, H. G., Dettinger, M. D. and Cayan, D. R. (2008), 'Downscaling with constructed analogues: Daily precipitation and temperature fields over the United States'. California Energy Commission, PIER Energy-Related Environmental Research. CEC-500-2007-123.

Hiebl, J. and Frei, C. (2016), 'Daily temperature grids for Austria since 1961 – concept, creation and applicability', *Theor. Appl. Climatol.* **124**(1–2), 161–178.

Hobaek Haff, I., Frigessi, A. and Maraun, D. (2015), 'How well do regional climate models simulate the spatial structure of precipitation? An application of pair-copula constructions', *J. Geophys. Res.* **120**(7), 2624–2646.

Hofstätter, M., Ganekind, M. and Hiebl, J. (2013), 'GPARD-6: A new 60-year gridded precipitation dataset for Austria based on daily rain gauge measurements'. Conference contribution: DACH 2013, Deutsh-Osterreichisch-Schweizerische Meteorologen-Tagung. Innsbruck, Austria.

Hofstra, N., New, M. and McSweeney, C. (2010), 'The influence of interpolation and station network density on the distributions and trends of climate variables in daily gridded data', *Clim. Dynam.* **35**, 841–858.

Holden, P. B. and Edwards, N. R. (2010), 'Dimensionally reduced emulation of an AOGCM for application to integrated assessment modelling', *Geophys. Res. Lett.* **37**(21).

Holton, J. R. and Hakim, G. J. (2013), *An introduction to dynamic meteorology*, 5 edn, Elsevier.

Hoppe, R., Wesselink, A. and Cairns, R. (2013), 'Lost in the problem: the role of boundary organisations in the governance of climate change', *WIREs Clim. Change* **4**(4), 283–300.

Hoskins, B. J. and Karoly, D. J. (1981), 'The steady linear response of a spherical atmosphere to thermal and orographic forcing', *J. Atmos. Sci.* **38**(6), 1179–1196.

Hoskins, B. J. and Valdes, P. J. (1990), 'On the existence of storm-tracks', *J. Atmos. Sci.* **47**(15), 1854–1864.

Hotelling, H. (1933), 'Analysis of a complex of statistical variables into principal components', *J. Educ. Psychol.* **24**(6), 417.

Hotelling, H. (1935), 'The most predictable criterion', *J. Educ. Psychol.* **26**(2), 139.

Howcroft, J. G. (1966), 'Fine-mesh limited-area forecasting model', Technical report, U.S. Air Weather Service, Scott Air Force Base.

Hu, Y., Maskey, S. and Uhlenbrook, S. (2013), 'Downscaling daily precipitation over the Yellow River source region in China: a comparison of three statistical downscaling methods', *Theor. Appl. Climatol.* **112**(3–4), 447–460.

Hughes, J. P. and Guttorp, P. (1994), 'A class of stochastic models for relating synoptic atmospheric patterns to regional hydrologic phenomena', *Wat. Resour. Res.* **30**(5), 1535–1546.

Hughes, J. P., Guttorp, P. and Charles, S. P. (1999), 'A non-homogeneous hidden Markov model for precipitation occurrence', *J. Roy. Stat. Soc. C* **48**(1), 15–30.

Hulme, M. (2007), 'The appliance of science', *The Guardian*. www.theguardian.com/society/2007/mar/14/scienceofclimatechange.climatechange.

Hulme, M., Jenkins, G. J., Lu, X., Turnpenny, J. R., Mitchell, T. D., Jones, R. G., Lowe, J., Murphy, J. M., Hassell, D., Boorman, P., McDonald, R. and Hill, S. (2002), 'Climate Change Scenarios for the United Kingdom. The UKCIP02 Scientific Report', Technical report, Tyndall Centre for Climate Change Research, School of Environmental Sciences, University of East Anglia, Norwich, UK.

Huntingford, C. and Cox, P. M. (2000), 'An analogue model to derive additional climate change scenarios from existing GCM simulations', *Clim. Dynam.* **16**(8), 575–586.

Huth, R. (1999), 'Statistical downscaling in central Europe: evaluation of methods and potential predictors', *Clim. Res.* **13**, 91–101, doi: 10.3354/cr013091.

Huth, R. (2002), 'Statistical downscaling of daily temperature in Central Europe', *J. Climate* **15**, 1731–1742.

Huth, R. (2004), 'Sensitivity of local daily temperature change estimates to the selection of downscaling models and predictors', *J. Climate* **17**(3), 640–652.

Huth, R. (2005), 'Downscaling of humidity variables: a search for suitable predictors and predictands', *Int. J. Climatol.* **25**(2), 243–250.

Huth, R., Kliegrova, S. and Metelka, L. (2008), 'Non-linearity in statistical downscaling: does it bring an improvement for daily temperature in Europe?', *Int. J. Climatol.* **28**(4), 465–477.

Huth, R., Miksovsky, J., Stepanek, P., Belda, M., Farda, A., Chladova, Z. and Pisoft, P. (2015), 'Comparative validation of statistical and dynamical downscaling models on a dense grid in central Europe: temperature', *Theor. Appl. Climatol.* **120**(3–4), 533–553.

Hyndman, R. J. and Grunwald, G. K. (2000), 'Applications: Generalized additive modelling of mixed distribution Markov models with application to Melbourne's rainfall', *Aust. N. Z. J. Stat.* **42**(2), 145–158.

Ineson, S. and Scaife, A. A. (2009), 'The role of the stratosphere in the European climate response to El Niño', *Nat. Geosci.* **2**(1), 32–36.

IPCC (1988), 'Report of the first session of the WMO/UNEP Intergovernmental Panel on Climate Change (IPCC)', *World Climate Programme Publications Series*. Geneva, 9–11 November.

Isotta, F. A., Frei, C., Weilguni, V., Perčec Tadić, M., Lassegues, P., Rudolf, B., Pavan, V., Cacciamani, C., Antolini, G., Ratto, S. M., Munari, M., Micheletti, S., Bonati, V., Lussana, C., Panettieri, C. R. E., Marigo, G. and Vertacnik, G. (2014), 'The climate of daily precipitation in the Alps: development and analysis of a high-resolution grid dataset from pan-Alpine rain-gauge data', *Int. J. Climatol.* **34**(5), 1657–1675.

Jacob, D., Petersen, J., Eggert, B., Alias, A., Christensen, O. B., Bouwer, L. M., Braun, A., Colette, A., Déqué, M., Georgievski, G., Georgopoulou, E., Gobiet, A., Nikulin, G., Haensler, A., Hempelmann, N., Jones, C., Keuler, K., Kovats, S., Kröner, N., Kotlarski, S., Kriegsmann, A., Martin, E., E, E. v., Moseley, C., Pfeifer, S., Preuschmann, S., Radtke, K., Rechid, D., Rounsevell, M., Samuelsson, P., Somot, S., Soussana, J.-F., Teichmann, C., Valentini, R., Vautard, R. and Weber, B. (2014), 'EURO-CORDEX: New high-resolution climate change projections for European impact research', *Reg. Environ. Change* **14**, 563–578.

Jacobeit, J., Hertig, E., Seubert, S. and Lutz, K. (2014), 'Statistical downscaling for climate change projections in the Mediterranean region: methods and results', *Reg. Environ. Change* **14**, 1891–1906.

Jacobeit, J., Wanner, H., Luterbacher, J., Beck, C., Philipp, A. and Sturm, K. (2003), 'Atmospheric circulation variability in the North-Atlantic-European area since the mid-seventeenth century', *Clim. Dynam.* **20**(4), 341–352.

Jakob, C. (2014), 'Going back to basics', *Nat. Clim. Change* **4**(12), 1042–1045.

Joe, H. (1996), *Distributions with fixed marginals and related topics*, IMS, Hayward, CA, chapter 'Families of m-variate distributions with given Margins and m(m-1)/2 Dependence Parameters'.

Jolliffe, I. T. (1986), *Principal component analysis*, Springer.

Jolliffe, I. T. (1990), 'Principal component analysis: a beginner's guide I. Introduction and application', *Weather* **45**(10), 375–382.

Jolliffe, I. T. (1993), 'Principal component analysis: A beginner's guide II. Pitfalls, myths and extensions', *Weather* **48**(8), 246–253.

Jolliffe, I. T. (1995), 'Rotation of principal components: choice of normalization constraints', *J. Appl. Stat.* **22**(1), 29–35.

Jolliffe, I. T. and Stephenson, D. B., eds. (2003), *Forecast verication: a practitioner's guide in atmospheric science*, Wiley.

Jones, D. A., Wang, W. and Fawcett, R. (2009*a*), 'High-quality spatial climate data-sets for Australia', *Austral. Meteorol. Oceanogr. J.* **58**(4), 233.

Jones, P. D., Harpham, C. and Briffa, K. R. (2013), 'Lamb weather types derived from reanalysis products', *Int. J. Climatol.* **33**(5), 1129–1139.

Jones, P. D., Kilsby, C. G., Harpham, C., Glenis, V. and Burton, A. (2009*b*), 'UK Climate Projections science report: Projections of future daily climate for the UK from the Weather Generator', Technical report, University of Newcastle, UK.

Jones, P. D., Lister, D. H., Osborn, T. J., Harpham, C., Salmon, M. and Morice, C. P. (2012), 'Hemispheric and large-scale land-surface air temperature variations: An extensive revision and an update to 2010', *J. Geophys. Res.* **117**(D5).

Jones, P. D. and Moberg, A. (2003), 'Hemispheric and large-scale surface air temperature variations: An extensive revision and an update to 2001', *J. Climate* **16**(2), 206–223.

Jordan, M. C. (1874), 'Mémoires sur les formes bilinéaires', *J. de Mathémathiques Pures et Appliquées* **19**, 35–54.

Joyce, R. J., Janowiak, J. E., Arkin, P. A. and Xie, P. (2004), 'CMORPH: A method that produces global precipitation estimates from passive microwave and infrared data at high spatial and temporal resolution', *J. Hydrometeorol.* **5**(3), 487–503.

Jung, T., Miller, M. J., Palmer, T. N., Towers, P., Wedi, N., Achuthavarier, D., Adams, J. M., Altshuler, E. L., Cash, B. A., Kinter III, J. L., Marx, L., Stan, C. and Hodges, K. I. (2012), 'High-resolution global climate simulations with the ECMWF model in Project Athena: Experimental design, model climate, and seasonal forecast skill', *J. Climate* **25**(9), 3155–3172.

Jury, M. W., Prein, A. F., Truhetz, H. and Gobiet, A. (2015), 'Evaluation of CMIP5 models in the context of dynamical downscaling over Europe', *J. Climate* **28**(14), 5575–5582.

Kaczmarska, J., Isham, V. and Onof, C. (2014), 'Point process models for fine-resolution rainfall', *Hydrol. Sci. J.* **59**(11), 1972–1991.

Kallache, M., Vrac, M. and Michelangeli, P.-A. (2011), 'Nonstationary probabilistic downscaling of extreme precipitation', *J. Geophys. Res.* **116**, D05113.

Kalnay, E. (2003), *Atmospheric modeling, data assimilation and predictability*, Cambridge University Press.

Kalnay, E., Kanamitsu, M., Kistler, R., Collins, W., Deaven, D., Gandin, L., Iredell, M., Saha, S., White, G., Woollen, J., Zhu, Y., Chelliah, M., Ebisuzaki, W., Higgins, W., Janowiak, J., Mo, K. C., Ropelewski, C., Wang, J., Leetmaa, A., Reynolds, R., Jenne, R. and Joseph, D. (1996), 'The NCEP/NCAR reanalysis project', *Bull. Amer. Meteorol. Soc.* **77**, 437–471.

Karl, T. R., Diamond, H. J., Bojinski, S., Butler, J. H., Dolman, H., Haeberli, W., Harrison, D. E., Nyong, A., Rösner, S., Seiz, G., Trenberth, K., Westermeyer, W. and Zillman, J. (2010), 'Observation needs for climate information, prediction and application: Capabilities of existing and future observing systems', *Proc. Environ. Sci.* **1**, 192–205.

Karl, T. R., Wang, W.-C., Schlesinger, M. E., Knight, R. W. and Portman, D. (1990), 'A method of relating general circulation model simulated climate to the observed local climate. Part I: seasonal statistics', *J. Climate* **3**, 1053–1079.

Katz, R. W. (1977), 'Precipitation as a chain-dependent process', *J. Appl. Meteorol.* **16**(7), 671–676.

Katz, R. W. and Parlange, M. B. (1993), 'Effects of an index of atmospheric circulation on stochastic properties of precipitation', *Wat. Resour. Res.* **29**(7), 2335–2344.

Katz, R. W. and Parlange, M. B. (1998), 'Overdispersion phenomenon in stochastic modeling of precipitation', *J. Climate* **11**(4), 591–601.

Kay, A. L. and Jones, R. G. (2012), 'Comparison of the use of alternative UKCP09 products for modelling the impacts of climate change on flood frequency', *Clim. Change* **114**(2), 211–230.

Keeley, S. P. E., Sutton, R. T. and Shaffrey, L. C. (2012), 'The impact of North Atlantic sea surface temperature errors on the simulation of North Atlantic European region climate', *Q. J. R. Meteorol. Soc.* **138**, 1774–1783.

Keeling, C. D. (1960), 'The concentration and isotopic abundances of carbon dioxide in the atmosphere', *Tellus* **12**(2), 200–2003.

Keenlyside, N. S., Latif, M., Jungclaus, J., Kornblueh, L. and Roeckner, E. (2008), 'Advancing decadal-scale climate prediction in the North Atlantic sector', *Nature* **453**(7191), 84–88.

Keller, D. E., Fischer, A. M., Liniger, M. A., Appenzeller, C. and Knutti, R. (2016), 'Testing a weather generator for downscaling climate change projections over Switzerland', *Int. J. Climatol.*

Keller, D., Fischer, A. M., Frei, C., Liniger, M. A., Appenzeller, C. and Knutti, R. (2015), 'Implementation and validation of a Wilks-type multi-site daily precipitation generator over a typical Alpine river catchment', *Hydrol. Earth Syst. Sci.* **19**, 2163–2177.

Kendon, E. J., Roberts, N. M., Fowler, H. J., Roberts, M. J., Chan, S. C. and Senior, C. A. (2014), 'Heavier summer downpours with climate change revealed by weather forecast resolution model', *Nat. Clim. Change* **4**, 570–576.

Kendon, E. J., Rowell, D. P. and Jones, R. G. (2010), 'Mechanisms and reliability of future projected changes in daily precipitation', *Clim. Dynam.* **35**(2–3), 489–509.

Kennedy, M. C. and O'Hagan, A. (2001), 'Bayesian calibration of computer models', *J. Roy. Stat. Soc. B* **63**(3), 425–464.

Kerr, R. A. (2011*a*), 'Time to adapt to a warming world, but where's the science?', *Science* **334**(6059), 1052–1053.

Kerr, R. A. (2011*b*), 'Vital details of global warming are eluding forecasters', *Science* **334**(6053), 173–174.

Kettle, H. and Thompson, R. (2004), 'Statistical downscaling in European mountains: verification of reconstructed air temperature', *Clim. Res.* **26**(2), 97–112.

Kida, H., Koide, T., Sasaki, H. and Chiba, M. (1991), 'A new approach for coupling a limited area model to a GCM for regional climate simulations', *J. Meteorol. Soc. Jap.* **69**(6), 723–728.

Kilsby, C. G., Jones, P. D., Burton, A., Ford, A. C., Fowler, H. J., Harpham, C., James, P., Smith, A. and Wilby, R. L. (2007), 'A daily weather generator for use in climate change studies', *Env. Mod. Soft.* **22**, 1705–1719.

Kim, J.-W., J.-T. Chang, Baker, N. L., Wilks, D. S. and Gates, W. L. (1984), 'The statistical problem of climate inversion: determination of the relationship between local and large-scale climate', *Mon. Wea. Rev.* **112**, 2069–2077.

Kim, J., Waliser, D. E., Mattmann, C. A., Goodale, C. E., Hart, A. F., Zimdars, P. A., Crichton, D. J., Nikulin, C. J. G., Hewitson, B., Jack, C., Lennard, C. and Favre, A. (2014), 'Evaluation of the CORDEX-Africa multi-RCM hindcast: systematic model errors', *Clim. Dynam.* **42**(5–6), 1189–1202.

Kistler, R., Kalnay, E., Collins, W., Saha, S., White, G., Woollen, J., Chelliah, M., Ebisuzaki, W., Kanamitsu, M., Kousky, V., van den Dool, H., Jenne, R. and Fiorino, M. (2001), 'The NCEP–NCAR 50-year reanalysis: monthly means CD–ROM and documentation', *Bull. Amer. Meteorol. Soc.* **82**(2), 247–267.

Klein Tank, A. M. G., Wijngaard, J. B., Können, G. P., Böhm, R., Demarée, G., Gocheva, A., Mileta, M., Pashiardis, S., Hejkrlik, L., Kern-Hansen, C., Heino, R., Bessemoulin, P., Müller-Westermeier, G., Tzanakou, M., Szalai, S., Pálsdóttir, T., Fitzgerald, D., Rubin, S., Capaldo, M., Maugeri, M., Leitass, A., Bukantis, A., Aberfeld, R., van Engelen, A. F. V., Forland, E., Mietus, M., Coelho, F., Mares, C., Razuvaev, V., Nieplova, E., Cegnar, T., López, J. A., Dahlström, B., Moberg, A., Kirchhofer, W., Ceylan, A., Pachaliuk, O., Alexander, L. V. and Petrovic, P. (2002), 'Daily dataset of 20th-century surface air temperature and precipitation series for the European Climate Assessment', *Int. J. Climatol.* **22**(12), 1441–1453.

Klein Tank, A. M. G., Zwiers, F. W. and Zhang, X. (2009), *Guidelines on analysis of extremes in a changing climate in support of informed decisions for adaptation, Climate Data and Monitoring WCDMP-No. 72*, World Meteorological Organisation.

Klein, W. H. (1948), 'Winter precipitation from the 700-millibar circulation', *Bull. Amer. Meteorol. Soc.* **9**, 439–453.

Klein, W. H. and Glahn, H. R. (1974), 'Forecasting local weather by means of model output statistics', *Bull. Amer. Meteorol. Soc.* **55**(10), 1217–1227.

Klein, W. H., Lewis, B. M. and Enger, I. (1959), 'Objective prediction of five-day mean temperatures during winter', *J. Meteorol.* **16**, 672–682.

Knight, J. R., Folland, C. K. and Scaife, A. A. (2006), 'Climate impacts of the Atlantic multi-decadal oscillation', *Geophys. Res. Lett.* **33**(17).

Kobayash, S., Yukinari, O., Harada, Y., Ebita, A., Moriya, M., Onoda, H., Onogi, K., Kamahori, H., Kobayashi, C., Endo, H., Miyaoka, K. and Takahashi, K. (2015), 'The JRA-55 reanalysis: general specifications and basic characteristics', *J. Meteorol. Soc. Japan II* **93**(1), 5–48.

Koenker, R. (2005), *Quantile regression*, number 38, Cambridge University Press.

Kohonen, T. (1998), 'The self-organizing map', *Neurocomputing* **21**(1–3), 1–6.

Kotlarski, S., Keuler, K., Christensen, O. B., Colette, A., Déqué, M., Gobiet, A., Goergen, K., Jacob, D., Lüthi, D., van Meijgaard, E., Nikulin, G., Schär, C., Teichmann, C., Vautard, R., Warrach-Sagi, K. and Wulfmeyer, V. (2014), 'Regional climate modelling on European scales: a joint standard evaluation of the EURO-CORDEX RCM ensemble', *Geosci. Model. Dev.* **7**, 1297–1333.

Krueger, O., Schenk, F., Feser, F. and Weisse, R. (2013), 'Inconsistencies between long-term trends in storminess derived from the 20CR reanalysis and observations', *J. Climate* **26**(3), 868–874.

Kruizinga, S. and Murphy, A. H. (1983), 'Use of an analogue procedure to formulate objective probabilistic temperature forecasts in the netherlands', *Mon. Wea. Rev.* **111**(11), 2244–2254.

Kukla, G., Gavin, J. and Karl, T. R. (1986), 'Urban warming', *J. Clim. Appl. Meteorol.* **25**(9), 1265–1270.

Kundzewicz, Z. W. and Stakhiv, E. Z. (2010), 'Are climate models "ready for prime time" in water resources management applications, or is more research needed?', *Hydrol. Sci. J.* **55**, 1085–1089.

Kuo, H.-L. (1965), 'On formation and intensification of tropical cyclones through latent heat release by cumulus convection', *J. Atmos. Sci.* **22**(1), 40–63.

Kutzbach, J. E. (1967), 'Empirical eigenvectors of sea-level pressure, surface temperature and precipitation complexes over North America', *J. Appl. Meteorol.* **6**(5), 791–802.

Kysely, J. and Plavcova, E. (2010), 'A critical remark on the applicability of E-OBS European gridded temperature data set for validating control climate simulations', *J. Geophys. Res.* **115**, D23118.

Lackmann, G. M. (2015), 'Hurricane Sandy before 1900 and after 2100', *Bull. Amer. Meteorol. Soc.* **96**(4), 547–560.

Lall, U. and Sharma, A. (1996), 'A nearest neighbor bootstrap for resampling hydrologic time series', *Wat. Resour. Res.* **32**(3), 679–693.

Lamb, H. H. (1972), *British Isles weather types and a register of the daily sequence of circulation patterns 1861–1971*, HMSO London No. 116, Meteorol. Off. Geophys. Mem.

Laprise, R. (2008), 'Regional climate modelling', *J. Comp. Phys.* **227**(7), 3641–3666.

Laprise, R. (2014), 'Comment on the added value to global model projections of climate change by dynamical downscaling: a case study over the continental US using the GISS-ModelE2 and WRF models" by Racherla et al.', *J Geophys. Res.* **119**(7), 3877–3881.

Latif, M. and Park, W. (2012), *The future of the world's climate*, Elsevier, Amsterdam, The Netherlands, chapter 'Climatic Variability on Decadal to Century Time-Scales', pp. 167–195.

Leadbetter, M. R., Lindgren, G. and Rootzen, H. (1983), *Extremes and related properties of random sequences and processes*, Springer Series in Statistics, Springer.

Leander, R. and Buishand, T. A. (2007), 'Resampling of regional climate model output for the simulation of extreme river flows', *J. Hydro.* **332**(3), 487–496.

Legates, D. R. (1991), 'The effect of domain shape on principal components analyses', *Int. J. Climatol.* **11**(2), 135–146.

Leloup, J., Lengaigne, M. and Boulanger, J.-P. (2008), 'Twentieth century ENSO characteristics in the IPCC database', *Clim. Dynam.* **30**(2–3), 277–291.

Lempert, R. J., Popper, S. W. and Banks, S. C. (2003), *Shaping the next one hundred years. New methods for quantitative, long-term policy analysis*, RAND.

Lempert, R., Nakicenovic, N., Sarewitz, D. and Schlesinger, M. (2004), 'Characterising climate-change uncertainties for decision-makers', *Clim. Change* **65**, 1–9.

Lenderink, G., Buishand, A. and van Deursen, W. (2007), 'Estimates of future discharges of the river Rhine using two scenario methodologies: direct versus delta approach', *Hydrol. Earth Syst. Sci.* **11**(3), 1145–1159.

Leonard, M., Westra, S., Phatak, A., Lambert, M., van den Hurk, B., McInnes, K., Risbey, J., Schuster, S., Jacob, D. and Stafford-Smith, M. (2014), 'A compound event framework for understanding extreme impacts', *WIREs Clim. Change* **4**(1), 113–128.

Li, C., Sinha, E., Horton, D. E., Diffenbaugh, N. S. and Michalak, A. M. (2014), 'Joint bias correction of temperature and precipitation in climate model simulations', *J. Geophys. Res.* **119**(23), 13153–13162.

Li, G. and Xie, S.-P. (2014), 'Tropical biases in CMIP5 multimodel ensemble: the excessive equatorial Pacific cold tongue and double ITCZ problems', *J. Climate* **27**(4), 1765–1780.

Li, H., Sheffield, J. and Wood, E. F. (2010), 'Bias correction of monthly precipitation and temperature fields from Intergovernmental Panel on Climate Change AR4 models using equidistant quantile matching', *J. Geophys. Res.* **115**, D10101.

Limpasuvan, V., Thompson, D. W. J. and Hartmann, D. L. (2004), 'The life cycle of the Northern Hemisphere sudden stratospheric warmings', *J. Climate* **17**(13), 2584–2596.

Liu, P. et al. (2009), 'An MJO simulated by the NICAM at 14- and 7-km resolutions', *Mon. Wea. Rev.* **137**, 3254–3268.

Livneh, B., Bohn, T. J., Pierce, D. W., Munoz-Arriola, F., Nijssen, B., Vose, R., Cayan, D. R. and Brekke, L. (2015), 'A spatially comprehensive, hydrometeorological data set for Mexico, the US, and Southern Canada 1950–2013', *Scientific Data* **2**, 150042.

Lorenz, E. N. (1956), 'Empirical orthogonal functions and statistical weather prediction', Technical report, Massachusetts Institute of Technology, Dept. of Meteorology.

Lorenz, E. N. (1969), 'Atmospheric predictability as revealed by naturally occuring analogs', *J. Atmos. Sci.* **26**, 639–646.

Lorenz, P. and Jacob, D. (2005), 'Influence of regional scale information on the global circulation: A two-way nesting climate simulation', *Geophys. Res. Lett.* **32**, L18706.

Lorenz, P. and Jacob, D. (2010), 'Validation of temperature trends in the ENSEMBLES regional climate model runs driven by ERA40', *Clim. Res.* **44**, 167–177.

Lovejoy, S. and Schertzer, D. (2010), 'Towards a new synthesis for atmospheric dynamics: space-time cascades', *Atmos. Res.* **96**(1), 1–52.

Lu, J., Vecchi, G. A. and Reichler, T. (2007), 'Expansion of the Hadley cell under global warming', *Geophys. Res. Lett.* **34**, L06805.

Luca, A. D., de Elía, R. and Laprise, R. (2012), 'Potential for added value in precipitation simulated by high-resolution nested regional climate models and observations', *Clim. Dynam.* **38**(5–6), 1229–1247.

Lynch, A. H. and Cassado, J. J. (2006), *Applied atmospheric dynamics*, Wiley.

MacQueen, J. (1967), 'Some methods for classification and analysis of multivariate observations', *in Proceedings of the fifth Berkeley symposium on mathematical statistics and probability*, Vol. 1, Univ. of Calif. Press, pp. 281–297.

Manabe, S., Smagorinsky, J. and Strickler, R. F. (1965), 'Simulated climatology of a general circulation model with a hydrologic cycle', *Mon. Wea. Rev.* **93**(12), 769–798.

Manabe, S. and Wetherald, R. T. (1967), 'Thermal equilibrium of the atmosphere with a given distribution of relative humidity', *J. Atmos. Sci.* **24**(3), 241–259.

Manabe, S. and Wetherald, R. T. (1975), 'The effects of doubling the CO_2 concentration on the climate of a general circulation model', *J. Atmos. Sci.* **32**(1), 3–15.

Mantua, N. J. and Hare, S. R. (2002), 'The Pacific decadal oscillation', *J. Ocean.* **58**(1), 35–44.

Mantua, N. J., Hare, S. R., Zhang, Y., Wallace, J. M. and Francis, R. C. (1997), 'A Pacific interdecadal climate oscillation with impacts on salmon production', *Bull. Amer. Meteorol. Soc.* **78**(6), 1069–1079.

Maraun, D. (2012), 'Nonstationarities of regional climate model biases in European seasonal mean temperature and precipitation sums', *Geophys. Res. Lett.* **39**, L06706.

Maraun, D. (2013a), 'Bias correction, quantile mapping and downscaling: revisiting the inflation issue', *J. Climate* **26**, 2137–2143.

Maraun, D. (2013b), 'When will trends in European mean and heavy daily precipitation emerge?', *Env. Res. Lett.* **8**, 014004.

Maraun, D. (2014), 'Reply to comment on "Bias correction, quantile mapping and downscaling: revisiting the inflation issue"', *J. Climate* **27**, 1821–1825.

Maraun, D. (2016), 'Bias correcting climate change simulations - a critical review', *Curr. Clim. Change Rep.* **2**(4), 211–220.

Maraun, D., Huth, R., Gutiérrez, J. M., San Martín, D., Dubrovsky, M., Fischer, A., Hertig, E., Soares, P. M. M., Bartholy, J., Pongrácz, R., Widmann, M., Casado, M. J., Ramos, P. & Bedia, J. (2017a), 'The VALUE perfect predictor experiment: evaluation of temporal variability', *Int. J. Climatol.*, DOI 10.1002/joc.5222, published online.

Maraun, D., Osborn, T. J. and Gillett, N. P. (2008), 'United Kingdom daily precipitation intensity: improved early data, error estimates and an update from 2000 to 2006', *Int. J. Climatol.* **28**(6), 833–842. DOI 10. 1002/joc. 1672.

Maraun, D., Osborn, T. J. and Rust, H. W. (2012), 'The influence of synptic airflow on UK daily precipitation extremes. Part II: regional climate model and E-OBS data validation', *Clim. Dynam.* **39**, 287–301.

Maraun, D., Rust, H. W. and Osborn, T. J. (2009), 'The annual cycle of heavy precipitation across the UK: a model based on extreme value statistics', *Int. J. Climatol.* **29**, 1731–1744. DOI 10.1002/joc.1811.

Maraun, D., Rust, H. W. and Osborn, T. J. (2010*a*), 'Synoptic airflow and UK daily precipitation extremes. Development and validation of a vector generalised model', *Extremes* **13**, 133–153.

Maraun, D., Shepherd, T. G., Widmann, M., Zappa, G., Walton, D., Gutierrez, J. M., Hagemann, S., Richter, I., Soares, P. M. M., Hall, A. and Mearns, L. (2017*b*), 'Towards process-informed bias correction of climate change simulations', *Nat. Clim. Change*, online first, DOI 10.1038/nclimate3418.

Maraun, D., Wetterhall, F., Ireson, A. M., Chandler, R. E., Kendon, E. J., Widmann, M., Brienen, S., Rust, H. W., Sauter, T., Themeßl, M., Venema, V. K. C., Chun, K. P., Goodess, C. M., Jones, R. G., Onof, C., Vrac, M. and Thiele-Eich, I. (2010*b*), 'Precipitation downscaling under climate change. Recent developments to bridge the gap between dynamical models and the end user', *Rev. Geophys.* **48**, RG3003.

Maraun, D. and Widmann, M. (2015), 'The representation of location by a regional climate model in complex terrain', *Hydrol. Earth Syst. Sci.* **19**, 3449–3456.

Maraun, D. and Widmann, M. (2017), 'Cross-validation of bias corrected climate simulations is misleading', *Hydrol. Earth Syst. Sci.* in prep.

Maraun, D., Widmann, M., Gutierrez, J. M., Kotlarski, S., Chandler, R. E., Hertig, E., Wibig, J., Huth, R. and Wilcke, R. A. I. (2015), 'VALUE: A framework to validate downscaling approaches for climate change studies', *Earth's Future* **3**, 1–14.

Masato, G., Hoskins, B. J. and Woollings, T. (2013), 'Winter and summer Northern Hemisphere blocking in CMIP5 models', *J. Climate* **26**, 7044–7059.

Mason, S. J. (2004), 'Simulating climate over western North America using stochastic weather generators', *Clim. Change* **62**(1), 155–187.

Masson-Delmotte, V., Schulz, M., Abe-Ouchi, A., Beer, J., Ganopolski, A., González Rouco, J. F., Jansen, E., Lambeck, K., Luterbacher, J., Naish, T., Osborn, T., Otto-Bliesner, B., Quinn, T., Ramesh, R., Rojas, M., Shao, X. and Timmermann, A. (2013), *Climate Change 2013: The Physical Science Basis. Contribution of Working Group I to the Fifth Assessment Report of the Intergovernmental Panel on Climate Change*, Cambridge University Press, Cambridge, United Kingdom, and New York, NY, USA, chapter 'Information from Paleoclimate Archives'.

Mastrandrea, M. D., Heller, N. E., Root, T. L. and Schneider, S. H. (2010), 'Bridging the gap: linking climate-impacts research with adaptation planning and management', *Clim. Change* **100**(1), 87–101.

Matsikaris, A., Widmann, M. and Jungclaus, J. (2016), 'Assimilating continental mean temperatures to reconstruct the climate of the late pre-industrial period', *Clim. Dynam.* **46**(11–12), 3547–3566.

Matulla, C., Zhang, X., Wang, X. L., Wang, J., Zorita, E., Wagner, S. and Von Storch, H. (2008), 'Influence of similarity measures on the performance of the analog method for downscaling daily precipitation', *Clim. Dynam.* **30**(2–3), 133–144.

Matyasovszky, I., Bogardi, I., Bardossy, A. and Duckstein, L. (1993), 'Space-time precipitation reflecting climate change', *Hydrol. Sci. J.* **38**(6), 539–558.

Maurer, E. P. (2007), 'Fine-resolution climate projections enhance regional climate change impact studies', *EOS* **88**(47), 504.

Maurer, E. P., Brekke, L., Pruitt, T., Thrasher, B., Long, J., Duffy, P., Dettinger, M., Cayan, D. and Arnold, J. (2014), 'An enhanced archive facilitating climate impact and adaptation analysis', *Bull. Amer. Meteorol. Soc.* **95**(7), 1011–1019.

Maurer, E. P. and Hidalgo, H. G. (2008), 'Utility of daily vs. monthly large-scale climate data: an intercomparison of two statistical downscaling methods', *Hydrol. Earth Syst. Sci.* **12**, 551–563.

Maurer, E. P., Hidalgo, H. G., Das, T., Dettinger, M. D. and Cayan, D. R. (2010), 'The utility of daily large-scale climate data in the assessment of climate change impacts on daily streamflow in California', *Hydrol. and Earth System Sci.* **14**(6), 1125–1138.

Maurer, E. P., Wood, A. W., Adam, J. C., Lettenmaier, D. P. and Nijssen, B. (2002), 'A long-term hydrologically-based data set of land surface fluxes and states for the conterminous United States', *J. Climate* **15**(22), 3237–3251.

McBean, G., McCarthy, J., Browning, K., Morel, P. and Rasool, I. (1990), *Climate Change. The IPCC Scientific Assessment. Report prepared for Intergovernmental Panel on Climate Change by Working Group I*, Cambridge University Press, Cambridge, United Kingdom, and New York, NY, USA, chapter 'Narrowing the Uncertainties: A Scientific Action Plan for Improved Prediction of Global Climate Change'.

McCabe, G. J., Palecki, M. A. and Betancourt, J. L. (2004), 'Pacific and Atlantic Ocean influences on multidecadal drought frequency in the United States', *Proc. Nat. Acad. Sci.* **101**(12), 4136–4141.

McCullagh, P. and Nelder, J. A. (1983), *Generalized linear Models*, Chapman and Hall, London.

McGuffie, K. and Henderson-Sellers, A. (2005), *A climate modelling primer*, John Wiley & Sons.

Mearns, L. O., Arritt, R., Biner, S., Bukovsky, M. S., McGinnis, S., Sain, S., Caya, D., Correia, J., Flory, D., Gutowski, W., Takle, E. S., Jones, R., Leung, R., Moufouma-Okia, W., McDaniel, L., Nunes, A. M. B., Qian, Y., Roads, J., Sloan, L. and Snyder, M. (2012), 'The North American Regional Climate Change Assessment Program. Overview of Phase I Results', *Bull. Amer. Meteorol. Soc.* **93**, 1337–1362.

Mearns, L. O., Gutowski, W. J., Jones, R., Leung, L.-Y., McGinnis, S., Nunes, A. M. B. and Qian, Y. (2009), 'A regional climate change assessment program for North America', *EOS* **90**(36), 311–312.

Mearns, L. O., Katz, R. W. and Schneider, S. H. (1984), 'Extreme high-temperature events: changes in their probabilities with changes in mean temperature', *J. Clim. Appl. Meteorol.* **23**(12), 1601–1613.

Meehl, G. A., Covey, C., Delworth, T., Latif, M., McAvaney, B., Mitchell, J., Stouffer, R. and Taylor, K. (2007a), 'The WCRP CMIP3 multi-model dataset: a new era in climate change research', *Bull. Amer. Meteorol. Soc.* **88**, 1383–1394.

Meehl, G. A., Goddard, L., Murphy, J., Stouffer, R. J., Boer, G., Danabasoglu, G., Dixon, K., Giorgetta, M. A., Greene, A. M., Hawkins, E., Hegerl, G., Karoly, D., Keenlyside, N., Kimoto, M., Kirtman, B., Navarra, A., Pulwarty, R., Smith, D., Stammer, D. and Stockdale, T. (2009), 'Decadal prediction. Can it be skillfull?', *Bull. Amer. Meteorol. Soc.* **90**, 1467–1485.

Meehl, G. A., Stocker, T. F., Collins, W. D., Friedlingstein, P., Gaye, A. T., Gregory, J. M., Kitoh, A., Knutti, R., Murphy, J. M., Noda, A., Raper, S. C. B., Watterson, I. G., Weaver, A. J. and Zhao, Z.-C. (2007b), *Climate Change 2007: The Physical Science Basis. Contribution of Working Group I to the Fourth Assessment Report of the Intergovernmental Panel on Climate Change*, Cambridge University Press, Cambridge, United Kingdom, and New York, NY, USA, chapter 'Global Climate Projections'.

Meredith, E. P., Maraun, D., Semenov, V. A. and Park, W. (2015*a*), 'Evidence for added value of convection permitting models for studying changes in extreme precipitation', *J. Geophys. Res. Atmos.* **120**, 12,500–12,513.

Meredith, E. P., Semenov, V. A., Maraun, D., Park, W. and Chernokulsky, A. V. (2015*b*), 'Crucial role of Black Sea warming in amplifying the 2012 Krymsk precipitation extreme', *Nat. Geosci.* **8**(8), 615–619.

Merlis, T. M. (2015), 'Direct weakening of tropical circulations from masked CO_2 radiative forcing', *Proc. Nat. Acad. Sci.* **112**(43), 13167–13171.

Mestre, O., Gruber, C., Prieur, C., Caussinus, H. and Jourdain, S. (2011), 'SPLIDHOM: A method for homogenization of daily temperature observations', *J. Appl. Meteorol. Climatol.* **50**(11), 2343–2358.

Mezghani, A. and Hingray, B. (2009), 'A combined downscaling-disaggregation weather generator for stochastic generation of multisite hourly weather variables over complex terrain: development and multi-scale validation for the Upper Rhone River basin', *J. Hydrol.* **377**(3), 245–260.

Michelangeli, P.-A., Vrac, M. and Loukos, H. (2009), 'Probabilistic downscaling approaches: application to wind cumulative distribution functions', *Geophys. Res. Lett.* **36**(11).

Minobe, S., Kuwano-Yoshida, A., Komori, N., Xie, S.-P. and Small, R. J. (2008), 'Influence of the Gulf Stream on the troposphere', *Nature* **452**(7184), 206–209.

Mitchell, J. F. B., Johns, T. C., Eagles, M., Ingram, W. J. and Davis, R. A. (1999), 'Towards the construction of climate change scenarios', *Clim. Change* **41**(3), 547–581.

Moron, V., Robertson, A. W., Ward, M. N. and Ndiaye, O. (2008), 'Weather types and rainfall over Senegal. Part I: Observational analysis', *J. Climate* **21**(2), 266–287.

Mountain Research Initiative EDW Working Group (2015), 'Elevation-dependent warming in mountain regions of the world', *Nat. Clim. Change* **5**(5), 424–430.

Murphy, J. M., Sexton, D. M. H., Jenkins, G. J., Booth, B. B. B., Brown, C. C., Clark, R. T., Collins, M., Harris, G. R., Kendon, E. J., Betts, R. A., Brown, S. J., Humphrey, K. A., McCarthy, M. P., McDonald, R. E., Stephens, A., Wallace, C., Warren, R., Wilby, R. and Wood, R. A. (2009), 'UK climate projections science report: Climate change projections', Technical report, Met Office Hadley Centre, Exeter UK.

Myhre, G., Shindell, D., Bréon, F.-M., Collins, W., Fuglestvedt, J., Huang, J., Koch, D., Lamarque, J.-F., Lee, D., Mendoza, B., Nakajima, T., Robock, A., Stephens, G., Takemura, T. and Zhang, H. (2013), *Climate Change 2013: The Physical Science Basis. Contribution of Working Group I to the Fifth Assessment Report of the Intergovernmental Panel on Climate Change*, Cambridge University Press, Cambridge, United Kingdom, and New York, NY, USA, chapter 'Anthropogenic and Natural Radiative Forcing'.

Nakamura, H., Sampe, T., Goto, A., Ohfuchi, W. and Xie, S.-P. (2008), 'On the importance of mid-latitude oceanic frontal zones for the mean state and dominant variability in the tropospheric circulation', *Geophys. Res. Lett.* **35**, L15709.

Nakicenovic, N. and Swart, R., eds (2000), *Special report on emissions scenarios: a special report of Working Group III of the Intergovernmental Panel on Climate Change*, Cambridge University Press.

Nature (2010), 'Validation required', *Nature* **463**(7283), 849–849.

Neelin, J. D. (2010), *Climate change and climate modeling*, Cambridge University Press.

New, M., Hulme, M. and Jones, P. (1999), 'Representing twentieth-century space–time climate variability. Part I: Development of a 1961–90 mean monthly terrestrial climatology', *J. Climate* **12**(3), 829–856.

New, M., Hulme, M. and Jones, P. (2000), 'Representing twentieth-century space–time climate variability. Part II: development of 1901–96 monthly grids of terrestrial surface climate', *J. Climate* **13**(13), 2217–2238.

North, G. R., Bell, T. L., Cahalan, R. F. and Moeng, F. J. (1982), 'Sampling errors in the estimation of empirical orthogonal functions', *Mon. Wea. Rev.* **110**(7), 699–706.

Northrop, P. (1998), 'A clustered spatial-temporal model of rainfall', *Proc. Roy. Soc. A* **454**(1975), 1875–1888.

Novella, N. S. and Thiaw, W. M. (2013), 'African rainfall climatology version 2 for famine early warning systems', *J. Appl. Meteorol. Climatol.* **52**(3), 588–606.

Obukhov, A. M. (1947), 'Statistically homogeneous fields on a sphere', *Usp. Mat. Nauk* **2**(2), 196–198.

Obukhov, A. M. (1954), 'Statistical description of continuous fields', *Transactions of the Geophysical International Academy Nauk USSR* **24**(24), 3–42.

Obukhov, A. M. (1960), 'The statistically orthogonal expansion of empirical functions', *Bulletin of the Academy of Sciences of the USSR. Geophysics Series (English Transl.)* **1**, 288–291.

O'Hagan, T. (2004), 'Dicing with the unknown', *Significance* **1**(3), 132–133.

O'Hare, G., Sweeney, J. and Wilby, R. (2014), *Weather, climate and climate change: human perspectives*, Routledge.

Olsson, J. (1998), 'Evaluation of a scaling cascade model for temporal rain-fall disaggregation', *Hydrol. Earth Syst. Sci.* **2**(1), 19–30.

Olsson, J., Uvo, C. and Jinno, K. (2001), 'Statistical atmospheric downscaling of short-term extreme rainfall by neural networks', *Phys. Chem. Earth B* **26**(9), 695–700.

Onof, C., Chandler, R. E., Kakou, A., Northrop, P., Wheater, H. S. and Isham, V. (2000), 'Rainfall modelling using Poisson-cluster processes: a review of developments', *Stoc. Env. Res. Risk Assess.* **14**(6), 384–411.

Onof, C. and Wheater, H. S. (1994), 'Improvements to the modelling of British rainfall using a modified random parameter Bartlett-Lewis rectangular pulse model', *J. Hydrol.* **157**(1), 177–195.

Osborn, T. J. and Hulme, M. (1997), 'Development of a relationship between station and grid-box rainday frequencies for climate model evaluation', *J. Climate* **10**(8), 1885–1908.

Osborn, T. J., Wallace, C. J., Harris, I. C. and Melvin, T. M. (2016), 'Pattern scaling using ClimGen: monthly-resolution future climate scenarios including changes in the variability of precipitation', *Clim. Change* **134**, 353–369.

Over, T. M. and Gupta, V. K. (1996), 'A space-time theory of mesoscale rainfall using random cascades', *J. Geophys. Res.* **101**(D21), 26319–26331.

Palmer, T. N. (2013), 'Climate extremes and the role of dynamics', *Proc. Nat. Acad. Sci.* **110**, 5281–5282.

Panofsky, H. W. and Brier, G. W. (1968), *Some applications of statistics to meteorology*, The Pennsylvania State University Press.

Parker, D. E. (2010), 'Urban heat island effects on estimates of observed climate change', *WIREs Clim. Change* **1**(1), 123–133.

Paschalis, A., Molnar, P., Fatichi, S. and Burlando, B. (2013), 'A stochastic model for high-resolution space-time precipitation simulation', *Wat. Resour. Res.* **49**(12), 8400–8417.

Paul, S., Liu, C. M., Chen, J. M. and Lin, S. H. (2008), 'Development of a statistical downscaling model for projecting monthly rainfall over east asia from a general circulation model output', *J. Geophys. Res.* **113**(D15).

Pearson, K. (1901), 'On lines and planes of closest fit to systems of points in space', *Philosophical Magazine* **2**(11), 559–572.

Peixoto, J. P. and Oort, A. H. (1992), *Physics of climate*, American Institute of Physics.

Penalba, O. C., Rivera, J. A. and Pántano, V. C. (2014), 'The CLARIS LPB database: constructing a long-term daily hydro-meteorological dataset for La Plata Basin, southern South America', *Geosci. Data J.* **1**(1), 20–29.

Perica, S. and Foufoula-Georgiou, E. (1996), 'Model for multiscale disaggregation of spatial rainfall based on coupling meteorological and scaling descriptions', *J. Geophys. Res* **101**, 26–347.

Philander, S. G. (1990), *El Niño, La Niña, and the Southern Oscillation*, Academic Press.

Philipp, A., Bartholy, J., Beck, C., Erpicum, M., Esteban, P., Fettweis, X., Huth, R., James, P., Jourdain, S., Kreienkamp, F. et al. (2010), 'Cost733cat - A database of weather and circulation type classifications', *Phys. Chem. Earth* **35**(9), 360–373.

Philipp, A., Beck, C., Huth, R. and Jacobeit, J. (2014), 'Development and comparison of circulation type classifications using the COST 733 dataset and software', *Int. J. Climatol.*

Philipp, A., Della-Marta, P.-M., Jacobeit, J., Fereday, D. R., Jones, P. D., Moberg, A. and Wanner, H. (2007), 'Long-term variability of daily North Atlantic-European pressure patterns since 1850 classified by simulated annealing clustering', *J. Climate* **20**(16), 4065–4095.

Piani, C. and Haerter, J. O. (2012), 'Two dimensional bias correction of temperature and precipitation copulas in climate models', *Geophys. Res. Lett.* **39**(20), L20401.

Piani, C., Haerter, J. O. and Coppola, E. (2010*a*), 'Statistical bias correction for daily precipitation in regional climate models over Europe', *Theor. Appl. Climatol.* **99**(1–2), 187–192.

Piani, C., Weedon, G. P., Best, M., Gomes, S. M., Viterbo, P., Hagemann, S. and Haerter, J. O. (2010*b*), 'Statistical bias correction of global simulated daily precipitation and temperature for the application of hydrological models', *J. Hydrol.* **395**, 199–215.

Pielke Jr., R. A. (2007), *The honest broker: making sense of science in policy and politics*, Cambridge University Press.

Pielke, R. A. and Wilby, R. L. (2012), 'Regional climate downscaling: What's the point?', *EOS* **93**(5), 52–53.

Pielke, R., Beven, K., Brasseur, G., Calvert, J., Chahine, M., Dickerson, R. R., Entekhabi, D., Foufoula-Georgiou, E., Gupta, H., Gupta, V., Krajewski, W., Krieder, E. P., Lau, W. K. M., Mc Donnell, J., Rossow, W., Schaake, J., Smith, J., Sorooshian, S. and Wood., E. (2009), 'Climate change: the need to consider human forcings besides greenhouse gases', *EOS* **90**(45), 413–413.

Pierce, D. W., Cayan, D. R., Maurer, E. P., Abatzoglou, J. T. and Hegewisch, K. C. (2015), 'Improved bias correction techniques for hydrological simulations of climate change', *J. Hydrometeorol.* **16**(6), 2421–2442.

Pierce, D. W., Cayan, D. R. and Thrasher, B. L. (2014), 'Statistical downscaling using localized constructed analogs (LOCA)', *J. Hydrometeorol.* **15**(6), 2558–2585.

Pingel, S., ed. (2012), *Toward a climate services enterprise*, Brussels, Belgium. Conference Report.

Pinto, J. G., Neuhaus, C. P., Leckebusch, G. C., Reyers, M. and Kerschgens, M. (2010), 'Estimation of wind storm impacts over western germany under future climate conditions using a statistical–dynamical downscaling approach', *Tellus A* **62**(2), 188–201.

Planton, S., ed. (2013), *Climate Change 2013: The Physical Science Basis. Contribution of Working Group I to the Fifth Assessment Report of the Intergovernmental Panel on Climate Change*, Cambridge University Press, Cambridge, United Kingdom, and New York, NY, USA, chapter 'Annex III: Glossary'.

Plaut, G. and Simonnet, E. (2001), 'Large-scale circulation classification, weather regimes, and local climate over France, the Alps and Western Europe', *Clim. Res.* **17**(3), 303–324.

Poli, P., Hersbach, H., Dee, D. P., Berrisford, P., Simmons, A. J., Vitart, F., Laloyaux, P., Tan, D. G. H., Peubey, C., Thépaut, J.-N., Yannick, T., Hólm, E. V., Bonavita, M., Isaksen, L. and Fisher, M. (2016), 'ERA-20C: An atmospheric reanalysis of the twentieth century', *J. Climate* **29**, 4083–4097.

Prein, A. F., Gobiet, A., Suklitsch, M., Truhetz, H., Awan, N. K., Keuler, K. and Georgievski, G. (2013), 'Added value of convection permitting seasonal simulations', *Clim. Dynam.* **41**, 2655–2677.

Prein, A. F., Langhans, W., Fosser, G., Ferrone, A., Ban, N., Goergen, K., Keller, M., Tölle, M., Gutjahr, O., Feser, F., Brisson, E., Kollet, S., Schmidli, J., van Lipzig, N. P. M. and Leung, R. (2015), 'A review on regional convection-permitting climate modeling: demonstrations, prospects, and challenges', *Rev. Geophys.* **53**(2), 323–361.

Prein, A. F., Rasmussen, R. M., Ikeda, K., Liu, C., Clark, M. P. and Holland, G. J. (2017), 'The future intensification of hourly precipitation extremes', *Nat. Clim. Change* **7**, 48–52.

Preisendorfer, R. W. and Barnett, T. P. (1977), Significance tests for empirical orthogonal functions, *in Fifth Conf. on Probability and Statistics in Atmos. Sci., Las Vegas, NV. American Meteorol. Soc*, Vol. 169, p. 172.

Preisendorfer, R. W. and Mobley, C. D. (1988), *Principal component analysis in meteorology and oceanography*, Developments in atmospheric science, Elsevier.

Preisendorfer, R. W., Zwiers, F. W. and Barnett, T. P. (1981), 'Foundations of principal component selection rules', *SIO Reference Series 81–4 May 1981*.

prepdata (n.d.), 'Partnership for Resilience and Preparedness (PREP)', www.prepdata.org.

Prudhomme, C., Reynard, N. and Crooks, S. (2002), 'Downscaling of global climate models for flood frequency analysis: where are we now?', *Hydrol. Proc.* **16**(6), 1137–1150. Sp. Iss. SI.

Prudhomme, C., Wilby, R. L., Crooks, S., Kay, A. L. and Reynard, N. S. (2010), 'Scenario-neutral approach to climate change impact studies: application to flood risk', *J. Hydrol.* **390**, 198–209.

Pruppacher, H. R., Klett, J. D. and Wang, P. K. (1998), *Microphysics of clouds and precipitation*, Taylor & Francis.

Pryor, S. C., Schoof, J. T. and Barthelmie, R. J. (2005), 'Empirical downscaling of wind speed probability distributions', *J. Geophys. Res.* **110**, D19109.

Racherla, P. N., Shindell, D. T. and Faluvegi, G. S. (2012), 'The added value to global model projections of climate change by dynamical downscaling: a case study over the continental US using the GISS-ModelE2 and WRF models', *J. Geophys. Res.* **117**(D20).

Racsko, P., Szeidl, L. and Semenov, M. (1991), 'A serial approach to local stochastic weather models', *Ecol. Mod.* **57**(1–2), 27–41.

Radanovics, S., Vidal, J.-P., Sauquet, E., Daoud, A. B. and Bontron, G. (2013), 'Optimising predictor domains for spatially coherent precipitation downscaling', *Hydrol. Earth Syst. Sci.* **17**, 4189–4208.

Räisänen, J. and Räty, O. (2013), 'Projections of daily mean temperature variability in the future: cross-validation tests with ENSEMBLES regional climate simulations', *Clim. Dynam.* **41**, 1553–1568.

Rajczak, J., Kotlarski, S. and Schär, C. (2016), 'Does quantile mapping of simulated precipitation correct for biases in transition probabilities and spell lengths?', *J. Climate* **29**, 1605–1615.

Randall, D. (2015), *An introduction to the global circulation of the atmosphere*, Princeton University Press.

Räty, O., Räisänen, J. and Ylhäisi, J. S. (2014), 'Evaluation of delta change and bias correction methods for future daily precipitation: intermodel cross-validation using ENSEMBLES simulations', *Clim. Dynam.* **42**(9–10), 2287–2303.

Raymond, D. J. and Emanuel, K. A. (1993), The Kuo cumulus parameterization, *in The representation of cumulus convection in numerical models*, Springer, pp. 145–147.

Rebora, N., Ferraris, L., von Hardenberg, J. and Provenzale, A. (2006), 'RainFARM: rainfall downscaling by a filtered autoregressive model', *J. Hydrometeorol.* **7**(4), 724–738.

Reyers, M., Pinto, J. G. and Moemken, J. (2015), 'Statistical–dynamical downscaling for wind energy potentials: evaluation and applications to decadal hindcasts and climate change projections', *Int. J. Climatol.* **35**(2), 229–244.

Richardson, C. W. (1981), 'Stochastic simulation of daily precipitation, temperature, and solar radiation', *Wat. Resour. Res.* **17**(1).

Richardson, C. W. and Wright, D. A. (1984), 'WGEN: A model for generating daily weather variables', Report No. 8, Agricultural Research Service, US Department of Agriculture, Washington, DC.

Richman, M. (1987), 'Rotation of principal components: a reply', *J. Climatol.* **7**(5), 511–520.

Richman, M. B. (1986), 'Rotation of principal components', *J. Climatol.* **6**(3), 293–335.

Richter, I. (2015), 'Climate model biases in the eastern tropical oceans: causes, impacts and ways forward', *WIRES: Clim. Change* **6**(3), 345–358.

Richter, I., Chang, P., Xu, Z., Doi, T., Kataoka, T., Nagura, M., Oettli, P., de Szoeke, S. and Tozuka, T. (2016), *Indo-Pacific climate variability and predictability*, World Scientific, chapter 'An Overview of Coupled GCM Performance in the Tropics'.

Richter, I. and S.-P. Xie (2008), 'On the origin of equatorial Atlantic biases in coupled general circulation models', *Clim. Dynam.* **31**(5), 587–598.

Richter, I., Xie, S.-P., Behera, S. K., Doi, T. and Masumoto, Y. (2014), 'Equatorial Atlantic variability and its relation to mean state biases in CMIP5', *Clim. Dynam.* **42**(1–2), 171–188.

Rienecker, M. M., Suarez, M. J., Gelaro, R., Todling, R., Bacmeister, J., Liu, E., Bosilovich, M. G., Schubert, S. D., Takacs, L., Kim, G.-K., Bloom, S., Chen, J., Collins, D., Conaty, A., da Silva, A., Gu, W., Joiner, J., Koster, R. D., Lucchesi, R., Molo, A., Owens, T., Pawson, S., Pegion, P., Redder, C. R., Reichle, R., Robertson, F. R., Ruddick, A. G., Sienkiewicz, M. and Woollen, J. (2011), 'MERRA: NASA's modern-era retrospective analysis for research and applications', *J. Climate* **24**(14), 3624–3648.

Rockel, B. (2015), 'The regional downscaling approach: a brief history and recent advances', *Curr. Clim. Change. Rep.* **1**(1), 22–29.

Rodriguez-Iturbe, I., Cox, D. R. and Isham, V. (1987), 'Some models for rainfall based on stochastic point processes', *in Proc. Roy. Soc. A*, Vol. 410, pp. 269–288.

Rodriguez-Iturbe, I., Cox, D. R. and Isham, V. (1988), 'A point process model for rainfall: further developments', *in Proc. Roy. Soc. A*, Vol. 417, The Royal Society, pp. 283–298.

Roe, G. H. (2005), 'Orographic precipitation', *Ann. Rev. Earth Planet. Sci.* **33**, 645–671.

Roehrig, R., Bouniol, D., Guichard, F., Hourdin, H. and Redelsperger, J.-L. (2013), 'The present and future of the West African monsoon: a process-oriented assessment of CMIP5 simulations along the AMMA transect', *J. Climate* **26**(17), 6471–6505.

Roessler, O., Fischer, A. M., Huebener, H., Maraun, D., Benestad, R. E., Christodoulides, P., Soares, P. M. M., Cardoso, R. M., Pagé, C., Kanamaru, H., Kreienkamp, F. and Vlachogiannis, D. (2017), 'Challenges to link climate change data provision and user needs - perspective from the COST-Action VALUE', *Int. J. Climatol.* DOI: 10.1002/joc.5060.

Rogers, R. R. and Yau, M. K. (1996), *A short course in cloud physics*, Elsevier.

Rosenzweig, C. (1985), 'Potential CO_2-induced climate effects on North American wheat-producing regions', *Clim. Change* **7**, 367–389.

Rummukainen, M. (1997), 'Methods of statistical downscaling of GCM simulations. Reports Meteorology and Climatology 80', Technical report, Swedish Meteorological and Hydrological Institute, SE-601 76 Norrköping, Sweden.

Rummukainen, M. (2010), 'State-of-the-art with regional climate models', *WIREs Clim. Change* **1**, 82–96. DOI: 10.1002/wcc.8.

Rust, H. W., Kruschke, T., Dobler, A., Fischer, M. and Ulbrich, U. (2015), 'Discontinuous daily temperatures in the WATCH forcing datasets', *J. Hydrometeorol.* **16**(1), 465–472.

Rust, H. W., Vrac, M., Lengaigne, M. and Sultan, B. (2010), 'Quantifying differences in circulation patterns based on probabilistic models: IPCC AR4 multimodel comparison for the North Atlantic', *J. Climate* **23**(24), 6573–6589.

Rust, H. W., Vrac, M., Sultan, B. and Lengaigne, M. (2013), 'Mapping weather-type influence on Senegal precipitation based on a spatial–temporal statistical model', *J. Climate* **26**(20), 8189–8209.

Saji, N. H., Goswami, B. N., Vinayachandran, P. N. and Yamagata, T. (1999), 'A dipole mode in the tropical Indian Ocean', *Nature* **401**(6751), 360–363.

Salameh, T., Drobinski, P., Vrac, M. and Naveau, P. (2009), 'Statistical downscaling of near-surface wind over complex terrain in southern France', *Meteorol. Atmos. Phys.* **103**(1), 253–265.

Salathé, E. P. (2005), 'Downscaling simulations of future global climate with application to hydro-logic modelling', *Int. J. Climatol.* **25**(4), 419–436.

Salathé, E. P., Steed, R., Mass, C. F. and Zahn, P. H. (2008), 'A high-resolution climate model for the U.S. Pacific Northwest: mesoscale feedbacks and local responses to climate change', *J. Climate* **21**, 5708–5726.

San-Martín, D., Manzanas, R., Brands, S., Herrera, S. and Gutiérrez, J. M. (2017), 'Reassessing model uncertainty for regional projections of precipitation with an ensemble of statistical downscaling methods', *J. Climate* **30**(1), 203–223.

Santer, B. (1985), 'The use of general circulation models in climate impact analysis – a preliminary study of the impacts of CO_2-induced climatic change on West European agriculture', *Clim. Change* **7**, 71–93.

Satoh, M. (2013), *Atmospheric circulation dynamics and general circulation models*, Springer Science & Business Media.

Satoh, M., Matsuno, T., Tomita, H., Miura, H., Nasuno, T. and Iga, S. (2008), 'Nonhydrostatic icosahedral atmospheric model (NICAM) for global cloud-resolving simulations', *J. Comput. Phys.* **227**, 3486–3514.

Saunders, I. R. and Byrne, J. M. (1999), 'Using synoptic surface and geopotential height fields for generating grid-scale precipitation', *Int. J. Climatol.* **19**(11), 1165–1176.

Sauter, T. and Venema, V. (2011), 'Natural three-dimensional predictor domains for statistical precipitation downscaling', *J. Climate* **24**(23), 6132–6145.

Scaife, A. A., Copsey, D., Gordon, C., Harris, C., Hinton, T., Keeley, S., O'Neill, A., Roberts, M. and Williams, K. (2011), 'Improved Atlantic winter blocking in a climate model', *Geophys. Res. Lett.* **38**, L23703.

Schamm, K., Ziese, M., Raykova, K., Becker, A., Finger, P., Meyer-Christoffer, A. and Schneider, U. (2015), 'GPCC Full Data Daily Version 1.0 at 1.0°: Daily Land-Surface Precipitation from Rain-Gauges built on GTS-based and Historic Data', DOI: 10.5676/DWD_GPCC/FD_D_V1_100.

Schär, C., Frei, C., Lüthi, D. and Davies, H. C. (1996), 'Surrogate climate-change scenarios for regional climate models', *Geophys. Res. Lett.* **23**(6), 669–672.

Schär, C., Lüthi, D., Beyerle, U. and Heise, E. (1999), 'The soil-precipitation feedback: A process study with a regional climate model', *J. Climate* **12**(3), 722–741.

Schertzer, D. and Lovejoy, S. (1987), 'Physical modeling and analysis of rain and clouds by anisotropic scaling multiplicative processes', *J. Geophys. Res.* **92**(D8), 9693–9714.

Schiermeier, Q. (2010), 'The real holes in climate science', *Nature* **463**(7279), 284–288.

Schlesinger, M. E. and Ramankutty, N. (1994), 'An oscillation in the global climate system of period 65–70 years', *Nature* **367**(6465), 723–726.

Schmidli, J., Frei, C. and Vidale, P. L. (2006), 'Downscaling from GCM precipitation: A benchmark for dynamical and statistical downscaling methods', *Int. J. Climatol.* **26**, 679–689.

Schmidli, J., Goodess, C. M., Frei, C., Haylock, M. R., Hundecha, Y., Ribalaygua, J. and Schmith, T. (2007), 'Statistical and dynamical downscaling of precipitation: an evaluation and comparison of scenarios for the European Alps', *J. Geophys. Res.* **112**(D4).

Schmith, T. (2008), 'Stationarity of regression relationships: Application to empirical downscaling', *J. Climate* **21**(17), 4529–4537.

Schneider, S. H. (2001), 'What is "Dangerous" climate change?', *Nature* **411**(6833), 17–19.

Schneider, U., Fuchs, T., Meyer-Christoffer, A. and Rudolf, B. (2008), 'Global precipitation analysis products of the GPCC', Global Precipitation Climatology Centre (GPCC), Deutscher Wetterdienst, Offenbach a. M., Germany, November.

Schölzel, C. and Friederichs, P. (2008), 'Multivariate non-normally distributed random variables in climate research – introduction to the copula approach', *Nonlin. Proc. Geophys.* **15**, 761–772.

Schölzel, C. and Hense, A. (2011), 'Probabilistic assessment of regional climate change in Southwest Germany by ensemble dressing', *Clim. Dynam.* **36**(9), 2003–2014.

Schoof, J. T. (2013), 'Statistical downscaling in climatology', *Geogr. Comp.* **7**(4), 249–265.

Schoof, J. T. and Pryor, S. C. (2001), 'Downscaling temperature and precipitation: A comparison of regression-based methods and artificial neural networks', *Int. J. Climatol.* **21**(7), 773–790.

Schrum, C., Hübner, U., Jacob, D. and Podzun, R. (2003), 'A coupled atmosphere/ice/ocean model for the North Sea and the Baltic Sea', *Clim. Dynam.* **21**(2), 131–151.

Schwarb, M. (2000), *The alpine precipitation climate*, PhD thesis, Swiss Federal Institute of Technology Zurich.

Schwarz, G. (1978), 'Estimating the dimension of a model', *Ann. Statist.* **6**, 461–464.

Schwarz, H. E. (1966), *Climate, Climatic change, and water supply*, National Academy of Sciences, chapter 'Climatic Change and Water Supply: How Sensitive is the Northeast?', pp. 111–120.

Semenov, M. A. and Barrow, E. M. (1997), 'Use of a stochastic weather generator in the development of climate change scenarios', *Clim. Change* **35**(4), 397–414.

Semenov, M. A., Brooks, R. J., Barrow, E. M. and Richardson, C. W. (1998), 'Comparison of the WGEN and LARS-WG stochastic weather generators for diverse climates', *Clim. Res.* **10**(2), 95–107.

Seneviratne, S. I., Corti, T., Davin, E. L., Hirschi, M., Jaeger, E. B., Lehner, I., Orlowsky, B. and Teuling, A. J. (2010), 'Investigating soil moisture – climate interactions in a changing climate: a review', *Earth Sci. Rev.* **99**(3), 125–161.

Seo, K.-H., Frierson, D. M. W. and Son, J.-H. (2014), 'A mechanism for future changes in Hadley circulation strength in CMIP5 climate change simulations', *Geophys. Res. Lett.* **41**(14), 5251–5258.

Sheffield, J., Goteti, G. and Wood, E. F. (2006), 'Development of a 50-year high-resolution global dataset of meteorological forcings for land surface modeling', *J. Climate* **19**(13), 3088–3111.

Shepard, D. S. (1984), 'Computer mapping: The SYMAP interpolation algorithm', *in Spatial statistics and models*, Springer, pp. 133–145.

Shepherd, T. G. (2014), 'Atmospheric circulation as a source of uncertainty in climate change projections', *Nat. Geosci.* **7**, 703–708.

Shindell, D., Racherla, P. and Milly, G. (2014), 'Reply to comment by Laprise on "The added value to global model projections of climate change by dynamical downscaling: a case study over the continental US using the GISS-ModelE2 and WRF models"', *J. Geophys. Res.* **119**(7), 3882–3885.

Sillmann, J., Kharin, V. V., Zhang, X., Zwiers, F. W. and Bronaugh, D. (2013), 'Climate extremes indices in the CMIP5 multimodel ensemble: Part 1. Model evaluation in the present climate', *J. Geophys. Res.* **118**(4), 1716–1733.

Simmons, A. J., Jones, P. D., da Costa Bechtold, V., Beljaars, A. C. M., Kållberg, P. W., Saarine, S., Uppala, S. M., Viterbo, P. and Wedi, N. (2004), 'Comparison of trends and low-frequency variability in CRU, ERA-40, and NCEP/NCAR analyses of surface air temperature', *J. Geophys. Res.* **109**(D24).

Simmons, A. J., Willett, K. M., Jones, P. D., Thorne, P. W. and Dee, D. P. (2010), 'Low-frequency variations in surface atmospheric humidity, temperature, and precipitation: Inferences from reanalyses and monthly gridded observational data sets', *J. Geophys. Res.* **115**(D1).

Sklar, A. (1959), 'Fonctions de répartition à n dimensions et leurs marges', *Publ. Inst. Stat. Univ. Paris* **8**.

Slingo, J., Inness, P., Neale, R., Woolnough, S. and Yang, G. (2003), 'Scale interactions on diurnal to seasonal timescales and their relevance to model systematic errors', *Ann. Geophys.* **46**(1), 139–155.

Smith, D. M., Cusack, S., Colman, A. W., Folland, C. K., Harris, G. R. and Murphy, J. M. (2007), 'Improved surface temperature prediction for the coming decade from a global climate model', *Science* **317**(5839), 796–799.

Smith, L. A. (2002), 'What might we learn from climate forecasts?', *Proc. Nat. Acad. Sci.* **99**(suppl 1), 2487–2492.

Smith, R. L. (1990), 'Regional estimation from spatially dependent data', Technical report.

Soares, P. et al. (2017), 'Process based evaluation of the VALUE perfect predictor experiment of statistical downscaling methods', *Int. J. Climatol., subm.*

Solman, S. A., Sanchez, E., Samuelsson, P., da Rocha, R. P., Li, L., Marengo, J., Pessacg, N. L., Remedio, A. R. C., Chou, S., C., Berbery, H, Le Treut, H., de Castro, M. and Jacob, D. (2013), 'Evaluation of an ensemble of regional climate model simulations over South America driven by the ERA-Interim reanalysis: model performance and uncertainties', *Clim. Dynam.* **41**(5-6), 1139–1157.

Solomon, S., Qin, D., Manning, M., Chen, Z., Marquis, M., Averyt, K. B., Tignor, M. and Miller, H. L., eds. (2007), *Climate change 2007: The physical science basis: Working Group I contribution to the Fourth Assessment Report of the Intergovernmental Panel on Climate Change*, Cambridge University Press.

Somot, S., Sevault, F., Déqué, M. and Crépon, M. (2008), '21st century climate change scenario for the Mediterranean using a coupled atmosphere–ocean regional climate model', *Glob. Planet. Change* **63**(2), 112–126.

Spearman, C. (1904), '"General intelligence," objectively determined and measured', *The American Journal of Psychology* **15**(2), 201–292.

Sperber, K. R., Annamalai, H., Kang, I.-S., Kitoh, A., Moise, A., Turner, A., Wang, B. and Zhou, T. (2013), 'The Asian summer monsoon: an intercomparison of CMIP5 vs. CMIP3 simulations of the late 20th century', *Clim. Dynam.* **41**, 2711–2744.

Stainforth, D. A. (2016), personal communication.

Stainforth, D. A., Allen, M. R., Tredger, E. R. and Smith, L. A. (2007), 'Confidence, uncertainty and decision-support relevance in climate predictions', *Phil. Trans. R. Soc. A* **365**, 2145–2161.

Stehlik, J. and Bárdossy, A. (2002), 'Multivariate stochastic downscaling model for generating daily precipitation series based on atmospheric circulation', *J. Hydrol.* **256**(1–2), 120–141.

Steiger, N. J., Steig, E. J., Dee, S. G., Roe, G. H. and Hakim, G. J. (2017), 'Climate reconstruction using data assimilation of water isotope ratios from ice cores', *J. Geophys. Res.* **122**(3), 1545–1568.

Stensrud, D. J. (2009), *Parameterization schemes: keys to understanding numerical weather prediction models*, Cambridge University Press.

Stern, R. D. and Coe, R. (1984), 'A model fitting analysis of daily rainfall data', *J. Roy. Stat. Soc. A* **134**(1), 1–34.

Stevens, B. and Bony, S. (2013), 'What are climate models missing?', *Science* **340**(6136), 1053–1054.

Stocker, T. (2011), *Introduction to climate modelling*, Springer Science & Business Media.

Stocker, T. F., Dahe, Q., Plattner, G.-K. and Tignor, M. (2015), 'IPCC Workshop on Regional Climate Projections and their Use in Impacts and Risk Analysis Studies', https://www.ipcc.ch/pdf/supporting-material/RPW_WorkshopReport.pdf.

Stocker, T. F., Qin, D., Plattner, K., Tignor, M. M. B., Allen, S. K., Boschung, J., Nauels, A., Xia, Y., Bex, V. and Midgley, P. M., eds (2013), *Climate change 2013: the physical science basis: Working Group I contribution to the Fifth Assessment Report of the Intergovernmental Panel on Climate Change*, Cambridge University Press.

Stone, M. (1977), 'An asymptotic equivalence of choice of model by cross-validation and Akaike's criterion', *J. R. Stat. Soc. B* **39**, 44–47.

Stoner, A. M. K., Hayhoe, K., Yang, X. and Wuebbles, D. J. (2013), 'An asynchronous regional regression model for statistical downscaling of daily climate variables', *Int. J. Climatol.* **33**, 2473–2494.

Sun, F., Walton, D. B. and Hall, A. (2015), 'A hybrid dynamical–statistical downscaling technique. Part II: End-of-century warming projections predict a new climate state in the Los Angeles region', *J. Climate* **28**(12), 4618–4636.

Sutton, R. T. and Dong, B. (2012), 'Atlantic Ocean influence on a shift in European climate in the 1990s', *Nat. Geosci.* **5**, 788–792.

Switanek, M. B., Troch, P. A., Castro, C. L., Leuprecht, A., Chang, H.-I., Mukherjee, R. and Demaria, E. M. C. (2016), 'Scaled distribution mapping: a bias correction method that preserves raw climate model projected changes', *Hydrol. Earth Syst. Sci.* **accepted**.

Takle, E. S., Gutowski, W. J., Arritt, R. A., Pan, Z., Anderson, C. J., da Silva, R. R., Caya, D., Chen, S.-C., Christensen, J. H., Hong, S.-Y., Juang, H.-M. H., M., J. K. W., Lapenta, Laprise, R., Lopez, P., McGregor, J. and Roads, J. O. (1999), 'Project to Intercompare Regional Climate Simulations (PIRCS): Description and initial results', *J. Geophys. Res.* **104**, 19443–19461.

Tareghian, R. and Rasmussen, P. F. (2013), 'Statistical downscaling of precipitation using quantile regression', *J. Hydrol.* **487**, 122–135.

Taylor, K. E., Stouffer, R. J. and Meehl, G. A. (2012), 'An overview of CMIP5 and the experiment design', *Bull. Amer. Meteorol. Soc.* **93**, 485–498.

Tebaldi, C. and Knutti, R. (2007), 'The use of the multi-model ensemble in probabilistic climate projections', *Phil. Trans. R. Soc. A* **365**, 2053–2075.

Tebaldi, C., Smith, R. L., Nychka, D. and Mearns, L. O. (2005), 'Quantifying uncertainty in projections of regional climate change: a Bayesian approach to the analysis of multimodel ensembles', *J. Climate* **18**(10), 1524–1540.

Teutschbein, C. and Seibert, J. (2012), 'Bias correction of regional climate model simulations for hydrological climate-change impact studies: review and evaluation of different methods', *J. Hydrol.* **456**, 12–29.

Themeßl, M. J., Gobiet, A. and Heinrich, G. (2012), 'Empirical-statistical downscaling and error correction of regional climate models and its impact on the climate change signal', *Clim. Change* **112**, 449–468.

Themeßl, M. J., Gobiet, A. and Leuprecht, A. (2011), 'Empirical-statistical downscaling and error correction of daily precipitation from regional climate models', *Int. J. Climatol.* **31**, 1530–1544.

Thober, S., Mai, J., Zink, M. and Samaniego, L. (2014), 'Stochastic temporal disaggregation of monthly precipitation for regional gridded data sets', *Wat. Resour. Res.* **50**(11), 8714–8735.

Thompson, D. W. J., Baldwin, M. P. and Wallace, J. M. (2002), 'Stratospheric connection to Northern Hemisphere wintertime weather: implications for prediction', *J. Climate* **15**(12), 1421–1428.

Thorne, P. W. and Vose, R. S. (2010), 'Reanalysis suitable for characterizing long term trends. Are they really achievable?', *Bull. Amer. Meteorol. Soc.* **91**, 353–361.

Thorne, P. and Vose, R. S. (2011), 'Reply to comments on "Reanalyses suitable for characterizing long-term trends"', *Bull. Amer. Meteorol. Soc.* **92**(1), 70–72.

Thorne, P. W., Willett, K. M., Allan, R. J., Bojinski, S., Christy, J. R., Fox, N., Gilbert, S., Jolliffe, I., Kennedy, J. J., Kent, E., Klein Tank, A., Lawrimore, J., Parker, D. E., Rayner, N., Simmons, A., Song, L., Stott, P. A., and Trewin, B. (2011), 'Guiding the creation of a comprehensive surface temperature resource for twenty-first-century climate science', *Bull. Amer. Meteorol. Soc.* **92**(11), ES40–ES47.

Thrasher, B., Maurer, E. P., McKellar, C. and Duffy, P. (2012), 'Technical note: bias correcting climate model simulated daily temperature extremes with quantile mapping', *Hydrol. Earth Syst. Sci.* **16**(9), 3309–3314.

Tiedtke, M. (1989), 'A comprehensive mass flux scheme for cumulus parameterization in large-scale models', *Mon. Wea. Rev.* **117**(8), 1779–1800.

Tippett, M. K., DelSole, T., Mason, S. J. and Barnston, A. G. (2008), 'Regression-based methods for finding coupled patterns', *J. Climate* **21**(17), 4384–4398.

Tokmakian, R., Challenor, P. and Andrianakis, Y. (2012), 'On the use of emulators with extreme and highly nonlinear geophysical simulators', *J. Atmos. Ocean. Tecnol.* **29**(11), 1704–1715.

Toreti, A., Kuglitsch, F. G., Xoplaki, E., Luterbacher, J. and Wanner, H. (2010), 'A novel method for the homogenization of daily temperature series and its relevance for climate change analysis', *J. Climate* **23**(19), 5325–5331.

Tran, G. T., Oliver, K. I. C., Sóbester, A., Toal, D. J. J., Holden, P. B., Marsh, R., Challenor, P. and Edwards, N. R. (2016), 'Building a traceable climate model hierarchy with multi-level emulators', *Adv. Stat. Climatol. Meteorol. Ocean.* **2**(1), 17–37.

Trenberth, K. E. (1975), 'A quasi-biennial standing wave in the Southern Hemisphere and inter-relations with sea surface temperature', *Quart. J. Roy. Meteorol. Soc.* **101**(427), 55–74.

Trenberth, K. E. (1992), *Climate system Modeling*, Cambridge University Press.

Trewin, B. (2010), 'Exposure, instrumentation, and observing practice effects on land temperature measurements', *WIREs Clim. Change* **1**(4), 490–506.

Trewin, B. (2013), 'A daily homogenized temperature data set for Australia', *Int. J. Climatol.* **33**(6), 1510–1529.

Underwood, F. M. (2009), 'Describing long-term trends in precipitation using generalized additive models', *J. Hydrol.* **364**(3), 285–297.

UNFCCC (1997), 'Report of the subsidiary body for scientific and technological advice on the work of its seventh session', FCCC/SBSTA/1997/14, United Nations, Geneva, Switzerland, 10 November.

United Nations (1992), 'United Nations Framework Convention on Climate Change', United Nations, 1992. FCCC/INFORMAL/84 GE.05-62220 (E) 200705. Available at www.unfccc.int.

Uppala, S. M., Källberg, P. W., Simmons, A. J. and Adrae et al., U. (2005), 'The ERA-40 reanalysis', *Quart. J. Roy. Meteorol. Soc* **131**, 2961–3012.

US Army Corps of Engineers (1971), 'HEC-4 monthly streamflow simulation'. Hydrologic Engineering Center, Davis, California.

van Asselt, M. B. A., J. and Rotmans (2002), 'Uncertainty in integrated assessment modelling', *Clim. Change* **54**(1), 75–105.

van den Dool, H. M. (1994), 'Searching for analogues, how long must we wait?', *Tellus A* **46**(3), 314–324.

van der Linden, P. and Mitchell, J. F. B. (2009), 'ENSEMBLES: Climate Change and its Impacts: Summary of research and results from the ENSEMBLES project', Technical report, Met Office Hadley Centre.

van Haren, R., van Oldenborgh, G. J., Lenderink, G., Collins, M. and Hazeleger, W. (2013*a*), 'SST and circulation trend biases cause an underestimation of European precipitation trends', *Clim. Dynam.* **40**(1–2), 1–20.

van Haren, R., van Oldenborgh, G. J., Lenderink, G. and Hazeleger, W. (2013*b*), 'Evaluation of modeled changes in extreme precipitation in Europe and the Rhine basin', *Environ. Res. Lett.* **8**(1), 014053.

van Meijgaard, E., van Ulft, L. H., van de Berg, W. J., Bosveld, F. C., van den Hurk, B. J. J. M., Lenderink, G. and Siebesma, A. P. (2008), 'The KNMI regional atmospheric climate model RACMO version 2.1, Technical Report 302, Royal Dutch Meteorological Institute, KNMI, Postbus 201, 3730 AE, De Bilt, The Netherlands.

van Oldenborgh, G. J., Doblas Reyes, F.-J., Drijfhout, S. S. and Hawkins, E. (2013), 'Reliability of regional climate model trends', *Env. Res. Lett.* **8**(1), 014055.

van Oldenborgh, G. J., Drijfhout, S., van Ulden, A., Haarsma, R., Sterl, A., Severijns, C., Hazeleger, W. and Dijkstra, H. (2009), 'Western Europe is warming much faster than expected', *Clim. Past* **5**(1), 1–12.

van Vuuren, D. P., Edmonds, J., Kainuma, M., Riahi, K., Thomson, A., Hibbard, K., Hurtt, G., Kram, T., Krey, V., Lamarque, J.-F., Masui, T., Meinshausen, M., Nakicenovic, N., Smith, S. J. and Rose, S. K. (2011), 'The representative concentration pathways: an overview', *Clim. Change* **109**(1), 5–31.

Vannitsem, S. (2011), 'Bias correction and post-processing under climate change', *Nonlin. Proc. Geophys.* **18**, 911–924.

Vautard, R. (1990), 'Multiple weather regimes over the North Atlantic: analysis of precursors and successors', *Mon. Wea. Rev.* **118**(10), 2056–2081.

Vecchi, G. A., Soden, B. J., Wittenberg, A. T., Held, I. M., Leetmaa, A. and Harrison, M. J. (2006), 'Weakening of tropical Pacific atmospheric circulation due to anthropogenic forcing', *Nature* **441**(7089), 73–76.

Venema, V. K. C., Mestre, O., Aguilar, E., Auer, I., Guijarro, J. A., Domonkos, P., Vertacnik, G., Szentimrey, T., Stepanek, P., Zahradnicek, P., Viarre, J., Müller-Westermeier, G., Lakatos,

M., Williams, C. N., Menne, M. J., Lindau, R., Rasol, D., Rustemeier, E., Kolokythas, K., Marinova, T., Andresen, L., Acquaotta, F., Fratianni, S., Cheval, S., Klancar, M., Brunetti, M., Gruber, C., Prohom Duran, M., Likso, T., Esteban, P. and Brandsma, T. (2012), 'Benchmarking homogenization algorithms for monthly data', *Clim. Past* **8**, 89–115.

Volosciuk, C., Maraun, D., Semenov, V. A. and Park, W. (2015), 'Extreme precipitation in an atmosphere general circulation model: impact of horizontal and vertical model resolutions', *J. Climate* **28**(3), 1184–1205.

Volosciuk, C., Maraun, D., Vrac, M. and Widmann, M. (2017), 'A combined statistical bias correction and stochastic downscaling method for precipitation', *Hydrol. Earth Syst. Sci.* **21**(3), 1693–1719.

von Storch, H. (1999), 'On the use of "inflation" in statistical downscaling', *J. Climate* **12**(12), 3505–3506.

von Storch, H. and Hannoschöck, G. (1985), 'Statistical aspects of estimated principal vectors (EOFs) based on small sample sizes', *J. Clim Appl. Meteorol.* **24**(7), 716–724.

von Storch, H., Langenberg, H. and Feser, F. (2000), 'A spectral nudging technique for dynamical downscaling purposes', *Mon. Wea. Rev.* **128**, 3664–3673.

von Storch, H., Zorita, E. and Cubasch, U. (1993), 'Downscaling of global climate change estimates to regional scales: an application to Iberian rainfall in wintertime', *J. Climate* **6**(6), 1161–1171.

von Storch, H. and Zwiers, F. W. (1999), *Statistical analysis in climate research*, Cambridge University Press, Cambridge.

Vrac, M. and Friederichs, P. (2015), 'Multivariate-intervariable, spatial, and temporal-bias correction', *J. Climate* **28**(1), 218–237.

Vrac, M., Marbaix, P., Paillard, D. and Naveau, P. (2007*a*), 'Non-linear statistical downscaling of present and LGM precipitation and temperatures over Europe', *Clim. Past* **3**(4), 669–682.

Vrac, M. and Naveau, P. (2007), 'Stochastic downscaling of precipitation: From dry events to heavy rainfalls', *Wat. Resour. Res.* **43**(7), W07402.

Vrac, M., Stein, M. and Hayhoe, K. (2007*b*), 'Statistical downscaling of precipitation through nonhomogeneous stochastic weather typing', *Clim. Res.* **34**, 169–184.

Vrac, M., Stein, M. L., Hayhoe, K. and Liang, X. Z. (2007*c*), 'A general method for validating statistical downscaling methods under future climate change', *Geophys. Res. Lett.* **34**, L18701.

Waldron, K. M., Paegle, J. and Horel, J. D. (1996), 'Sensitivity of a spectrally filtered and nudged limited-area model to outer model options', *Mon. Wea. Rev.* **124**(3), 529–547.

Wallace, J. M. and Hobbs, P. V. (2006), *Atmospheric science. An introductory survey*, Academic Press.

Walton, D. B., Sun, F., Hall, A. and Capps, S. (2015), 'A hybrid dynamical–statistical downscaling technique. Part I: Development and validation of the technique', *J. Climate* **28**(12), 4597–4617.

Wang, C., Zhang, L., Lee, S.-K., Wu, L. and Mechoso, C. R. (2014), 'A global perspective on CMIP5 climate model biases', *Nat. Clim. Change* **4**, 201–205.

Ward, J. H. (1963), 'Hierarchical grouping to optimize an objective function', *J. Amer. Stat. Assoc.* **58**(301), 236–244.

Warszawski, L., Frieler, K., Huber, V., Piontek, F., Serdeczny, O. and Schewe, J. (2014), 'The Inter-Sectoral Impact Model Intercomparison Project (ISI-MIP): project framework', *Proc. Nat. Acad. Sci.* **111**(9), 3228–3232.

Washington, W. M. and Parkinson, C. L. (2005), *An introduction to three-dimensional climate modeling*, University Science Books.

WCRP WGRC (2014), 'WCRP WGRC Expert Meeting on Climate Information "Distillation"'. 29–31 October 2014, Santander, Spain.

Weedon, G. P., Balsamo, G., Bellouin, N., Gomes, S., Best, M. J. and Viterbo, P. (2014), 'The WFDEI meteorological forcing data set: WATCH Forcing Data methodology applied to ERA-Interim reanalysis data', *Wat. Resour. Res.* **50**(9), 7505–7514.

Weedon, G. P., Gomes, S., Viterbo, P., Shuttleworth, W. J., Blyth, E., Österle, H., Adam, J. C., Bellouin, N., Boucher, O. and Best, M. (2011), 'Creation of the WATCH forcing data and its use to assess global and regional reference crop evaporation over land during the twentieth century', *J. Hydrometeorol.* **12**(5), 823–848.

Wehrens, R. and Buydens, L. M. C. (2007), 'Self-and super-organizing maps in R: the Kohonen package', *J. Stat. Softw.* **21**(5), 1–19.

Wheater, H. S., Chandler, R. E., Onof, C. J., Isham, V. S., Bellone, E., Yang, C., Lekkas, D., Lourmas, G. and Segond, M.-L. (2005), 'Spatial-temporal rainfall modelling for flood risk estimation', *Stoch. Environ. Res. Risk Assess.* **19**, 403–416.

Wheater, H. S., Isham, V. S., Cox, D. R., Chandler, R. E., Kakou, A., Northrop, P. J., Oh, L., Onof, C. and Rodriguez-Iturbe, I. (2000), 'Spatial-temporal rainfall fields: modelling and statistical aspects', *Hydrol. Earth Syst. Sci.* **4**(4), 581–601.

White, R., Cooley, D., Derby, R. and Seaver, F. (1958), 'The development of efficient linear statistical operators for the prediction of sea-level pressure', *J. Meteorol.* **15**(5), 426–434.

White, R. H. and Toumi, R. (2013), 'The limitations of bias correcting regional climate model inputs', *Geophys. Res. Lett.* **40**(12), 2907–2912.

Whiteman, C. D. (2000), *Mountain meteorology: fundamentals and applications*, Oxford University Press.

Widmann, M. (2005), 'One-dimensional CCA and SVD, and their relationship to regression maps', *J. Climate* **18**(14), 2785–2792.

Widmann, M. and Bretherton, C. S. (2000), 'Validation of mesoscale precipitation in the NCEP reanalysis using a new gridcell dataset for the northwestern United States', *J. Climate* **13**(11), 1936–1950.

Widmann, M., Bretherton, C. S. and Salathe, E. P. (2003), 'Statistical precipitation downscaling over the northwestern United States using numerically simulated precipitation as a predictor', *J. Climate* **16**(5), 799–816.

Widmann, M. et al. (2017), 'Validation of spatial variability in downscaling results from the VALUE perfect predictor experiment', *Int. J. Climatol., subm.*

Widmann, M., Goosse, H., van der Schrier, G., Schnur, R. and Barkmeijer, J. (2010), 'Using data assimilation to study extratropical Northern Hemisphere climate over the last millennium', *Clim. Past* **6**(5), 627–644.

Widmann, M. and Schär, C. (1997), 'A principal component and long-term trend analysis of daily precipitation in Switzerland', *Int. J. Climatol.* **17**(12), 1333–1356.

Wigley, T. M. L., Jones, P. D., Briffa, K. R. and Smith, G. (1990), 'Obtaining subgrid scale information from coarse-resolution general circulation model output', *J. Geophys. Res.* **95**, 1943–1953.

Wigley, T. M. L., Jones, P. D. and Kelly, P. M. (1986), *The greenhouse effect. Climatic change and ecosystems*, John Wiley, New York, chapter 'Empirical Climate Studies'.

Wilby, R. L. (2010), 'Opinion: evaluating climate model outputs for hydrological applications', *Hydrol. Sci. J.* **55**, 1090–1093.

Wilby, R. L. (2017), *Climate change in practice*, Cambridge University Press.

Wilby, R. L., Charles, S. P., Zorita, E., Timbal, B., Whetton, P. and Mearns, L. O. (2004), 'Guidelines for use of climate scenarios developed from statistical downscaling methods', IPCC Task Group on Data and Scenario Support for Impact and Climate Analysis (TGICA).

Wilby, R. L., Dawson, C. W. and Barrow, E. M. (2002), 'SDSM – a decision support tool for the assessment of regional climate change impacts', *Env. Mod. Soft.* **17**(2), 145–157.

Wilby, R. L., Dawson, C. W., Murphy, C., O'Connor, P. and Hawkins, E. (2014), 'The Statistical DownScaling Model - Decision Centric (SDSM-DC): conceptual basis and applications', *Clim. Res.* **61**(3), 259–276.

Wilby, R. L. and Dessai, S. (2010), 'Robust adaptation to climate change', *Weather* **65**, 180–185.

Wilby, R. L., Friedhoff, M., Connell, R., Minikulov, N. and Leonidova, N. (2011), 'Tajikistan Pilot Programme for Climate Resilience (PPCR) Project A4 – Improving the Climate Resilience of Tajikistan's Hydropower Sector', Final Report.

Wilby, R. L., Hay, L. E. and Leavesley, G. H. (1999), 'A comparison of downscaled and raw GCM output: implications for climate change scenarios in the San Juan River basin, Colorado', *J. Hydrol.* **225**(1), 67–91.

Wilby, R. L., Hay, L. E., Gutowski, W. J., Arritt, R. W., Takle, E. S., Pan, T., Leavesley, G. H. and Clark, M. P. (2000), 'Hydrological responses to dynamically and statistically downscaled climate model output', *Geophys. Res. Lett.* **27**(8), 1199–1202.

Wilby, R. L., Tomlinson, O. J. and Dawson, C. W. (2003), 'Multi-site simulation of precipitation by conditional resampling', *Climate Res.* **23**(3), 183–194.

Wilby, R. L. and Wigley, T. M. L. (1997), 'Downscaling general circulation model output: a review of methods and limitations', *Prog. Phys. Geogr.* **21**, 530–548.

Wilby, R. L. and Wigley, T. M. L. (2000), 'Precipitation predictors for downscaling: observed and general circulation model relationships', *Int. J. Climatol.* **20**(6), 641–661.

Wilby, R. L., Wigley, T. M. L., Conway, D., Jones, P. D., Hewitson, B. C., Main, J. and Wilks, D. S. (1998), 'Statistical downscaling of general circulation model output: a comparison of methods', *Wat. Resour. Res.* **34**(11), 2995–3008.

Wilcke, R. A. I., Mendlik, T. and Gobiet, A. (2013), 'Multi-variable error correction of regional climate models', *Clim. Change* **120**(4), 871–887.

Wilks, D. S. (1988), 'Estimating the consequences of CO_2-induced climatic change on North American grain agriculture using general circulation model information', *Clim. Change* **13**, 19–42.

Wilks, D. S. (1998), 'Multisite generalization of a daily precipitation generation model', *J. Hydrol.* **210**, 178–191.

Wilks, D. S. (2006), *Statistical methods in the atmospheric sciences*, 2 edn, Academic Press/Elsevier.

Wilks, D. S. (2009), 'A gridded multisite weather generator and synchronization to observed weather data', *Wat. Resour. Res.* **45**(10).

Wilks, D. S. (2010), 'Use of stochastic weather generators for precipitation downscaling', *WIREs Clim. Change* **1**(6), 898–907.

Wilks, D. S. (2012), 'Stochastic weather generators for climate-change downscaling, Part II: Multivariable and spatially coherent multisite downscaling', *WIREs Clim. Change* **3**(3), 267–278.

Wilks, D. S. and Wilby, R. L. (1999), 'The weather generation game: a review of stochastic weather models', *Prog. Phys. Geogr.* **23**(3), 329–357.

Willems, P., Olsson, J., Arnbjerg-Nielsen, K., Beecham, S., Pathirana, A., Gregersen, I. B. and Madsen, H. (2012), *Impacts of climate change on rainfall extremes and urban drainage systems*, IWA Publishing.

Willems, P. and Vrac, M. (2011), 'Statistical precipitation downscaling for small-scale hydrological impact investigations of climate change', *J. Hydrol.* **402**(3), 193–205.

Willett, K. M., Jones, P. D., Gillett, N. P. and Thorne, P. W. (2008), 'Recent changes in surface humidity: development of the HadCRUH dataset', *J. Climate* **21**(20), 5364–5383.

Williams, K., Brown, A., Jakob, C., Best, M., Arribas, A., Bodas-Salcedo, A., Bony, S., Danabasoglu, G., Ebert, B., Gleckler, P., Donner, L., Miller, M., Petch, J., Scaife, A., Waliser, D. and Watanabe, M. (2013), '4th WGNE Workshop on Systematic Errors in Weather and Climate Models', Workshop Summary, UK Met Office, Exeter, UK.

Winkler, J. A., Palutikof, J. P., Andresen, J. A. and Goodess, C. M. (1997), 'The simulation of daily temperature time series from GCM output. Part II: Sensitivity analysis of an empirical transfer function methodology', *J. Climate* **10**(10), 2514–2532.

Wippermann, F. and Gross, G. (1981), 'On the construction of orographically influenced wind roses for given distributions of the large-scale wind', *Beiträge zur Physik der Atmosphäre* **54**(4), 492–501.

Wong, G., Maraun, D., Vrac, M., Widmann, M., Eden, J. and Kent, T. (2014), 'Stochastic model output statistics for bias correcting and downscaling precipitation including extremes', *J. Climate* **27**, 6940–6959.

Wood, A. W., Leung, L. R., Sridhar, V. and Lettenmaier, D. P. (2004), 'Hydrologic implications of dynamical and statistical approaches to downscaling climate model outputs', *Clim. Change* **62**(1–3), 189–216.

Wood, A. W., Maurer, E. P., Kumar, A. and Lettenmaier, D. P. (2002), 'Long-range experimental hydrologic forecasting for the eastern United States', *J. Geophys. Res. Atmos.* **107**(D20).

Woollings, T. (2010), 'Dynamical influences on European climate: an uncertain future', *Phil. Trans. R. Soc. A* **368**, 3733–3756.

Woollings, T., Gregory, J. M., Pinto, J. G., Reyers, M. and Brayshaw, D. J. (2012), 'Response of the North Atlantic storm track to climate change shaped by ocean-atmosphere coupling', *Nat. Geosci.* **5**(5), 313–317.

World Bank (2013), 'Turn Down the Heat: Climate Extremes, Regional Impacts, and the Case for Resilience', a report for the World Bank by the Potsdam Institute for Climate Impact Research and Climate Analytics.

Worldbank (n.d.), 'Climate change knowledge portal', http://sdwebx.worldbank.org/climate portal/.

Xie, P., Chen, M. and Shi, W. (2010), 'CPC unified gauge-based analysis of global daily precipitation', 24th Conference on Hydrology, Atlanta, USA.

Xu, C. (1999), 'From GCMs to river flow: a review of downscaling methods and hydrologic modelling approaches', *Prog. Phys. Geogr.* **23**(2), 229.

Yang, C., Chandler, R. E. and Isham, V. S. (2005), 'Spatial-temporal rainfall simulation using generalized linear models', *Wat. Resour. Res.* **41**, W11415.

Yang, G.-Y. and Slingo, J. (2001), 'The diurnal cycle in the tropics', *Mon. Wea. Rev.* **129**(4), 784–801.

Yarnal, B., Comrie, A. C., Frakes, B. and Brown, D. P. (2001), 'Developments and prospects in synoptic climatology', *Int. J. Climatol.* **21**(15), 1923–1950.

Yatagai, A., Kamiguchi, K., Arakawa, O., Hamada, A., Yasutomi, N. and Kitoh, A. (2012), 'APHRODITE: Constructing a long-term daily gridded precipitation dataset for Asia based on a dense network of rain gauges', *Bull. Amer. Meteorol. Soc.* **93**(9), 1401–1415.

Yates, D., Gangopadhyay, S., Rajagopalan, B. and Strzepek, K. (2003), 'A technique for generating regional climate scenarios using a nearest-neighbor algorithm', *Wat. Resour. Res.* **39**(7).

Yee, T. W. and Wild, C. J. (1996), 'Vector generalized additive models', *J. R. Stat. Soc.* **B 58**, 481–493.

Yip, S., Ferro, C. A. T., Stephenson, D. B. and Hawkins, E. (2011), 'A simple, coherent framework for partitioning uncertainty in climate predictions', *J. Climate* **24**(17), 4634–4643.

Young, K. C. (1994), 'A multivariate chain model for simulating climatic parameters from daily data', *J. Appl. Meteorol.* **33**(6), 661–671.

Zappa, G., Shaffrey, L. C. and Hodges, K. I. (2013), 'The ability of CMIP5 models to simulate North Atlantic extratropical cyclones', *J. Climate* **26**, 5379–5396.

Zhou, Z.-Q. and Xie, S.-P. (2015), 'Effects of climatological model biases on the projection of tropical climate change', *J. Climate* **28**(24), 9909–9917.

Zorita, E., Kharin, V. and von Storch, H. (1992), 'The atmospheric circulation and sea surface temperature in the North Atlantic area in winter: their interaction and relevance for Iberian precipitation', *J. Climate* **5**(10), 1097–1108.

Zorita, E. and von Storch, H. (1997), 'A survey of statistical downscaling techniques', Technical report, GKSS report 97/E/20, GKSS Research Center: Geesthacht.

Zorita, E. and von Storch, H. (1999), 'The analog method as a simple statistical downscaling technique: comparison with more complicated methods', *J. Climate* **12**(8), 2474–2489.

Zuidema, P., Chang, P., Medeiros, B., Kirtman, B. P., Mechoso, R., Schneider, E. K., Toniazzo, T., Richter, I., Small, R. J., Bellomo, K., Brandt, P., de Szoeke, S., Farrar, J. T., Jung, E., Kato, S., Li, M., Patricola, C., Wang, Z., Wood, R. and Xu, Z. (2016), 'Challenges and prospects for reducing coupled climate model SST biases in the eastern tropical Atlantic and Pacific oceans: the US CLIVAR Eastern Tropical Oceans Synthesis Working Group', *Bull. Amer. Meteorol. Soc.* **97**, 2305–2327.

Index